Mathematics: Concepts, Theories and Applied Principles

Mathematics: Concepts, Theories and Applied Principles

Editor: Frank West

New York

Published by NY Research Press
118-35 Queens Blvd., Suite 400,
Forest Hills, NY 11375, USA
www.nyresearchpress.com

Mathematics: Concepts, Theories and Applied Principles
Edited by Frank West

International Standard Book Number: 978-1-63238-559-8 (Hardback)

Cataloging-in-Publication Data

Mathematics : concepts, theories and applied principles / edited by Frank West.
p. cm.
Includes bibliographical references and index.
ISBN 978-1-63238-559-8
1. Mathematics. 2. Applied mathematics. I. West, Frank.
QA36 .M38 2017
510--dc23

Printed in the United States of America.

Contents

Preface

This book on mathematics deals with the concepts and theories that are derived with a view to group and compute numbers. Mathematics is an essential requirement for almost all sciences that require abstraction for its functioning. Topics included in this book seek to advance the level of research that has been seen in this field in the past decade. This book, with its detailed analyses and data, will prove immensely beneficial to professionals and students involved in this area at various levels. Case studies as well as conceptual theorems allow the readers to get a holistic understanding of the areas of study that fall under mathematics. This text is appropriate for students seeking detailed information in this area as well as for experts.

All of the data presented henceforth, was collaborated in the wake of recent advancements in the field. The aim of this book is to present the diversified developments from across the globe in a comprehensible manner. The opinions expressed in each chapter belong solely to the contributing authors. Their interpretations of the topics are the integral part of this book, which I have carefully compiled for a better understanding of the readers.

At the end, I would like to thank all those who dedicated their time and efforts for the successful completion of this book. I also wish to convey my gratitude towards my friends and family who supported me at every step.

Editor

1

A mathematical analysis of an exchange-traded horse race betting fund with deterministic payoff betting strategy for institutional investment to challenge EMH

Craig George Leslie Hopf[1] and Gurudeo Anand Tularam[1*]

*Corresponding author: Gurudeo Anand Tularam, Mathematics and Statistics, Griffith Sciences (ENV), Environmental Futures Research Institute (EFRI), Griffith University, Brisbane, Australia

E-mail: a.tularam@griffith.edu.au

Reviewing editor: Quanxi Shao, CSIRO, Australia

Abstract: This paper's primary alternative hypothesis is H_a: profitable exchange-traded horserace betting fund with deterministic payoff exists for acceptable institutional portfolio return—risk. The primary hypothesis challenges the semi-strong efficient market hypothesis applied to horse race wagering. An optimal deterministic betting model (DBM) is derived from the existing stochastic model fundamentals, mathematical pooling principles, and new theorem. The exchange-traded betting fund (ETBF) is derived from force of interest first principles. An ETBF driven by DBM processes conjointly defines the research's betting strategy. Alpha is excess return above financial benchmark, and invokes betting strategy alpha that is composed of model alpha and fund alpha. The results and analysis from statistical testing of a global stratified data sample of three hundred galloper horse races accepted at the ninety-five percent confidence-level positive betting strategy alpha, to endorse an exchange-traded horse race betting fund with deterministic payoff into financial market.

Subjects: Applied Mathematics; Mathematics & Statistics; Science

Keywords: complex number system optimization; deterministic betting model; exchange-traded betting fund; portfolio betting inputs

1. Preliminary

The Bank for International Settlements (2007) examined institutional investors, global savings, and asset allocation, and remarked that one important investment decision of insurance companies and pension funds is the selection of an optimal asset mix that best reflects their risk—reward trade-offs. There is little corporate consideration presently toward horserace betting as a mainstream or alternative asset for portfolio input, to support the efficient market hypothesis (EMH) toward horserace

ABOUT THE AUTHOR

Craig George Leslie Hopf (BSc, MPhil) is a research associate and Gurudeo Anand Tularam (BSc, PhD) is a senior lecturer in Mathematics and Statistics at Griffith University in Brisbane Australia. Their joint research contribution is toward betting input modeling for legitimate portfolio investment and stock market benefit.

PUBLIC INTEREST STATEMENT

An exchange-traded horserace betting fund with deterministic payoff betting strategy is developed from new and existing mathematical principles and theory. The results from sample testing statistically support a profitable betting strategy to challenge the existing semi-strong efficient market hypothesis (EMH) toward horse racing.

betting (Bank for International Settlements, 2007; Roca, Wong, and Tularam (2010); Tularam, Roca, & Wong, 2010). One conclusion drawn from this hypothesis is that traders are unable to achieve consistently superior returns from either technical parametric or fundamental nonparametric betting strategies. Correspondingly, there is extensive research on optimal stochastic horserace betting systems that forecast horserace outcomes based on both "technical" and "fundamental" analysis techniques to challenge the semi-strong EMH applied to horseracing (Edelman, 2007). The concept of an efficient betting market incorporates forecasting future asset prices and future expected return remarked Snyder (1978). Essentially, the horserace bet as suggested by Koning and Van Velzen (2009) is a binary option asset that generates a defined return from the betting market; as argument against the EMH. The questioning of the credibility of the EMH suggests that a profitable deterministic horserace betting strategy may be achievable. The profoundness of a deterministic model is that every set of variable states is uniquely determined by statistics in the model and by sets of previous states of these variables wrote Yang (2008).

2. Rank-order permutation

The foundation for the existing literature written and research conducted on horserace betting models is largely the rank-order permutation of stochastic processes (Critchlow, Fligner, & Verducci, 1991). The rank for the outcome of an event expressed by a rater is the result of ranking of a random variable from an underlying parametric or nonparametric distribution (D'Elia, 2003). The number of rank-order arrangements in an n racer field is $n!$, and the subsequent number of arrangements in an n racer field with i associations is $(n-i)!$, where i denotes the number of associations. The order of finish for a field size of n racers is represented by the general permutation $\mathrm{R} = \left(\mathrm{R}_{1,\ldots,j}, \mathrm{R}_{k,\ldots,l}, \ldots, \mathrm{R}_{s,\ldots,n}\right)$; $\mathrm{R}_{1,\ldots,j}$: first to jth ranked racers; $j, k, l, s, n \in [1, \ldots, n]$ for the n racer field. Permutation group $R_n = \{\mathrm{R}\}, n \varepsilon I^+$, represents the set of complete and partial rank-order permutations for the n racer field, and $p(\mathrm{R})$ the parametric or nonparametric distribution on permutation R. Plackett's (1975) first-order permutation is represented by $\mathrm{R} = (\mathrm{R}_1, \mathrm{R}_2, \ldots, \mathrm{R}_n)$, whereby R_1 represents the first placed racer and R_n is the last placed racer from the field. $\mathrm{R} = \left(\mathrm{R}_{1,2}, \mathrm{R}_{2,3}, \ldots, \mathrm{R}_{i,i+1}, \ldots, \mathrm{R}_{n-1,n}\right)$ illustrates Plackett's second-order permutation for the n racer field ($\mathrm{R}_{i,i+1}$: paired racers). The stochastic technical horserace betting models of parametric distribution revised in this paper utilize the bettor win or place real odds vector, $\mathcal{O} = \left(O_{i1,\ldots,j}, O_{jk,\ldots,l}, \ldots, O_{is,\ldots,n}\right); O_{i1,\ldots,j}$: first j racer's odds event i; $\mathcal{O} \in R^n; i, j, k, l, s, \ldots, n \in [1, \ldots, n]$, to calculate permutation probability.

3. Permutation probability of parametric distribution

In this paper, $\mathrm{R}_{i,\ldots,j}$ and $X_{\mathrm{R}_{i,\ldots,j}}$ represent independent, non-identically distributed discrete and continuous random variables, respectively. $\mathrm{R}_{i,\ldots,j}$ represents the rank order for racer i (and associations) with probability density $F\left(\mathrm{R}_{i,\ldots,j}; \mathcal{O}\right)$; and $X_{\mathrm{R}_{i,\ldots,j}}$ represents the time taken for the ith rank racer (and associations), with probability distribution $F\left(X_{\mathrm{R}_{i,\ldots,j}}; \alpha_{i,\ldots,j}\right)$. Accordingly, R_1 and X_{R_1} designate the first ranked racer and time taken by first ranked racer; and subsequently, $\mathrm{R}_{1,2}$ and $X_{\mathrm{R}_{1,2}}$ represent the first and second ranked racer pair, and the time taken by the winning racer pair, respectively. Stochastic technical model theory has developed from the L-Decomposable and Plackett's logistic models of distribution $F\left(\mathrm{R}_{i,\ldots,j}; \mathcal{O}\right)$, to models (and their derivatives) based upon the normal, gamma and exponential parametric distributions $F\left(X_{\mathrm{R}_{i,\ldots,j}}; \alpha_{i,\ldots,j}\right)$. Random variables $\mathrm{R}_{i,\ldots,j}$ and $X_{\mathrm{R}_{i,\ldots,j}}$ are statistically independent when the joint probability density function equals the marginal densities product; i.e. $f\left(X_{\mathrm{R}_{1,\ldots,j}}, X_{\mathrm{R}_{k,\ldots,l}}, \ldots, X_{\mathrm{R}_{s,\ldots,n}}\right) = f\left(X_{\mathrm{R}_{1,\ldots,j}}; \alpha_{1,\ldots,j}\right) f\left(X_{\mathrm{R}_{k,\ldots,l}}; \alpha_{k,\ldots,l}\right), \ldots, f\left(X_{\mathrm{R}_{s,\ldots,n}}; \alpha_{s,\ldots,n}\right)$.

The permutation probability calculated from the conditional product of the choice probabilities is outlined in Equation 3.1;

$$p(\mathrm{R}) = p\left(\mathrm{R}_{1,\ldots,j}, \mathrm{R}_{k,\ldots,l}, \ldots, \mathrm{R}_{s,\ldots,n}\right)$$

$$\equiv \frac{p\left(\mathrm{R}_{1,\ldots,j}\right) p\left(\mathrm{R}_{k,\ldots,l} | \mathrm{R}_{1,\ldots,j}\right), \ldots, p\left(\mathrm{R}_{s,\ldots,n} | \mathrm{R}_{1,\ldots,j}, \mathrm{R}_{k,\ldots,l}, \ldots, \mathrm{R}_{m,\ldots,r}\right)}{\left[1 - p\left(\mathrm{R}_{1,\ldots,j}\right)\right]\left[1 - p\left(\mathrm{R}_{1,\ldots,j}\right) - p\left(\mathrm{R}_{k,\ldots,l}\right)\right], \ldots, p\left(\mathrm{R}_{s,\ldots,n}\right)},$$

$p(R_{k,\dots,l}|R_{1,\dots,j})$: probability k^{th} to l^{th} ranked racers from remaining "n − k + 1" racers,

$$p(R_{k,\dots,l}) = \frac{1}{O_{k,\dots,l}}; O_{k,\dots,l}: k\text{th to } l\text{th ranked odds.} \tag{3.1}$$

The ranking processes of the L-Decomposable models of Harville (1973), Luce (1959), and Stern (1990) all determine the permutation probability from the individual racer win (or place) odds of r-1 objects ($r \le n$) as shown by Equation 3.2;

$$p(R) = p(R_1, R_2, \dots, R_n) = \frac{p(R_1)p(R_2|R_1), \dots, p(R_n)}{(1-p(R_1))(1-p(R_1)) - p(R_2)), \dots, p(R_n)} = \prod_{i=1}^{n-1}\left[\frac{p(R_i)}{\sum_{j\varepsilon B_i} p(R_j)}\right],$$

$$B_i = \{R_i, \dots, R_n\} \text{ is the set remaining at stage } i\text{, with } \sum_{i=1}^{n} p(R_i) = 1, \text{ and } p(R_i) = \frac{1}{O_i}. \tag{3.2}$$

Plackett's (1975) logistic model expresses the $n!-1$ independent permutation probabilities from n objects in terms of a finite set of $n!-1$ parameters of hierarchical structure $K, \lambda_i, \lambda_{ij}, \lambda_{ijk}, \dots, \lambda_{ijk,\dots,n}$. All the parameters of the logistic model are logarithms of probability win ratios, except K (Plackett, 1975), and the hierarchical structure is outlined in Equation 3.3.

$$\lambda_{ijk,\dots,ls} = \log\left(\frac{p(R_{ijk,\dots,ls})p(R_{jk,\dots,lt})}{p(R_{ijk,\dots,lt})p(R_{jk,\dots,ls})}\right); p(R_{ijk,\dots,ls}) = \frac{1}{O_{ijk,\dots,ls}}; i,j,k,\dots,l,s,t \in [1,\dots,n],$$

$$K = \log(p(R_n)); R_n\text{:last placed racer}\{\text{zero stage}\},$$

$$\lambda_i = \log\left(\frac{p(R_i)}{p(R_n)}\right); i \in [1,\dots,n-1]\{\text{first stage}\},$$

$$\lambda_{ij} = \log\left(\frac{p(R_{ij})p(R_s)}{p(R_{is})p(R_j)}\right); i \text{ fixed};$$

$$s = \max 1,\dots,n, \text{exclude } i; j \in [1,\dots,n], \text{exclude } i,s\{\text{second stage}\},$$

$$\lambda_{ijk} = \log\left(\frac{p(R_{ijk})p(R_{jt})}{p(R_{ijt})p(R_{jk})}\right); i,j \text{ fixed};$$

$$t = \max 1,\dots,n, \text{exclude } i,j; k \in [1,\dots,n], \text{exclude } i,j,t\{\text{third stage}\}. \tag{3.3}$$

When $\lambda_{ij}, \lambda_{ijk}, \dots, \lambda_{ijk,\dots,n}$ equal zero, Plackett's first-order model is the L-decomposable model; and the permutation probability from a second-order logistic model for n racer field is illustrated in Equation 3.4 (Dansie, 1983; Plackett, 1975).

$$p(R) = p\left[\left\{R_{1,2} < R_{i,j}, \forall i,j \ne 1,2\right\} \text{ and } \left\{R_{2,3} < R_{k,l}|R_{1,2}, \forall k,l \ne 2,3\right\}\right],$$

$$p(R) = p(R_{1,2}, R_{2,3}, R_{3,4}, \dots, R_{n-1,n}) = \frac{p(R_{1,2})p(R_{2,3}|R_{1,2}), \dots, p(R_{n-1,n}|R_{1,2}, R_{2,3}, \dots, R_{n-2,n-1})}{[1-p(R_{1,2})][1-p(R_{1,2}) - p(R_{2,3})], \dots, p(R_{n-1,n})}. \tag{3.4}$$

The parametric distributions of the normal random variable $f(X_{R_i}; \alpha_i, \sigma^2)$, the gamma random variable $f(X_{R_i}; \alpha_i, r)$, and the exponential random variable $f(X_{R_i}; \alpha_i, r = 1)$ have been proposed for horserace betting modeling. The normal rank-order model is a class of rank model that is a function of a single independent variable of parametric distribution $N(X_{R_{i,\dots,j}}; \alpha_{i,\dots,j})$ with joint pdf

$\prod_{i=1}^{n} f\left(X_{R_{i,\ldots,j}} - \alpha_{i,\ldots,j}\right)$($\alpha_{i,\ldots,j}$: average time ith to jth racers). The permutation probability from the normal rank model is represented by the multivariate integral,

$$p(\mathrm{R}) = p\left(\mathrm{R}_{1,\ldots,j}, \mathrm{R}_{k,\ldots,l}, \ldots, \mathrm{R}_{s,\ldots,n}\right) \equiv p\left(X_{R_{1,\ldots,j}} < X_{R_{k,\ldots,l}} < \ldots < X_{R_{s,\ldots,n}}\right),$$

$$p(\mathrm{R}) = \int_0^\infty f\left(X_{R_{1,\ldots,j}} - \alpha_{1,\ldots,j}\right) \int_{X_{R_{1,\ldots,j}}}^\infty f\left(X_{R_{k,\ldots,l}} - \alpha_{k,\ldots,l}\right) \cdots \int_{X_{R_{m,\ldots,r}}}^\infty f\left(X_{R_{s,\ldots,n}} - \alpha_{s,\ldots,n}\right)$$
$$dX_{R_{1,\ldots,j}}, \ldots, dX_{R_{s,\ldots,n}} \left(\sigma^2 = 1\right). \tag{3.5}$$

The probability of racer i first for X_R (expected win time) is expressed in Equation 3.6;

$$p(\mathrm{R}_i = 1) = \int_0^{X_R} f\left(X_{R_i}; \alpha_i\right) \prod_{j \neq i = 1}^{n-1} [1 - F\left(X_{R_j} - \alpha_j\right)] dX_{R_i}, \tag{3.6}$$

Henery (1983) derived the normal rank approximate model from first-degree Taylor expansion to calculate the permutation probability shown in Equation 3.7 which has been modified for associations.

$$p(\mathrm{R}) \approx \frac{1}{(n-i)!} + \frac{\sum_{1,\ldots,j}^{s,\ldots,n} \alpha_{a,\ldots,b}\mu_{a,\ldots,b;n}}{(n-i)!} \left(\sum \alpha_{a,\ldots,b} = 0\right), \tag{3.7}$$

The win, place, and trifecta probability approximations are illustrated by the following forms

$$p(\mathrm{R}_i = 1) \approx \frac{1}{n} + \frac{\alpha_i \mu_{1;n}}{(n-1)},$$

$$p\left(\mathrm{R}_j = k\right) \approx \frac{1}{n} + \frac{\alpha_j \mu_{k;n}}{(n-1)},$$

$$p(\mathrm{R}_1\mathrm{R}_2\mathrm{R}_3) \approx \frac{1}{n_{(3)}} \left[1 + \sum_{i=1}^{3} \alpha_i \mu_{i;n} + \frac{\sum_{j=1}^{3} \alpha_j \sum_{j=1}^{3} \mu_{j;n}}{(n-3)}\right]. \tag{3.8}$$

The gamma rank-order model class is a function of bivariate independent variables with gamma distribution $\Gamma(X_{R_i}; \alpha_i, r)$ and joint pdf $\prod_{i=1}^{n} \left[\frac{\alpha_i^r X_{R_i}^{r-1} \exp(-\alpha_i X_{R_i})}{\Gamma(r)}\right]$ (α: rate; r: distance). The gamma rank probability permutation by Stern (1990) is in Equation 3.9 which has been modified for associations;

$$p\left(\mathrm{R}_{1,\ldots,j}, \mathrm{R}_{k,\ldots,l}, \ldots, \mathrm{R}_{s,\ldots,n}\right) \approx \int_0^{X_{R_{k,\ldots,l}}} \left(\frac{\alpha_{1,\ldots,j}^r X_{R_{1,\ldots,j}}^{r-1} \exp\left(-\alpha_{1,\ldots,j} X_{R_{1,\ldots,j}}\right)}{\Gamma(r)}\right.$$
$$\int_0^{X_{R_{m,r}}} \left(\frac{\alpha_{k,\ldots,l}^r X_{R_{k,\ldots,l}}^{r-1} \exp\left(-\alpha_{k,\ldots,l} X_{R_{k,\ldots,l}}\right)}{\Gamma(r)}\right. \cdots \tag{3.9}$$
$$\int_0^\infty \left(\frac{\alpha_{s,\ldots,n}^r X_{R_{s,\ldots,n}}^{r-1} \exp\left(-\alpha_{s,\ldots,n} X_{R_{s,\ldots,n}}\right)}{\Gamma(r)}\right. dX_{R_{1,\ldots,j}}, \ldots, dX_{R_{s,\ldots,n}},$$

and the racer probability to win by Henery (1983) is expressed in Equation 3.10;

$$p(\mathrm{R}_i = 1) = \int_0^\infty f(X_{R_i}; \alpha_i, r) \prod_{j \neq i = 1}^{n-1} [1 - F(X_{R_j}; \alpha_j, r)] dX_{R_i}, \tag{3.10}$$

The gamma density function is represented by

$$f\left(X_{R_{i,\dots,j}};\alpha_{i,\dots,j},r\right)=\left(\frac{\alpha_{i,\dots,j}^{r}X_{R_{i,\dots,j}}^{r-1}\exp\left(-\alpha_{i,\dots,j}X_{R_{i,\dots,j}}\right)}{\Gamma(r)}\right)\equiv\exp\left\{-\alpha_{i,\dots,j}X_{R_{i,\dots,j}}+C\left(X_{R_{i,\dots,j}}\right)+D\left(\alpha_{i,\dots,j}\right)\right\},\tag{3.11}$$

from which a first-degree Taylor expansion approximation for the model in Equation 3.12 is derived by Henery (1983) and modified to allow for associations accordingly:

$$p(\mathrm{R})\approx p_0(\mathrm{R})+\sum_{1,\dots,k}^{s,\dots,n}\frac{\partial p(\mathrm{R})}{\partial\alpha_{i,\dots,j}}\left(\alpha_{i,\dots,j}-\alpha_0\right),$$

$$p_0(\mathrm{R})=\frac{1}{(n-i)!}\left(\forall\alpha_{a,\dots,b}=\alpha_0\right),$$

$$\frac{\partial p(\mathrm{R})}{\partial\alpha_{i,\dots,j}}=\int\int,\dots,\int\frac{\partial}{\partial\alpha_{i,\dots,j}}\ln f(X_{R_{i,\dots,j}};\alpha_{i,\dots,j},r)\prod_{1,\dots,k}^{s,\dots,n}f\left(X_{R_{i,\dots,j}};\alpha_{i,\dots,j},r\right)dX_{R_{i,\dots,j}},$$

$$\frac{\partial p(\mathrm{R})}{\partial\alpha_{i,\dots,j}}=\frac{1}{(n-i)!}\left(-\mu_{i,\dots,j;n}+D'(\alpha_{i,\dots,j})\right),$$

$$p(\mathrm{R})\approx\frac{1}{(n-i)!}+\frac{\sum_{1,\dots,k}^{s,\dots,n}\left\{r-\mu_{i,\dots,j;n}\right\}\left\{\alpha_{i,\dots,j}-1\right\}}{(n-i)!}\left(D\left(\alpha_{i,\dots,j}\right)=\ln\alpha_{i,\dots,j}^{r},\alpha_{i,\dots,j}\to1\right).\tag{3.12}$$

According to Henery (1983) a gamma rank probability approximation for the kth placed racer is expressed below in Equation 3.13;

$$p(\mathrm{R}_i=k)\approx\frac{1}{n}+\frac{\{r-\mu_{k;n}\}\{\alpha_i-1\}}{(n-1)},\tag{3.13}$$

The exponential model with density function $f\left(X_{R_{i,\dots,j}};\alpha_{i,\dots,j}\right)=\alpha_{i,\dots,j}\exp\left(-\alpha_{i,\dots,j}X_{R_{i,\dots,j}}\right)$ denotes the gamma rank-order model with shape parameter $r=1$. The exponential model is equivalent to the logistic model (Dansie, 1983; Stern, 1990); and the derivation of Plackett's ordered logistic models from the gamma rank-order model ($r=1$) using first principles is detailed in Equation 3.14;

$$f\left(X_{R_{i,\dots,j}};\alpha_{i,\dots,j},r=1\right)=\alpha_{i,\dots,j}\exp\left(-\alpha_{i,\dots,j}X_{R_{i,\dots,j}}\right),$$

$$p\left(X_{R_{1,\dots,j}}=\min X_{R_{a,\dots,b}}\right)=\int_0^\infty\int_{X_{R_{1,\dots,j}}}^\infty,\dots,\int_{X_{R_{1,\dots,j}}}^\infty\left[\prod_{1,\dots,k}^{s,\dots,n}\alpha_{a,\dots,b}e^{-\alpha_{a,\dots,b}X_{R_{a,\dots,b}}}\right]dX_{R_{1,\dots,j}}dX_{R_{k,\dots,l}},\dots,dX_{R_{s,\dots,n}},$$
$$=\int_0^\infty p\left(X_{R_{a,\dots,b}}>X_{R_{1,\dots,j}}|X_{R_{1,\dots,j}}\right)\alpha_{1,\dots,j}e^{-\alpha_{1,\dots,j}X_{R_{1,\dots,j}}}dX_{R_{1,\dots,j}},$$

$$p\left(X_{R_{1,\dots,j}}=\min X_{R_{a,\dots,b}}\right)=\int_0^\infty\alpha_{1,\dots,j}e^{-\sum\alpha_{a,\dots,b}X_{R_{1,\dots,j}}}dX_{R_{1,\dots,j}}=\frac{\alpha_{1,\dots,j}}{\sum_{1,\dots,j}^{s,\dots,n}\alpha_{a,\dots,b}},$$

$$\therefore p(\mathrm{R})=p\left(X_{R_{1,\dots,j}}<X_{R_{k,\dots,l}}<\dots<X_{R_{s,\dots,n}}\right)=\prod_{1,\dots,j}^{s,\dots,n}\frac{p(\mathrm{R}_{i,\dots,k})}{\sum_1^n p(\mathrm{R}_{a,\dots,b})};a,b,i,j,k,l,s,\dots,n\in[1,\dots,n].\tag{3.14}$$

A model developed from the both the logistic first-order and gamma models is the discount model; the discount model includes a discount factor λ_k^r(r: shape coefficient, k: position) to discount the anticipated diminished performance by a racer with a decrease in its placing suggests Lo, Bacon-Shone, and Busche (1995). The log odds ratio assumption that racer i beats racer j for kth place being a discounted function of racer i defeating racer j for the win; $\log\left(\mathrm{R}_i, \mathrm{R}_j | k\right) = \log\left(\mathrm{R}_i, \mathrm{R}_j | 1\right)^{\lambda_k^r} \equiv \lambda_k^r \log\left(\mathrm{R}_i, \mathrm{R}_j | 1\right)$ (Lo et al., 1995). Both a trifecta probability approximation and permutation probability approximation (that allows for associations) using the discount model technique is illustrated in Equation 3.15.

$$\log\left(p\left(\mathrm{R}_i\mathrm{R}_j\mathrm{R}_k\right)\right) = \log\left[p(\mathrm{R}_i)p\left(\mathrm{R}_j, \mathrm{R}_k | 2\right)p\left(\mathrm{R}_k, \mathrm{R}_s | 3\right)\right],$$

$$\log\left(p\left(\mathrm{R}_i\mathrm{R}_j\mathrm{R}_k\right)\right) = \log\left[p(\mathrm{R}_i)p\left(\left(\mathrm{R}_j, \mathrm{R}_k | 1\right)^{\lambda_2^r}\right)p\left(\left(\mathrm{R}_k, \mathrm{R}_s | 1\right)^{\lambda_3^r}\right)\right],$$

$$p\left(\mathrm{R}_i\mathrm{R}_j\mathrm{R}_k\right) \approx p(\mathrm{R}_i)\frac{\left[\left(p\left(\mathrm{R}_j\right)\right)^{\lambda_2^r}\right]}{\left[\sum_{s\neq i}\left(p\left(\mathrm{R}_s\right)\right)^{\lambda_2^r}\right]}\frac{\left[\left(p(\mathrm{R}_k)\right)^{\lambda_3^r}\right]}{\left[\sum_{t\neq i,j}\left(p\left(\mathrm{R}_t\right)\right)^{\lambda_3^r}\right]},$$

$$p\left(\mathrm{R}_{1,\dots,j}, \mathrm{R}_{k,\dots,l}, \dots, \mathrm{R}_{s,\dots,n}\right) \approx p\left(\mathrm{R}_{1,\dots,j}\right) \times \frac{\left[\left(p(\mathrm{R}_{k,\dots,l})\right)^{\lambda_{k,\dots,l}^r}\right]}{\left[\sum_{a,\dots,b\neq 1,\dots,j}\left(p\left(\mathrm{R}_{n,\dots,p}\right)\right)^{\lambda_{k,\dots,l}^r}\right]} \times \dots \times \frac{\left[\left(p(\mathrm{R}_{s,\dots,n})\right)^{\lambda_{s,\dots,n}^r}\right]}{\left[\sum_{t,\dots,w\neq k,\dots,l}\left(p\left(\mathrm{R}_{t,\dots,w}\right)\right)^{\lambda_{s,\dots,n}^r}\right]}. \tag{3.15}$$

The discount model is a function of the win probabilities of the n racers in the event, unlike the L-Decomposable model which is a function of the win probabilities of r−1 racers ($r \leq n$). These stochastic technical rank-order models use permutation conditional probability to determine expected outcomes for typical horserace betting products, such as win (R_1), exacta and quinella $\left(\mathrm{R}_1, \mathrm{R}_2\right)$, trifecta ($\mathrm{R}_1, \mathrm{R}_2, \mathrm{R}_3$), and first four events.

These permutation probabilities and the odds real vector $\mathcal{O}$ become arguments to optimize the expected payoff function for an event's betting products, and to determine the optimal bet vector $b = \left(b_{i1,\dots,j}, b_{jk,\dots,l}, \dots, b_{is,\dots,n}\right); b_{ik,\dots,l}$: bet on racers "$k$" to "$l$" event "$i$" $b \in R^n, j \in [1, \dots, n]$.

One approach to determine the optimal win and place bets on r racers ($r \leq n$) for a race is the extension of the Kelly criterion (Edelman, 2007) of maximizing the expected logarithmic return for the race, as a function of the market bettor odds and probability forecasts for win, exacta, quinella, trifecta, and additional permutation products as outlined in Equation 3.16.

$$\underset{\{b_{k,\dots,l}\}}{\text{Maximize}}\; EX(R) = f\left(\mathrm{R}_n; \mathcal{O}, b, p_{k,\dots,l}\right) = \sum_{1,\dots,j}^{s,\dots,n} p_{k,\dots,l} \log\left[b_{k,\dots,l}O_{k,\dots,l} - \left(\sum_{1,\dots,j}^{s,\dots,n} b_{k,\dots,l}\right)\right],$$

subject to: $\mathcal{O}, b \in R^n \forall O_{k,\dots,l}, b_{k,\dots,l} \in R; j, k, l, s, \dots, n \in [1, \dots, n]$,

$EX(R)$: expected logarithm return,

$p_{k,\dots,l}$: permutation probability forecasts for win, quinella, …,

$O_{k,\dots,l}$: bettor odds racers k to l; $b_{k,\dots,l}$: optimal bet for kth to lth racers. (3.16)

When $r = n$, wagering on the set of partial or complete permutations is analogous to optimal field betting over the entire racer field.

4. Permutation set pooling odds payoff properties and theorem over C^n

The zero return-to-nil risk pooling property of the gross bettor odds applies to the partial and complete set of rank-order permutations over R^n as illustrated by Equation 4.1 accordingly;

$$\sum \frac{1}{QO_{k,\ldots,l}} = X(>1),$$

$$\frac{1}{Q}\sum \frac{1}{O_{k,\ldots,l}} = X \therefore Q = \frac{1}{X}, Q = 1 - t; t\text{: track commission},$$

$$\sum \frac{1}{O_{k,\ldots,l}} = 1,$$

$$QO_{k,\ldots,l}\text{: net bettor product odds(less track commission)},$$

$$O_{k,\ldots,l}\text{: gross bettor odds(pre − commission)}; Q\text{: payback portion.} \tag{4.1}$$

Equation 4.2 extends the pooling property over C^n, and "arbitrage" is achievable from field betting of permutation wager products such as win $(\mathbb{R}_1)$, quinella $(\mathbb{R}_1, \mathbb{R}_2)$ and first four bet $(\mathbb{R}_1, \mathbb{R}_2, \mathbb{R}_3, \mathbb{R}_4)$, with trading the market bettor odds.

$$\underset{\{b_{k,\ldots,l}\}}{\text{Maximize}} \; R_{k,\ldots,l} = f(\mathbb{R}_n; \mathcal{O}, b) = |b \cdot \mathcal{O}| - \left(\sum_{1,\ldots,j}^{s,\ldots,n} b_{k,\ldots,l}\right) \geq c; R_{k,\ldots,l}, c \geq 0, R_{k,\ldots,l} = f(\mathbb{R}_n; \mathcal{O}, b)$$

$$= \sum_{1,\ldots,j}^{s,\ldots,n} |b_{k,\ldots,l} O_{k,\ldots,l}| - \left(\sum_{1,\ldots,j}^{s,\ldots,n} b_{k,\ldots,l}\right),$$

$$\text{subject to: } \left|O_{k,\ldots,l} \times b_{k,\ldots,l}\right| = \left|O_{s,\ldots,r} \times b_{s,\ldots,r}\right|, \forall k, \ldots, l \neq s, \ldots, r,$$

$$\mathcal{O} \in R^n \,\forall\, O_{k,\ldots,l} \in R; b \in C^n \forall\, b_{k,\ldots,l} = (x_{k,\ldots,l}, y_{k,\ldots,l}) \in C; j, k, l, s, \ldots n \in [1, \ldots, n]. \tag{4.2}$$

To establish optimal solution from the DBM of Equation 4.2, the multiple system optimization (MSO) theorem over C^n (Hopf & Tularam, 2014) is outlined in Equation 4.3, and the proof is provided in Appendix A1.

THEOREM 4.1 *Multiple system optimization (MSO) over an "n" finite series of complex systems generates a constant real component over each consecutive system.*

$$\{\max/\min\} Z_{1,\ldots,n}(C^n_{1,\ldots,n}; \odot) = \mathfrak{R}^n,$$

$$Z_{1,\ldots,n}(C^n_1 C^n_2, \ldots, C^n_n)\text{: "n" multiple system complex function},$$

$$C^n\text{: complex vector input}; \odot\text{: set of operators},$$

$$\hat{Z}_{1,\ldots,n}\left(\hat{C}^n_{1,\ldots,n}\right)\text{: complex output}; Re\left(\hat{Z}_{1,\ldots,n}\right) = R^n. \tag{4.3}$$

From the argument of MSO, the optimal solution for field betting on an individual horserace is a constant net payoff across the entire field; and with wagering on multiple, consecutive races, the DBM attempts to lock in arbitrage and a constant optimal payoff over the race series. The working of the DBM is demonstrated in Appendix A2.

Assume the unit investment period for the fund to be the day's racing, and $\delta_{R_{i,\ldots,j}} = i^{(365)}$ *denotes the generated racer force of interest; i.e. nominal rate of return per annum convertible daily. During the day's racing, an ETBF may invest on a single race, or invest over consecutive races. In the case of investing on a race series, the value of the fund at day's end is provided in Equation 4.4 as follows;*

$$X_1 = X_0\left[\prod_{i=1}^{n}(1+r_i)\right],$$

$$1+\frac{\delta_{R_{i,\ldots,j}}}{365} = \prod_{i=1}^{n}(1+r_i);r_i\text{: race } i \text{ return};i,\ldots,j\in[1,\ldots,n];n \text{ races.} \tag{4.4}$$

Subsequently, the value of the fund at interval end, after the ETBF makes investment over a series of races over consecutive days is calculated by Equation 4.5.

$$X_t = X_0\left[1+\frac{\delta_{R_{1,i,\ldots,j}}}{365}\right]\left[1+\frac{\delta_{R_{2,i,\ldots,j}}}{365}\right],\ldots,\left[1+\frac{\delta_{R_{t,i,\ldots,j}}}{365}\right] = X_0\left[\prod_{t=1}^{n}\left[1+\frac{\delta_{R_{t,i,\ldots,j}}}{365}\right]\right],$$

$$X_0\text{: initial investment; } X_t\text{: final wealth time } t. \tag{4.5}$$

Equation 4.6 converts from the force of interest to the required nominal rate of return per annum convertible pthly;

$$i^{(p)} = p\left[\left(1+\frac{\delta_{R_{1,i,\ldots,j}}}{365}\right),\ldots,\left(1+\frac{\delta_{R_{t,i,\ldots,j}}}{365}\right)-1\right],$$

$$i^{(p)} = p\left[\prod_{t=1}^{n}\left[1+\frac{\delta_{R_{t,i,\ldots,j}}}{365}\right]-1\right],$$

$$i^{(p)}\text{: nominal rate per annum convertible pthly; } \delta_{R_{t,i,\ldots,j}}\text{: force of interest day } t. \tag{4.6}$$

5. Pooling betting strategy

Alpha return in industry is defined as the excess investment return over a benchmark (King, 2007). Betting strategy alpha is invoked and assumed to be composed of model alpha and ETF alpha. An ETF may be trading at a premium to its benchmark even though the two financial models share identical model alpha (State Street Global Advisors, n.d.). α_R denotes betting strategy alpha to represent the excess return (per race) above the financial benchmark, from exchange traded betting fund with deterministic payoff betting strategy; $\alpha_R = \text{Payoff}_{\text{ETBF|DBM}} - \text{Payoff}_{\text{Benchmark}}$(excess return above benchmark).

The primary hypothesis statement tested at $\alpha = 5\%$ is $\mu_{\alpha_R} > 0$, an exchange traded betting fund with deterministic payoff as betting strategy can achieve abnormal returns above financial benchmarks. The sample mean betting strategy alpha is assumed to be normally distributed, and $\overline{X}_{\alpha_R} \sim N\left(\mu, \frac{\sigma^2}{n}\right)$ represents sample mean betting strategy payoff excess (per race) from the financial benchmark payoff.

The distribution for $\overline{X}_{\alpha_R}$ is derived from the sum of the two assumed normally distributed independent random variables $\overline{X}_{\alpha_{R_1}}$ and $\overline{X}_{\alpha_{R_2}}$. $\overline{X}_{\alpha_{R_1}} \sim N\left(\mu_1, \frac{\sigma^2_1}{n_1}\right)$ and $\overline{X}_{\alpha_{R_2}} \sim N\left(\mu_2, \frac{\sigma^2_2}{n_2}\right)$ denote sample means model alpha and fund alpha respectively. Essentially, nested within the primary hypothesis are two sub-hypotheses to be tested, H_{1a}: $\mu_{@R_1} > 0$ and H_{2a}: $\mu_{@R_2} > 0$; $\overline{X}_{@R_1}$ represents sample mean DBM payoff excess (per race) from an existing stochastic betting model benchmark, and $\overline{X}_{@R_2}$ represents the sample mean ETBF payoff excess (per race) from the ETF SPDR S&P/ASX 200 fund benchmark.

6. Sample betting strategy alpha hypothesis testing and results

A stratified sample of three-hundred consecutive galloper races, $\sum_{t=1}^{6} n_t = 300$, conveniently commencing from the start of the calendar year on 01 January 2012, from global racetracks of Australasia (Australia (173) and New Zealand (45)), Asia (Singapore (10) and Hong Kong (10)), South Africa (20) and the United Kingdom (42), provides the data for testing the performance of the DBM and ETBF race payoffs, against their respective benchmarks. The online totalizer market prices and race information were taken from the market-makers http://www.sportsbet.com.au/results/horse_racing (day one), and https://tatts.com/tattsbet (days two to six), with exposure predominantly in the Australasian region. Appropriately, horseraces from the Australasian region contributed seventy-three percent of test sample; and the Asian, African, and United Kingdom regions accounted for the remaining twenty-seven percent of the sample. A stratified random sampling technique is employed within these regions (strata) on the assumption that the random sample statistics generated will be a true reflection for the population parameters and is not subject to sampling error. A simple random sampling technique to select a sample size of fifty races from these regions was conducted daily for the six-day interval. The daily analysis of a randomly selected fifty races over six consecutive days simulates the continuous trading on global betting markets from the southern to the northern hemisphere; a premise for financial markets trading comparison.

Initially, regional daily and weekly average payoff (%) per race and accumulative payoff statistics, for "field win wagering", are calculated using the DBM and SBM benchmark techniques for the sampling period from 01 January 2012 to 06 January 2012 (Hopf, 2014). The workings of the DBM and SBM techniques are illustrated in Appendix A2. The regional weekly average payoff (%) per race statistics using the DBM $\left(\bar{X}_{DBM}\right)$ and SBM $\left(\bar{X}_{SBM}\right)$ techniques are provided in Table 1. The cross tabulation of Table 2 summarizes the regional daily and weekly accumulative payoffs reported from the DBM technique for sample. Paired payoff results are reported according to two approaches; (a) nil trade strategy where maximum exposure from an individual racer losing is 100%; and (b) trade strategy where the maximum risk of loss from any individual racer is capped at a maximum of 10%.

UK race series results for day one report \$1 investment from ETBF accumulated to (a) \$1.44 (nil trade) and (b) \$1.83 (trade strategy). UK race series accumulative results for the sampling period report \$1 investment from ETBF diminished to (a) $\$1.22 \times 10^{-4}$ (19.3% compounding loss per race from consecutive betting on the 42 race series). However, \$1 investment from ETBF accumulated to (b) \$4.81 (3.8% compounding profit per race from trading market odds). Similarly, day one's race series results display \$1 investment from ETBF diminished to (a) $\$3.91 \times 10^{-6}$ (22.05% compounding loss per race from consecutive betting on the 50 race series); in comparison, \$1 investment from ETBF accumulated to (b) \$18.65 (6.03% compounding profit per race from trading market odds using DBM).The sample statistics for DBM and no trade intervention disclosed a constant accumulative investment loss of $Я = -25.2\%$ from betting consecutively on three-hundred races, and average daily investment loss on each race of $\overline{Я} = -19\%$ calculated from the five regions and six-day time period. Essentially, results from trade intervention with global market-makers to maximize loss from a race to ten percent of the investment, reported a credible sample compounding race return of $Я = 3.05\%$; \$1 invested by ETBF accumulated to a prodigious \$8180 from consecutive betting on the three-hundred race sample over the six-day period using the DBM method with hedging; to legitimize the hypotheses testings conducted.

Sample payoff results from the DBM technique are benchmarked against payoff results recorded using the normal rank approximate method. Although dated, both Ali (1998) and Lo et al. (1995) tests showed that the normal rank-order model forecasted horserace rank probabilities more

Table 1. Model regional mean alpha (per race)—daily and weekly averages

Region	Day	$\bar{X}_{DBM}$	$\bar{X}_{SBM}$	$\bar{X}_{®_i} = \bar{X}_{DBM} - \bar{X}_{SBM}$	$\overline{X}_{®R_1}$	$s_{®R_1}$
United Kingdom	1	(a) 11.2; (b) 15.04	−16.82	(a) 28.02; (b) 31.86	(a) 2.98; (b) 19.60	(a) 31.06; (b) 33.81
South Africa		n/a	n/a	n/a		
Asia		(a) −4.78; (b) 6.97	37.52	(a) −42.3; (b) −30.55		
Australia		(a) −11.73; (b) 6.38	−27.37	(a) 15.64; (b) 33.75		
New Zealand		(a) −30.28; (b) 2.52	−40.83	(a) 10.55; (b) 43.35		
United Kingdom	2	(a) −17.01; (b) 4.32	−21.8	(a) 4.79; (b) 26.12	(a) 21; (b) 42.53	(a) 26.47; (b) 21.51
South Africa		(a) −2.55 9; (b) 8.61	−62.8	(a) 60.25; (b) 71.41		
Asia		n/a	n/a	n/a		
Australia		(a) −18.90; (b) 1.88	−24.25	(a) 5.35; (b) 26.13		
New Zealand		(a) −35.63; (b) −2.81	−49.25	(a) 13.62; (b) 46.44		
United Kingdom	3	(a) −26.4 9; (b) −0.8	62.16	(a) −88.56; (b) −62.96	(a) −19.84; (b) −8.6	(a) 47.15; (b) 37
South Africa		7.67	−8.65	16.32		
Asia		n/a	n/a	n/a		
Australia		(a) −2.53; (b) 6.7	−6.35	(a) 3.82; (b) 13.05		
New Zealand		(a) −5; (b) 5.13	5.92	(a) −10.92; (b) −0.79		
United Kingdom	4	(a) −17.16; (b) −1.72	−21.1	(a) 3.94; (b) 19.38	(a) 5.49; (b) 21.27	(a) 11.45; (b) 3.18
South Africa		(a) 2.87; (b) 8.41	−9.3	(a) 12.17; (b) 17.71		
Asia		5.68	−15.67	21.35		
Australia		(a) −29.95; (b) −0.43	−26.6	(a) −3.35; (b) 26.17		
New Zealand		(a) −28.76; (b) −0.36	−22.1	(a) −6.66; (b) 21.74		
United Kingdom	5	(a) −8.77; (b) 5.02	29	(a) −37.77; (b) −23.98	(a) 2.89; (b) 12.85	(a) 32.68; (b) 33.97
South Africa		10.77	16.6	−5.83		
Asia		n/a	n/a	n/a		
Australia		(a) −9.84; (b) 4.15	−26.1	(a) 16.26; (b) 30.25		
New Zealand		(a) −8.23; (b) 3.85	−47.12	(a) 38.89; (b) 50.97		
United Kingdom	6	(a) −7.26; (b) 3.59	−21.93	(a) 14.67; (b) 25.52	(a) 5.79; (b) 21.64	(a) 6.41; (b) 16.25
South Africa		n/a	n/a	n/a		
Asia		(a) −37.66; (b) −3.06	−41.52	(a) 3.86; (b) 38.46		
Australia		(a) −17.01; (b) 0.93	−22.18	(a) 5.17; (b) 23.11		
New Zealand		5.01	5.56	−0.55		
Regional averages	*T = 6 days*	*(a) −10.9; (b) 4.24*	*2.1*	*(a) −12.5; (b) 2.66*	*(a) 3; (b) 18*	*(a) 13; (b) 11*
		(a) 4.69; (b) 8.87	*−16*	*(a) 20.69; (b) 24.87*		
		(a) −12.25; (b) 3.2	*−6.56*	*(a) −5.69; (b) 9.76*		
		(a) −15; (b) 3.27	*−22.14*	*(a) 7.14; (b) 25.41*		
		(a) −17.15; (b) 2.22	*−24.64*	*(a) 7.49; (b) 26.86*		

(a) Maximum risk 100%, (b) Maximum risk 10%.

accurately than the gamma rank or L-decomposable models. Uniform unit bets on the field for the normal rank approximate model resulted in the net expected win return per \$1 unit bet from a winning racer equaling the negative of the tote track take (t); and the net expected field win payoff is $E(\text{payoff}) = -(n + t - 1)$; n:field size, t:track take.

The sample statistics using the SBM technique in Table 3 recorded a constant, compounding race loss of $Я = 37.4\%$ from consecutive wagering on the three-hundred race sample that equates to an

Table 2. DBM accumulative payoff cross tabulation

Region	Day 1	Day 2	Day 3	Day 4	Day 5	Day 6	$\prod_{t=1}^{6}(1+Я_t)$	Я (%)	$\bar{Я}$ (%)
United Kingdom, $n = 42$	(a) 1.44; (b) 1.83	(a) 0.031; (b) 1.42	(a) 0.125; (b) 0.93	(a) 0.289; (b) 0.905	(a) 0.129; (b) 1.936	(a) 0.588; (b) 1.135	(a) 1.22×10^{-4}; (b) 4.81	(a) −19.3; (b) 3.8	(a) −17.47; (b) 3.67
South Africa, $n = 20$		(a) 0.671; (b) 1.41	1.796	(a) 0.530; (b) 1.421	1.224		(a) 0.782; (b) 4.40	(a) −1.2; (b) 7.7	(a) −0.34; (b) 8.15
Asia, $n = 20$	(a) 0.215; (b) 1.87			1.314		(a) 0.03; (b) 0.84	(a) 3.48×10^{-3}; (b) 2.064	(a) −21.2; (b) 3.7	(a) −19.68; (b) 2.88
Australia, $n = 173$	(a) 6.65×10^{-5}; (b) 5.27	(a) 3.15×10^{-6}; (b) 1.542	(a) 0.126; (b) 3.794	(a) 3.024×10^{-7}; (b) 0.753	(a) 1.17×10^{-3}; (b) 2.876	(a) 1.35×10^{-5}; (b) 1.183	(a) 1.26×10^{-25}; (b) 78.628	(a) −88; (b) 2.56	(a) −27.37; (b) 2.64
New Zealand, $n = 45$	(a) 0.19; (b) 1.033	(a) 0.038; (b) 0.85	(a) 0.0943; (b) 1.916	(a) 3.6×10^{-4}; (b) 0.926	(a) 0.392; (b) 1.193	1.275	(a) 1.225×10^{-7}; (b) 2.37	(a) −88.9; (b) 1.9	(a) −27.59; (b) 1.71
$\prod_{t=1}^{5}(1+Я_t)$	(a) 3.91×10^{-6}; (b) 18.65	(a) 2.49×10^{-9}; (b) 2.624	(a) 2.67×10^{-3}; (b) 12.14	(a) 2.19×10^{-11}; (b) 1.178	(a) 7.24×10^{-5}; (b) 8.13	(a) 3.036×10^{-7}; (b) 1.438	(a) 1.25×10^{-38}; (b) 8180	(a) −25.2; (b) 3.05	
Я (%)	(a) −22.05; (b) 6.03	(a) −32.71; (b) 1.95	(a) −11.18; (b) 5.12	(a) −38.79; (b) 0.33	(a) −17.36; (b) 4.28	(a) −25.93; (b) 0.73	(a) −25.2; (b) 3.05		
$\bar{Я}$ (%)	(a) −18.8; (b) 6.43	(a) −29.87; (b) 2.23	(a) −12.5; (b) 4.21	(a) −24.92; (b) 1.81	(a) −10.3; (b) 5.75	(a) −21.16; (b) 1.31			(a) −19; (b) 4

(a) Maximum risk 100% (b) Maximum risk 10%.

Table 3. SBM accumulative payoff cross tabulation

Region	Day 1	Day 2	Day 3	Day 4	Day 5	Day 6	$\prod_{t=1}^{6}(1+Я_t)$	$Я(\%)$	$\overline{Я}(\%)$
United Kingdom, $n = 42$	0.219	1.92×10^{-3}	0.388	0.023	0.016	0.12	7.20×10^{-9}	−36	−35.22
South Africa, $n = 20$		0.0049	0.198	0.282	0.882		2.41×10^{-4}	−34	−28.07
Asia, $n = 20$	0.19			0.292		0.012	6.66×10^{-4}	−30.6	−31.94
Australia, $n = 173$	1.365×10^{-8}	1.72×10^{-7}	3×10^{-4}	2.83×10^{-7}	4.21×10^{-7}	3.6×10^{-7}	3.02×10^{-38}	−39.3	−39.2
New Zealand, $n = 45$	0.084	0.019	0.023	2088×10^{-3}	0.038	0.959	2.79×10^{-9}	−35.4	−37.04
$\prod_{t=1}^{5}(1+Я_t)$	4.77×10^{-11}	3.07×10^{-14}	5.3×10^{-7}	1.12×10^{-12}	2.26×10^{-10}	4.97×10^{-10}	9.75×10^{-62}	−37.4	
$Я(\%)$	−37.83	−46.33	−25.1	−36.76	−35.87	−34.85	−37.4	-	-
$\overline{Я}(\%)$	−35.23	−51.81	−22.17	−37.1	−30.01	−33.62	-	-	−35

Table 4. Fund alpha (per race)—daily and week average

Results	Day 1	Day 2	Day 3	Day 4	Day 5	Day 6	Average
$Payoff\%_{ETBF}[Я(\%)]$	(a) −22.05; (b) 6.03	(a) −32.71; (b) 1.95	(a) −11.18; (b) 5.12	(a) −38.79; (b) 0.33	(a) −17.36; (b) 4.28	(a) −25.93; (b) 0.73	(a) −24.67; (b) 3.07
$Payoff\ \%_{SPDRETF}$	0.14	0.23	−0.12	−0.10	3.33×10^{-3}	0.136	0.048
$@R_2$	(b) 5.89	(b) 1.72	(b) 5.24	(b) 0.43	(b) 4.277	(b) 0.594	(b) 3.025

average daily regional loss per race from total sample of $\overline{Я} = -35\%$. Both the daily and regional race series recorded consecutive race losses of $Я < -20.0\%$.

The regional weekly average payoff (%) per race and accumulative payoff per race statistics provide the data to determine the alpha central tendency and dispersion measures. Table 1 displays the daily and weekly mean model alpha (per race) measures, $X_{@R_1}$ and $^{@R_1}$, for the regions; and Table 4 displays the daily and average weekly fund alpha race returns, $\overline{X}_{@R_2}$ and $_{@R_2}$. The daily returns for the

Table 5. Betting strategy sample statistics

Regional results	Day 1	Day 2	Day 3	Day 4	Day 5	Day 6	Average
$\overline{X}_{@R_1}$	19.60	42.53	-8.6	21.27	12.85	21.64	18
$\overline{X}_{@R_2}$	5.89	1.72	5.24	0.43	4.277	0.594	3.025
$\overline{X}_{@R} = \overline{X}_{@R_1} + \overline{X}_{@R_2}$	25.5	44.25	-3.36	21.7	17.127	22.234	21.025
$s_{ЯR_1}$	33.81	21.51	37	3.18	33.97	16.25	16.5
$s_{ЯR_2}$	-	-	-	-	-	-	2.4
$\frac{\bar{X}_{ЯR}}{s_{ЯR}}$	-	-	-	-	-	-	3.09
$\frac{s_{ЯR}}{\bar{X}_{ЯR}}$	-	-	-	-	-	-	0.32

SPDR S&P/ASX 200 fund benchmark are tabled in Appendix A3. The time interval between successive online races is forty minutes, and appropriately the daily returns reported during sampling period for the SPDR S&P/ASX 200 fund, which trades daily for six hours on the ASX, are converted into equivalent compounding return per forty minutes, to match the ETBF compounding race payoff.

Collectively, these betting strategy sample statistics (using trade hedging) are provided in Table 5 for testing of the secondary sub-hypotheses and primary hypothesis statements. From Table 5 credible Sharpe and coefficient of variation measures were recorded from sample of $\frac{\overline{X}_{@R}}{s_{@R}} = 3.09$ and $\frac{s_{@R}}{\overline{X}_{@R}} = 0.32$, respectively, for asset benchmark comparison.

The secondary hypotheses and primary hypothesis tests and outcomes conducted at ninety-five percent level of confidence are detailed accordingly;

$H_{10}: \mu_{@R_1} = 0,$

$H_{1a}: \mu_{@R_1} > 0,$

$$t = \frac{\overline{X}_{@R_1} - \mu_{@R_1}}{\frac{s_{@R_1}}{\sqrt{n_1}}} = \frac{18}{16.5/\sqrt{6}} = 2.67\,;\;\; n_1 = 6, df = 5, \alpha = 0.05\,,$$

$t_{5,0.05} = 2.015 < 2.67,$

∴ Rejection of H_{10}, and acceptance of alternative sub-hypothesis at $\alpha = 0.05$ that the DBM achieves abnormal profits from the normal rank approximate model benchmark.

$H_{20}: \mu_{@R_2} = 0,$

$H_{2a}: \mu_{@R_2} > 0,$

$$t = \frac{\overline{X}_{@R_2} - \mu_{@R_2}}{\frac{s_{@R_2}}{\sqrt{n_2}}} = \frac{3.025}{2.4/\sqrt{6}} = 3.09\,;\;\; n_2 = 6, df = 5, \alpha = 0.05\,,$$

$t_{5,0.05} = 2.015 < 3.09,$

∴ Rejection of H_{20} based upon data sample and acceptance of alternative sub-hypothesis at $\alpha = 0.05$ that the ETBF outperforms the SPDR S&P/ASX 200 fund; statistical evidence for a profitable ETBF product for the risk averse investor.

$H_0: \mu_{@R} = 0,$

$H_a: \mu_{@R} > 0,$

$$t = \frac{\left(\overline{X}_{@R_1} + \overline{X}_{@R_2}\right) - \left(\mu_{@R_1} + \mu_{@R_2}\right)}{\sqrt{\frac{s^2_{@R_1}}{n_1} + \frac{s^2_{@R_2}}{n_2}}} = \frac{21.025}{\sqrt{\frac{16.5^2}{6} + \frac{2.4^2}{6}}} = 3.09\,,$$

$t_{5,0.05} = 2.015 < 3.09,$

∴ The conclusion is rejection of H_0 at level of significance $\alpha = 0.05$ and statistical acceptance for the research primary hypothesis, and endorsement of an exchange-traded horserace betting fund with deterministic payoff for institutional investment consideration.

7. Conclusion and implication

A deterministic betting model for pool wagering is developed from the existing stochastic literature, mathematical pooling principles, and from the development of the MSO theorem over C^n space, to validate the optimal solution generated from the DBM.

The MSO theorem states that optimization over an *n* finite series of complex systems generates a constant real component over each consecutive system. The MSO theorem has application for all science, and one application is finding optimal solution for deterministic modeling of pool betting. The DBM applied to horse racing optimizes field wagering to determine a feasible, actual payoff—risk trade-off pre-race (or within race). Arbitrage for ETBF is achievable from DBM and market odds trading processes. The structure of the ETBF is a savings fund generating a continuous rate of return.

Testing of betting strategy alpha from race sample invokes a profitable investment strategy of acceptable return—risk for institutional investment, to challenge the existing semi-strong EMH toward horserace betting, and implication for positive benefit to financial markets.

Funding

The authors received no direct funding for this research.

Author details

Craig George Leslie Hopf[1]

E-mail: cglhopf@yahoo.com

Gurudeo Anand Tularam[1]

E-mail: a.tularam@griffith.edu.au

ORCID ID: http://orcid.org/0000-0002-7015-8589

[1] Mathematics and Statistics, Griffith Sciences (ENV), Environmental Futures Research Institute (EFRI), Griffith University, Brisbane, Australia.

References

Ali, M. M. (1998). Probability models on horse-race outcomes. *Journal of Applied Statistics, 25*, 221–229. http://dx.doi.org/10.1080/02664769823205

Bank for International Settlements. (2007). *Institutional investors, global savings and asset allocation.* Retrieved January 6, 2013, from http://www.bis.org/publ/cgfs27.pdf

Critchlow, D. E., Fligner, M. A., & Verducci, J. S. (1991). Probability models on rankings. *Journal of Mathematical Psychology, 35*, 294–318. http://dx.doi.org/10.1016/0022-2496(91)90050-4

Dansie, B. R. (1983). A note on permutation probabilities. *Journal of the Royal Statistical Society. Series B (Methodological), 45*, 22–24.

D'Elia, A. (2003). Modelling ranks using the inverse hypergeometric distribution. *Statistical Modelling, 3*, 65–78. http://dx.doi.org/10.1191/1471082X03st047oa

Edelman, D. (2007). Adapting support vector machine methods for horserace odds prediction. *Annals of Operations Research, 151*, 325–336. http://dx.doi.org/10.1007/s10479-006-0131-7

Harville, D. A. (1973). Assigning probabilities to the outcomes of multi-entry competitions. *Journal of the American Statistical Association, 68*, 312–316. http://dx.doi.org/10.1080/01621459.1973.10482425

Henery, R. J. (1983). Permutation probabilities for gamma random variables. *Journal of Applied Probability, 20*, 822–834. http://dx.doi.org/10.2307/3213593

Hopf, C. G. L. (2014). *Exchange traded horserace betting fund with deterministic payoff—a mathematical analysis of a profitable deterministic horserace betting model* (Unpublished master's thesis, Sup. Tularam, G. A). Griffith University, Brisbane.

Hopf, C. G. L., & Tularam, G. A. (2014). A mathematical analysis of the inclusion of institutional betting funds into stock market: The case of technical and fundamental payoff models in horserace betting. *Journal of Mathematics and Statistics, 10*, 390–400. http://dx.doi.org/10.3844/jmssp.2014.390.400

King, D. (2007). Portfolio optimisation and diversification. *Journal of Asset Management, 8*, 296–307. http://dx.doi.org/10.1057/palgrave.jam.2250082

Koning, R. H., & Van Velzen, B. (2009). Betting exchanges: The future of sports betting? *International Journal of Sport Finance, 4*, 42–62.

Lo, V. S., Bacon-Shone, J., & Busche, K. (1995). The application of ranking probability models to racetrack betting. *Management Science, 41*, 1048–1059. http://dx.doi.org/10.1287/mnsc.41.6.1048

Luce, R. D. (1959). *Individual choice behaviour*. New York, NY: Wiley.

Plackett, R. L. (1975). The analysis of permutations. *Journal of the Royal Statistical Society Series C, 24*, 193–202.

Roca, E., Wong, V. S. H., & Tularam, G. A. (2010). Are socially responsible investment markets worldwide integrated? *Accounting Research Journal, 23*, 281–301.

Snyder, W. W. (1978). Horse racing: Testing the efficient markets model. *The Journal of Finance, 33*, 1109–1118. http://dx.doi.org/10.1111/j.1540-6261.1978.tb02051.x

State Street Global Advisors. (n.d.). *All SPDR ETFS.* Retrieved December 1, 2013, from http://www.spdr.com.au/viewall/index.html

Stern, H. (1990). Models for distributions on permutations. *Journal of the American Statistical Association, 85*, 558–564. http://dx.doi.org/10.1080/01621459.1990.10476235

Tularam, G. A., Roca, E., & Wong, V. S. H. (2010). Investigation of socially responsible investment markets (SRI) using dynamic conditional correlation (DCC) method: Implications for diversification. *Journal of Mathematics and Statistics, 6*, 385–394.

Yang, X. S. (2008). *Mathematical modelling for earth sciences.* Edinburgh: Dunedin Academic Press. Retrieved from http://www.dunedinacademicpress.co.uk/display.asp?ISB=9781903765920&DS=Mathematical%20Modelling%20for%20Earth%20Sciences

Appendix A1. Multiple System Optimization Theorem

Multiple System Optimization (MSO) Theorem, Proof by Induction,

$$\{\max/\min\}Z_{1,\dots,n}(C^n_{1,\dots,n};\odot) = Я^n;Z(C^n;\odot),Z_{1,\dots,2}(C^n_1C^n_2;\odot)\{n \in I\}, \tag{A1}$$

$$\{\max/\min\}Z(C^n;\odot) = Re(\hat{Z}) = Я, \{n = 1\},$$

$$Z(C_1C_2,\dots,C_n)\text{: complex function,} \tag{A2}$$

$$C^n = (C_1C_2,\dots,C_n), C^n\text{: complex vector input; } \odot\text{ : set of operators,}$$

$$\hat{Z}(\hat{C}^n)\text{: complex output; } Re(\hat{Z}) = Я = \text{constant,}$$

Maximizing or minimizing on the space C^n of n-tuples of complex numbers must satisfy the n-dimensional Cauchy–Riemann equations in order for the complex function to be complex differentiable.

$$Z(C^n;\odot) = Z(C_1,C_2,\dots,C_n;\odot)|C_j = (x_j,y_j), j \in \{1,\dots,n\}, \tag{A3}$$

$$\frac{dZ}{dC} = \left[\frac{dZ}{dC_1},\frac{dZ}{dC_2},\dots,\frac{dZ}{dC_n}\right], \tag{A4}$$

$$\frac{dZ}{dC} = \left[\frac{\partial Z}{\partial x_1}\frac{\partial x_1}{\partial C_1},\frac{\partial Z}{\partial x_2}\frac{\partial x_2}{\partial C_2},\dots,\frac{\partial Z}{\partial x_n}\frac{\partial x_n}{\partial C_n}\right] + \left[\frac{\partial Z}{\partial y_1}\frac{\partial y_1}{\partial C_1},\frac{\partial Z}{\partial y_2}\frac{\partial y_2}{\partial C_2},\dots,\frac{\partial Z}{\partial y_n}\frac{\partial y_n}{\partial C_n}\right], \tag{A5}$$

$$\frac{dZ}{dC} = \frac{1}{2}\left[\frac{\partial Z}{\partial x_1} - i\frac{\partial Z}{\partial y_1},\frac{\partial Z}{\partial x_2} - i\frac{\partial Z}{\partial y_2},\dots,\frac{\partial Z}{\partial x_n} - i\frac{\partial Z}{\partial y_n}\right], \tag{A6}$$

$$\frac{\partial Z}{\partial x_1} = \frac{\partial Z}{\partial x_2} =,\dots,= \frac{\partial Z}{\partial x_n} = 0\{\text{Cauchy} - \text{Riemann}\}, \tag{A7}$$

$$Re(\hat{Z}) = Я = constant \tag{A8}$$

Mathematical Induction step,

$$Z_{1,\dots,k}(C^n_1C^n_2,\dots,C^n_k;\odot) = Я^k = \{Z_1(C^n;\odot)\}^k\{n = k\}, \tag{A 9}$$

$$Z_{1,\dots,k+1} = Z_{1,\dots,k}(C^n_1C^n_2,\dots,C^n_k;\odot)Z_1(C^n;\odot), = \{Z_1(C^n;\odot)\}^kZ(C^n;\odot) = Я^kЯ = Я^{k+1} = Z_{1,\dots,k+1}(C^n_1C^n_2,\dots,C^n_{k+1};\odot)\{\text{QED}\}. \tag{A10}$$

Appendix A2. DBM Workings

TrackInvest© Model – Field Win & Place Bet Payoff

Race Result: $R(R_1R_2R_3) = (1234)$,

Win Payoff: $Z(R) = Я = 30.30\%$,

(b_i) Place Payoff: $Z(R_1 \vee R_2 \vee R_3) = Я = 5.9\%$

Racer (R_i)	Market odds (O_i)	Bet (b_i)	Win $R(b_i)$	Net return $R_{net}(b_i)$	Payoff % $R_{net}(b_i)$	Racer (R_i)	Market odds (O_i)	Bet (b_i)	Place $R_{ijk}(b_i)$	Net return (max, min)	Payoff (max, min)
1	17.8	\$12000	\$213600	\$48600	(29.45%, 0)	1	3.4	\$21000	\$71400	\$13600, −\$43600	6.6%, −21.3%
2	20.8	\$10000	\$208000	\$43000	(26.06%, 0)	2	3.6	\$20000	\$72000	\$14200, −\$43000	6.9%, −21%
3	4.9	\$45000	\$220500	\$55500	(33.64%, 0)	3	1.6	\$45000	\$72000	\$14200, −\$43000	6.9%, −21%
4	16.6	\$13000	\$215800	\$50800	(30.79%, 0)	4	7.3	\$10000	\$73000	\$14200, −\$42000	6.9%, −20.5%
5	17.4	\$12000	\$208800	\$43800	(26.55%, 0)	5	5.3	\$14000	\$74200	\$14200, −\$40800	6.9%, −19.9%
6	*0*					*6*	*0*				
7	*0*					*7*	*0*				
8	4.7	\$47000	\$220900	\$55900	(33.88%, 0)	8	1.9	\$37000	\$70300	\$12500, −\$44700	6.1%, −21.8%
9	2.4	\$0	\$0	−\$165000	(0%, −100%)	9	1	\$20000	\$20000	−\$37800, −\$44700	−18.4%, −21.8%
10	13.5	\$16000	\$216000	\$51000	(30.91%, 0)	10	3.5	\$20000	\$70000	\$12200, −\$44700	6%, −21.8%
11	*0*					*11*	*0*				
12	21.5	\$10000	\$215000	\$50000	(30.30%, 0)	12	4	\$18000	\$72000	\$14200, −\$43000	6.9%, −21%

Source: 24/02/2012 Riccarton ZS1 (https://tatts.com/racing/racingsearch.aspx).

laximize $\text{Payoff}_{(n)}(R; O^n, b^n) = |b \cdot \mathbf{O}| - \sum^{n} b_i = \sum^{n} |b_i O_i| - \sum^{n} b_i$

$$|O_i b_i| = \left|O_j b_j\right| \forall i \neq j;\ \mathbf{O} = O^n \in R^n, O_i \geq 0, O_i \in R;\ \mathbf{b} = b^n \in C^n, b_i = (x_i, y_i),$$

$$\mathrm{R}(b_i) = |b_i O_i|;\ R_{\text{net}}(b_i) = \left|b_i O_i\right| - \sum_{i=1}^{n} b_i;\ \%R_{\text{net}}(b_i) = \left(\frac{\mathrm{Re}(R_{\text{net}}(b_i))}{\mathrm{Re}(\sum_{i=1}^{n} b_i)}, \frac{\mathrm{Im}(R_{\text{net}}(b_i))}{\mathrm{Im}(\sum_{i=1}^{n} b_i)}\right) \times 100\%.$$

The payoff function is represented by a complex couple that separates field net payoff from individual racer loss accordingly;

$$\text{Payoff}_{(n=9)}(R; O^n, b^n) = |b \cdot \mathbf{O}| - \sum_{i=1}^{9} b_i = \sum_{i=1}^{9} |b_i O_i| - \sum_{i=1}^{9} b_i,$$

$$\mathrm{R}(b_{12}) = |b_{12} \times O_{12}| = \$10000 \times 21.5 = \$215000,$$

$$\mathrm{R}_{\text{net}}(b_{12}) = |b_{12} O_{12}| - \sum_{i=1}^{n} b_i = \$215000 - \left(\$;165000, \$89583\tfrac{1}{3}\right) = \left(\$50000, -\$89583\tfrac{1}{3}\right),$$

$$\%\mathrm{R}_{\text{net}}(b_{12}) = \left(\frac{\mathrm{Re}(\mathrm{R}_{\text{net}}(b_i))}{\mathrm{Re}(\sum_{i=1}^{n} b_i)}, \frac{\mathrm{Im}(\mathrm{R}_{\text{net}}(b_i))}{\mathrm{Im}(\sum_{i=1}^{n} b_i)}\right) \times 100\%,$$

$$\%\mathrm{R}_{\text{net}}(b_{12}) = \left(\frac{50000}{165000}, \frac{-89583\frac{1}{3}}{89583\frac{1}{3}}\right) \times 100\% = (30.30\%, -100\%),$$

$$|O_{12}b_{12}| = |O_9 b_9|; \left|21.5 \times (10000, 0)\right| = |2.4 \times b_9|,$$

$$b_9 = \frac{\$215000}{2.4} = \$89583\frac{1}{3}; b_9 \in C = (x, y) = \left(0, 89583\frac{1}{3}\right),$$

$$\mathrm{R}_{\mathrm{net}}(b_9) = |b_9 O_9| - \sum_{i=1}^{9} b_i = |(2.4) \times \left(0, 89583\frac{1}{3}\right)| - \left(165000, 89583\frac{1}{3}\right),$$

$$\mathrm{R}_{\mathrm{net}}(b_9) = \left|(0, 215000)\right| - \left(165000 + 89583\frac{1}{3}i\right),$$

$$\mathrm{R}_{\mathrm{net}}(b_9) = \left(50000, -89583\frac{1}{3}\right); \%\mathrm{R}_{\mathrm{net}}(b_9) = (30.30\%, -100\%).$$

The feasible payoff solution is 30.30% (with nil trade and 100% exposure to shortest priced favorite winning). Trading the market odds reduces the racer exposure (such as hedging to 10% loss), and "arbitrage" optimal payoff solution of 3% is achievable from the bet taking win market odds of 4.7 for the shortest priced racer.

The payoff (%) calculation from SBM is accordingly; Payoff $= \frac{\$21.5-\$9}{\$9} \times 100\% \approx 140\%$.

Appendix A3. ETF and ETBF Daily Payoffs

Date	*STW*		*ETBF*	
	Adj close	**Return (%)**	**Day**	**Return (%)**
10/01/2012	36.47	1.22	6	43.8
09/01/2012	36.03	0.03	5	713
6/01/2012	36.02	−0.91	4	17.8
5/01/2012	36.35	−1.09	3	1114
04/01/2012	36.75	2.08	2	162.4
03/01/2012	36.00	1.24	1	1765
30/12/2011	35.56			

Source: http://au.finance.yahoo.com/q/hp?s; STW.AX: SPDR S&P/ASX 200 Fund; adj Close: close price adjusted for dividends and splits.

2

On analytical and numerical study of implicit fixed point iterations

Renu Chugh[1], Preety Malik[1] and Vivek Kumar[2*]

*Corresponding author: Vivek Kumar, Department of Mathematics, KLP College, Rewari, India.
E-mail: ratheevivek15@yahoo.com

Reviewing editor: Kok Lay Teo, Curtin University, Australia

Abstract: In this article, we define a new three-step implicit iteration and study its strong convergence, stability and data dependence. It is shown that the new three-step iteration has better rate of convergence than implicit and explicit Mann iterations as well as implicit Ishikawa-type iteration. Numerical example in support of validity of our results is provided.

Subjects: Science; Mathematics & Statistics; Applied Mathematics

Keywords: implicit iterations; strong convergence; convergence rate; stability; data dependence

AMS subject classifications: 47H10; 47H17; 47H15

1. Introduction

Implicit iterations are of great importance from numerical standpoint as they provide accurate approximation compared to explicit iterations. Computer-oriented programmes for the approximation of fixed point by using implicit iterations can reduce the computational cost of the fixed-point problem. Numerous papers have been published on convergence of explicit as well as implicit iterations in various spaces (Anh & Binh, 2004; Berinde, 2004; Berinde, 2011; Chidume & Shahzad, 2005; Chugh & Kumar, 2013; Ciric, Rafiq, Cakić, & Ume, 2009; Ćirić, Rafiq, Radenović, Rajović, & Ume, 2008; Khan, Fukhar-ud-din, & Khan, 2012; Rhoades, 1993; Shahzad & Zegeye, 2009). Data dependence of fixed points is a related and new issue which has been studied by many authors; see (Gursoy,

ABOUT THE AUTHORS

Renu Chugh is a professor in the Department of Mathematics, Maharshi Dayanand University, Rohtak, Haryana. There she teaches Functional Analysis, Topology and other topics subjects in postgraduate level. So far she has supervised 18 PhD students and 28 MPhil students. Her research interest focuses on nonlinear analysis and fuzzy mathematics. She has published her research contributions in some national and international journals.

Preety Malik is persuing her PhD under the supervision of Prof. Renu Chugh as a research scholar from MDU, Rohtak. Also, she has published her research papers in some national and international journals.

Vivek Kumar is an assistant professor in Department of Mathematics, KLP College, Rewari, where he teaches in undergraduate level. He has completed his PhD from MDU, Rohtak, Haryana. He has published many research papers in international journals.

PUBLIC INTEREST STATEMENT

In this paper, a new three-step implicit iteration is defined as:

$$x_n = W(x_{n-1}, Ty_n, \alpha_n)$$
$$y_n = W(z_n, Tz_n, \beta_n)$$
$$z_n = W(x_n, Tx_n, \gamma_n)$$

where α_n, β_n and γ_n are sequences in [0, 1].

For this three-step implicit iteration, strong convergence and stability results are proved in convex metric spaces. Also, we have done the comparison of rate of convergence of newly defined iteration with implicit and explicit Mann, and implicit and explicit Ishikawa-type iterations analytically and numerically. It is found that our newly defined implicit iteration has better rate of convergence than these iterations. Also, data-dependence result for new implicit iteration is proved in hyperbolic spaces.

Karakaya, & Rhoades, 2013; Khan, Kumar, & Hussain, 2014 and references therein). In computational mathematics, it is of theoretical and practical importance to compare the convergence rate of iterations and to find out, if possible, which one of them converges more rapidly to the fixed point. Recent works in this direction are (Chugh & Kumar, 2013; Ciric, Lee, & Rafiq, 2010; Hussian, Chugh, Kumar, & Rafiq, 2012; Khan et al., 2014; Kumar et al., 2013). Motivated by the works of Ciric (Ciric, 1971; Ciric, 1974; Ciric, 1977; Ciric et al., 2010; Ćirić & Nikolić, 2008a, 2008b; Ciric et al., 2009; Ćirić et al., 2008; Ciric, Ume, & Khan, 2003) and the fact that three-step iterations give better approximation than one-step and two-step iterations (Glowinski & Tallec, 1989), we define a new and more general three-step implicit iteration with higher convergence rate as compared to implicit Mann, explicit Mann and implicit Ishikawa iterations.

Let K be a nonempty convex subset of a convex metric space X and $T{:}K \to K$ be a given mapping. Then for $x_0 \in K$, we define the following implicit iteration:

$$\begin{aligned} x_n &= W(x_{n-1}, Ty_n, \alpha_n) \\ y_n &= W(z_n, Tz_n, \beta_n) \\ z_n &= W(x_n, Tx_n, \gamma_n) \end{aligned} \tag{1.1}$$

where $\{\alpha_n\}$, $\{\beta_n\}$;$\{\gamma_n\}$ are sequences in [0, 1].

Equivalence form of iteration (1.1) in linear space can be written as

$$\begin{aligned} x_n &= \alpha_n x_{n-1} + (1-\alpha_n)Ty_n \\ y_n &= \beta_n z_n + (1-\beta_n)Tz_n \\ z_n &= \gamma_n x_n + (1-\gamma_n)Tx_n \end{aligned} \tag{IN}$$

Putting $\gamma_n = 1$ in (IN), we get Ishikawa-type implicit iteration:

$$\begin{aligned} x_n &= \alpha_n x_{n-1} + (1-\alpha_n)Ty_n \\ y_n &= \beta_n x_n + (1-\beta_n)Tx_n \end{aligned} \tag{II}$$

Putting $\gamma_n = \beta_n = 1$ in (IN), we get well-known implicit Mann iteration (Ćirić et al., 2008; Ciric et al., 2003):

$$x_n = W(x_{n-1}, Tx_n, \alpha_n) = \alpha_n x_{n-1} + (1-\alpha_n)Tx_n \tag{IM}$$

Also, Mann iteration (Mann, 1953) is defined as :

$$x_{n+1} = (1-\alpha_n)x_n + \alpha_n Tx_n \tag{M}$$

Zamfirescu operators (Zamfirescu, 1972) are most general contractive-like operators which have been studied by several authors, satisfying the following condition: for each pair of points x, y in X, at least one of the following is true:

$$\begin{aligned} &\text{(i)} \quad d(Tx,Ty) \le p\,d(x,y) \\ &\text{(ii)} \quad d(Tx,Ty) \le q[d(x,Tx) + d(y,Ty)] \\ &\text{(iii)} \quad d(Tx,Ty) \le r[d(x,Ty) + d(y,Tx)] \end{aligned} \tag{1.2}$$

where p, q, r are nonnegative constants satisfying $0 \le p \le 1$, $0 \le q, r \le \frac{1}{2}$.

Z-operators are equivalent to the following contractive contraction:

$$d(Tx,Ty) \le c\max\{d(x,y), \{d(x,Tx) + d(y,Ty)\}/2, \{d(x,Ty) + d(y,Tx)\}/2\} \quad \forall x,y \in X, 0 < c < 1 \tag{1.3}$$

The contractive condition (1.3) implies

$$d(Tx, Ty) \leq 2ad(x, Tx) + ad(x, y), \forall x, y \in X \tag{1.4}$$

where $a = \max\left\{c, \frac{c}{2-c}\right\}$ (see Berinde, 2004).

Rhoades (1993) used the following more general contractive condition than (1.4): there exists $c \in [0, 1)$, such that

$$d(Tx, Ty) \leq c\max\left\{d(x, y), \{d(x, Tx) + d(y, Ty)\}/2, d(x, Ty), d(y, Tx)\right\} \quad \forall x, y \in X \tag{1.5}$$

Osilike (1995) used a more general contractive definition than those of Rhoades': there exists $a \in (0, 1)$, $L \geq 0$, such that

$$d(Tx, Ty) \leq Ld(x, Tx) + ad(x, y) \forall x, y \in X \tag{1.6}$$

We use the contractive condition due to Imoru and Olatinwo (Olatinwo & Imoru, 2008), which is more general than (1.6): there exists $a \in [0, 1)$ and a monotone-increasing function φ: $R^+ \rightarrow R^+$ with $\phi(0) = 0$, such that

$$d(Tx, Ty) \leq \phi(d(x, Tx)) + ad(x, y), \quad a \in [0, 1), \forall x, y \in X \tag{1.7}$$

Also, we use the following definitions and lemmas to achieve our main results.

Definition 1.1 (Takahashi, 1970) A map W:$X^2 \times [0, 1] \rightarrow X$ is a convex structure on X if

$$d(u, W(x, y, \lambda)) \leq \lambda d(u, x) + (1 - \lambda)d(u, y)$$

for all $x, y, u \in X$ and $\lambda \in [0, 1]$. A metric space (X, d) together with a convex structure W is known as convex metric space and denoted by (X, d, W). A nonempty subset C of a convex metric space is convex if $W(x, y, \lambda) \in C$ for all $x, y \in C$ and $\lambda \in [0, 1]$.

All normed spaces and their subsets are the examples of convex metric spaces. But there are many examples of convex metric spaces which are not embedded in any normed space (see Takahashi, 1970). Several authors extended this concept in many ways later, one such convex structure is hyperbolic space introduced by Kohlenbach (2004) as follows:

Definition 1.2 (Kohlenbach, 2004) A hyperbolic space (X, d, W) is a metric space (X, d) together with a convexity mapping W: $X^2 \times [0, 1] \rightarrow X$ satisfying

(W1) $d(z, W(x, y, \lambda)) \leq (1 - \lambda)\, d(z, x) + \lambda\, d(z, y)$

(W2) $d(W(x, y, \lambda_1), W(x, y, \lambda_2)) = |\lambda_1 - \lambda_2|\, d(x, y)$

(W3) $W(x, y, \lambda) = W(y, x, 1 - \lambda)$

(W4) $d(W(x, z, \lambda), W(y, w, \lambda)) \leq (1 - \lambda)d(x, y) + \lambda d(z, w)$ for all $x, y, z, w \in X$ and $\lambda, \lambda_1, \lambda_2 \in [0, 1]$.

Evidently, every hyperbolic space is a convex metric space but converse may not true. For example, if $X = R$, $W(x, y, \lambda) = \lambda x + (1 - \lambda) y$ and define $d(x, y) = \frac{|x-y|}{1+|x-y|}$ for $x, y \in R$, then (X, d, W) is a convex metric space but not a hyperbolic space.

The stability of explicit as well as implicit iterations has extensively been studied by various authors (Berinde, 2011; Khan et al., 2014; Olatinwo, 2011; Olatinwo & Imoru, 2008; Ostrowski, 1967; Timis, 2012) due to its increasing importance in computational mathematics, especially due to revolution in computer programming. The concept of T-stability in convex metric space setting was given by Olatinwo (Olatinwo, 2011):

Definition 1.3. (Olatinwo, 2011) Let (X, d, W) be a convex metric space and $T:X \to X$ a self-mapping.

Let $\{x_n\}_{n=0}^{\infty} \subset X$ be the sequence generated by an iterative scheme involving T, which is defined by

$$x_{n+1} = f_{T,\alpha_n}^{x_n}, \quad n = 0, 1, 2, \ldots \tag{1.8}$$

where $x_0 \in X$ is the initial approximation and $f_{T,\alpha_n}^{x_n}$ is some function having convex structure, such that $\alpha_n \in [0, 1]$. Suppose that $\{x_n\}$ converges to a fixed-point p of T. Let $\{y_n\}_{n=0}^{\infty} \subset X$ be an arbitrary sequence and set $\varepsilon_n = d(y_{n+1}, f_{T,\alpha_n}^{y_n})$. Then, the iteration (1.8) is said to be T-stable with respect to T if and only if $\lim_{n\to\infty} \varepsilon_n = 0$, implies $\lim_{n\to\infty} y_n = p$.

LEMMA 1.4 *(Berinde, 2004; Khan et al., 2014) If δ is a real number such that $0 \le \delta < 1$ and $\{\in_n\}_{n=0}^{\infty}$ is a sequence of positive numbers such that $\lim_{n\to\infty} \in_n = 0$, then for any sequence of positive numbers $\{u_n\}_{n=0}^{\infty}$ satisfying*

$$u_{n+1} \le \delta u_n + \in_n, n = 0, 1, 2, \ldots$$

we have $\lim_{n\to\infty} u_n = 0$.

Definition 1.5 (Berinde, 2004) Suppose $\{a_n\}$ and $\{b_n\}$ are two real convergent sequences with limits a and b, respectively. Then $\{a_n\}$ is said to converge faster than $\{b_n\}$ if

$$\lim_{n\to\infty} \left| \frac{a_n - a}{b_n - b} \right| = 0$$

Definition 1.6 (Berinde, 2004) Let $\{u_n\}$ and $\{v_n\}$ be two fixed-point iterations that converge to the same fixed point p on a normed space X, such that the error estimates

$$\|u_n - p\| \le a_n$$

and

$$\|v_n - p\| \le b_n$$

are available, where $\{a_n\}$ and $\{b_n\}$ are two sequences of positive numbers (converging to zero) .If $\{a_n\}$ converge faster than $\{b_n\}$, then we say that $\{u_n\}$ converge faster to p than $\{v_n\}$.

Definition 1.7 (Gursoy et al., 2013) Let T, T_1 be two operators on X. We say T_1 is approximate operator of T if for all $x \in X$ and for a fixed $\in > 0$, we have $d(Tx, T_1x) \le \in$.

LEMMA 1.8 *(Gursoy et al., 2013; Khan et al., 2014) Let $\{a_n\}_{n=0}^{\infty}$ be a nonnegative sequence for which there exists $n_0 \in N$, such that for all $n \ge n_0$, one has the following inequality:*

$$a_{n+1} \le (1 - r_n)\, a_n + r_n t_n$$

where $r_n \in (0, 1)$, for all $n \in N$, $\sum_{n=1}^{\infty} r_n = \infty$ and $t_n \ge 0 \ \forall n \in N$.

Then, $0 \le \limsup_{n\to\infty} a_n \le \limsup_{n\to\infty} t_n$.

Having introduced the implicit iteration (1.1), we use it to prove the results concerning convergence, stability and rate of convergence for contractive condition (1.7) in convex metric spaces. Furthermore, data-dependence result of the same iteration is proved in hyperbolic spaces.

2. Convergence and stability results for new implicit iteration in convex metric spaces

THEOREM 2.1 *Let K be a nonempty closed convex subset of a convex metric space X and T be a quasi-contractive operator satisfying (1.7) with $F(T) \neq \varphi$. Then, for $x_0 \in C$, the sequence $\{x_n\}$ defined by (1.1) with $\sum (1-\alpha_n) = \infty$, converges to the fixed point of T.*

Proof Using (1.1) and (1.7), we have for $p \in F(T)$,

$$\begin{aligned} d(x_n,p) &= d(W(x_{n-1},Ty_n,\alpha_n),p) \\ &\le \alpha_n d(x_{n-1},p) + (1-\alpha_n)d(Ty_n,p) \\ &\le \alpha_n d(x_{n-1},p) + (1-\alpha_n)ad(y_n,p) \end{aligned} \tag{2.1}$$

Now, we have the following estimates:

$$\begin{aligned} d(y_n,p) &= d(W(z_n,Tz_n,\beta_n),p) \\ &\le \beta_n d(z_n,p) + (1-\beta_n)d(Tz_n,p) \\ &\le \beta_n d(z_n,p) + (1-\beta_n)ad(z_n,p) \\ &= [\beta_n + a(1-\beta_n)]d(z_n,p) \end{aligned} \tag{2.2}$$

and

$$\begin{aligned} d(z_n,p) &= d(W(x_n,Tx_n,\gamma_n),p) \\ &\le \gamma_n d(x_n,p) + (1-\gamma_n)d(Tx_n,p) \\ &\le \gamma_n d(x_n,p) + (1-\gamma_n)ad(x_n,p) \\ &= [\gamma_n + a(1-\gamma_n)]d(x_n,p) \end{aligned} \tag{2.3}$$

Inequalities (2.1), (2.2) and (2.3) yield

$$d(x_n,p) \le \alpha_n d(x_{n-1},p) + (1-\alpha_n)a[\beta_n + a(1-\beta_n)][\gamma_n + a(1-\gamma_n)]d(x_n,p)$$

which further implies

$$\{1-(1-\alpha_n)a[\beta_n + a(1-\beta_n)][\gamma_n + a(1-\gamma_n)]\}d(x_n,p) \le \alpha_n\, d(x_{n-1},p)$$

and therefore

$$d(x_n,p) \le \frac{\alpha_n}{1-(1-\alpha_n)a[\beta_n + a(1-\beta_n)][\gamma_n + a(1-\gamma_n)]}d(x_{n-1},p) \tag{2.4}$$

Let $\frac{P_n}{Q_n} = \frac{\alpha_n}{1-(1-\alpha_n)a[\beta_n+a(1-\beta_n)][\gamma_n+a(1-\gamma_n)]}$

then

$$\begin{aligned} 1-\frac{P_n}{Q_n} &= \frac{1-(1-\alpha_n)a[\beta_n+a(1-\beta_n)][\gamma_n+a(1-\gamma_n)]-\alpha_n}{1-(1-\alpha_n)a[\beta_n+a(1-\beta_n)][\gamma_n+a(1-\gamma_n)]} \\ &\ge 1-(1-\alpha_n)a[\beta_n + a(1-\beta_n)][\gamma_n + a(1-\gamma_n)] + \alpha_n \end{aligned}$$

which further implies,

$$\begin{aligned} \frac{P_n}{Q_n} &\le (1-\alpha_n)a[\beta_n + a(1-\beta_n)][\gamma_n + a(1-\gamma_n)] + \alpha_n \\ &= (1-\alpha_n)a[\beta_n + a(1-\beta_n)][\gamma_n + a(1-\gamma_n)] + \alpha_n \\ &= (1-\alpha_n)a[1-(1-a)(1-\beta_n)][1-(1-a)(1-\gamma_n)] + \alpha_n \end{aligned} \tag{2.5}$$

$$\le (1-\alpha_n)a + \alpha_n$$

$$= 1-(1-\alpha_n)(1-a). \tag{2.6}$$

Using (2.6), (2.4) becomes

$$d(x_n,p) \le [1-(1-\alpha_n)(1-a)]d(x_{n-1},p)$$

$$\le \prod_{i=1}^{n}[1-(1-\alpha_i)(1-a)]d(x_0,p)$$

$$\le e^{-\sum_{i=1}^{n}(1-\alpha_i)(1-a)}d(x_0,p) \tag{2.7}$$

But $\sum_{i=1}^{n}(1-\alpha_i)=\infty$, hence (2.7) yields $\lim_{n\to\infty} d(x_n,p)=0$. Therefore, $\{x_n\}$ converges to p.

THEOREM 2.2 *Let K be a nonempty closed convex subset of a convex metric space X and T be a quasi-contractive operator satisfying (1.7) with $F(T) \neq \varphi$. Then, for $x_0 \in C$, the sequence $\{x_n\}$ defined by (1.1) with $\alpha_n \le \alpha < 1$, $\sum(1-\alpha_n)=\infty$, is T-stable.*

Proof Suppose that $\{p_n\}_{n=0}^{\infty} \subset K$ be an arbitrary sequence, $\varepsilon_n = d(p_n, W(p_{n-1}, Tq_n, \alpha_n))$, where $q_n = W(r_n, Tr_n, \beta_n)$, $r_n = W(p_n, Tp_n, \gamma_n)$ and let $\lim_{n\to\infty} \varepsilon_n = 0$.

Then, using (1.7), we have

$$\begin{aligned} d(p_n,p) &\le d(p_n, W(p_{n-1},Tq_n,\alpha_n)) + d(W(p_{n-1},Tq_n,\alpha_n),p) \\ &\le \varepsilon_n + \alpha_n d(p_{n-1},p) + (1-\alpha_n)d(Tq_n,p) \\ &\le \varepsilon_n + \alpha_n d(p_{n-1},p) + (1-\alpha_n)\varphi d(Tp,p) + (1-\alpha_n)ad(q_n,p) \\ &\le \varepsilon_n + \alpha_n d(p_{n-1},p) + (1-\alpha_n)a[\beta_n + a(1-\beta_n)][\gamma_n + a(1-\gamma_n)]d(p_n,p) \end{aligned} \tag{2.8}$$

which implies

$$\{1-(1-\alpha_n)a[\beta_n+a(1-\beta_n)][\gamma_n+a(1-\gamma_n)]\}d(p_n,p) \le \varepsilon_n + \alpha_n d(p_{n-1},p)$$

and therefore

$$\begin{aligned} d(p_n,p) \le\ & \frac{\alpha_n}{1-(1-\alpha_n)a[\beta_n+a(1-\beta_n)][\gamma_n+a(1-\gamma_n)]}d(p_{n-1},p) \\ &+\frac{\varepsilon_n}{1-(1-\alpha_n)a[\beta_n+a(1-\beta_n)][\gamma_n+a(1-\gamma_n)]} \end{aligned} \tag{2.9}$$

But from (2.6), we have

$$\frac{\alpha_n}{1-(1-\alpha_n)a[\beta_n+a(1-\beta_n)][\gamma_n+a(1-\gamma_n)]} \le 1-(1-\alpha_n)(1-a) \tag{2.10}$$

Hence (2.9) becomes

$$d(p_n,p) \le [1-(1-\alpha_n)(1-a)]d(p_{n-1},p) + \frac{\varepsilon_n}{1-(1-\alpha_n)a[\beta_n+a(1-\beta_n)][\gamma_n+a(1-\gamma_n)]} \tag{2.11}$$

Using $\alpha_n \le \alpha < 1$ and $a \in (0,1)$, we have

$$1-(1-\alpha_n)(1-a) < 1$$

Hence, using Lemma 1.4, (2.11) yields $\lim_{n\to\infty} p_n = p$

Conversely, if we let $\lim_{n\to\infty} p_n = p$ then using contractive condition (1.7), it is easy to see that $\lim_{n\to\infty} \varepsilon_n = 0$.

Therefore, the iteration (1.1) is T-stable.

Remark 2.3 As contractive condition (1.7) is more general than those of (1.2)–(1.6), the convergence and stability results for implicit iteration (IN) using contractive conditions (1.2)–(1.6) can be obtained as special cases.

Remark 2.4 As implicit Mann iteration (IM) and Ishikawa-type iteration (II) are special cases of new implicit iteration (1.1), results similar to Theorem 2.1 and Theorem 2.2 hold for implicit Mann iteration (IM) and Ishikawa-type iteration (II).

3. Rate of convergence for implicit iterations

THEOREM 3.1 *Let K be a nonempty closed convex subset of a convex metric space X and T be a quasi-contractive operators satisfying (1.7) with $F(T) \neq \varphi$. Then, for $x_0 \in C$, the sequence $\{x_n\}$ defined by (1.1) with $\sum (1-\alpha_n) = \infty$, converges faster than implicit Mann iteration (IM) as well as Ishikawa-type iteration (II) to the fixed-point of T.*

Proof For implicit Mann iteration (IM), we have

$$\begin{aligned} d(x_n, p) &\leq \alpha_n d(x_{n-1}, p) + (1-\alpha_n) d(Tx_n, p) \\ &\leq \alpha_n d(x_{n-1}, p) + (1-\alpha_n) a d(x_n, p) \end{aligned}$$

which further yield

$$[1-(1-\alpha_n)a] d(x_n, p) \leq \alpha_n d(x_{n-1}, p)$$

and so

$$d(x_n, p) \leq \frac{\alpha_n}{1-(1-\alpha_n)a} d(x_{n-1}, p) \tag{3.1}$$

If we take $\frac{\alpha_n}{1-(1-\alpha_n)a} = \frac{A_n}{B_n}$

then,

$$1 - \frac{A_n}{B_n} = 1 - \frac{\alpha_n}{1-(1-\alpha_n)a} = \frac{1-[(1-\alpha_n)a + \alpha_n]}{1-(1-\alpha_n)a} \geq 1 - [(1-\alpha_n)a + \alpha_n]$$

and hence

$$\frac{A_n}{B_n} \leq (1-\alpha_n)a + \alpha_n \tag{3.2}$$

Keeping in mind the Berinde's Definition 1.6, inequalities (2.6) and (3.3) yields fast convergence of three-step implicit iteration (IN) than implicit Mann iteration (IM).

Also, for explicit Mann iteration, we have

$$\begin{aligned} d(x_n, p) &= d(W(x_{n-1}, Tx_{n-1}, \alpha_n), p) \\ &\leq \alpha_n d(x_{n-1}, p) + (1-\alpha_n) d(Tx_{n-1}, p) \\ &\leq \alpha_n d(x_{n-1}, p) + (1-\alpha_n) a d(x_{n-1}, p) \\ &\leq [\alpha_n + (1-\alpha_n)a] d(x_{n-1}, p) \end{aligned} \tag{3.3}$$

Similarly, for implicit Ishikawa-type iteration (II), we have

$$d(x_n, p) \leq \{(1-\alpha_n)a[1-(1-a)(1-\beta_n)] + \alpha_n\} d(x_{n-1}, p) \tag{3.4}$$

Using (3.1), (3.2) and (3.3), we conclude that implicit Mann iteration converges faster than corresponding explicit Mann iteration. Also, from (2.5) and (3.4), it is obvious that new three-step implicit iteration converges faster than Ishikawa-type implicit iteration (II).

Example 3.2 Let $K = [0, 1]$, $T(x) = \frac{x}{4}$, $x \neq 0$ and $\alpha_n = \beta_n = \gamma_n = 1 - \frac{4}{\sqrt{n}}$, $n \geq 25$ and for $n = 1, 2, \ldots, 24$, $\alpha_n = \beta_n = \gamma_n = 0$, then for implicit Mann iteration, we have

$$\begin{aligned} x_n &= \alpha_n x_{n-1} + (1-\alpha_n)Tx_n \\ &= \left(1 - \frac{4}{\sqrt{n}}\right)x_{n-1} + \frac{4}{\sqrt{n}}\frac{x_n}{4} \end{aligned}$$

which further implies

$$x_n\left[1 - \frac{1}{\sqrt{n}}\right] = \left(1 - \frac{4}{\sqrt{n}}\right)x_{n-1}$$

and so

$$x_n = \frac{\sqrt{n}-4}{\sqrt{n}-1}x_{n-1} = \prod_{i=25}^{n}\left(\frac{\sqrt{i}-4}{\sqrt{i}-1}\right)x_0 \tag{3.5}$$

Also, for the new three-step iteration (IN), we have

$$\begin{aligned} z_n &= \alpha_n x_n + (1-\alpha_n)Tx_n \\ &= \left(1 - \frac{4}{\sqrt{n}}\right)x_n + \frac{4}{\sqrt{n}}\frac{x_n}{4} \\ &= \left(1 - \frac{3}{\sqrt{n}}\right)x_n \end{aligned}$$

$$y_n = \left(1 - \frac{3}{\sqrt{n}}\right)z_n = \left(1 - \frac{3}{\sqrt{n}}\right)^2 x_n$$

and so

$$\begin{aligned} x_n &= \left(1 - \frac{4}{\sqrt{n}}\right)x_{n-1} + \frac{4}{\sqrt{n}}\frac{y_n}{4} \\ &= \left(1 - \frac{4}{\sqrt{n}}\right)x_{n-1} + \frac{1}{\sqrt{n}}\left(1 - \frac{3}{\sqrt{n}}\right)^2 x_n \\ &= \left(1 - \frac{4}{\sqrt{n}}\right)x_{n-1} + \left(\frac{1}{\sqrt{n}} + \frac{9}{n^{3/2}} - \frac{6}{n}\right)x_n \end{aligned}$$

which further implies

$$x_n\left[1 - \left(\frac{1}{\sqrt{n}} + \frac{9}{n^{3/2}} - \frac{6}{n}\right)\right] = \frac{\sqrt{n}-4}{\sqrt{n}}x_{n-1}$$

and hence

$$x_n = \frac{n^{3/2} - 4n}{n^{3/2} - n + 6\sqrt{n} - 9}x_{n-1}$$

$$= \prod_{i=25}^{n}\left(\frac{i^{3/2} - 4i}{i^{3/2} - i + 6\sqrt{i} - 9}\right)x_0 \tag{3.6}$$

Also, for explicit Mann iteration, we have

$$x_n = \alpha_n x_{n-1} + (1-\alpha_n)Tx_{n-1} = \left(1-\frac{4}{\sqrt{n}}\right)x_{n-1} + \frac{4}{\sqrt{n}}\frac{x_{n-1}}{4} = \left(1-\frac{3}{\sqrt{n}}\right)x_{n-1} \quad (3.7)$$

For two-step Ishikawa-type implicit iteration, we have

$$y_n = \left(1-\frac{3}{\sqrt{n}}\right)x_n$$

and so

$$x_n = \left(1-\frac{4}{\sqrt{n}}\right)x_{n-1} + \left(\frac{1}{\sqrt{n}}\right)\left(1-\frac{3}{\sqrt{n}}\right)x_n$$

$$x_n\left(1-\frac{1}{\sqrt{n}}+\frac{3}{n}\right) = \left(1-\frac{4}{\sqrt{n}}\right)x_{n-1}$$

$$x_n = \frac{(\sqrt{n}-4)\sqrt{n}}{n-\sqrt{n}+3}x_{n-1} = \frac{n-4\sqrt{n}}{n-\sqrt{n}+3}x_{n-1} = \prod_{i=25}^{n}\left(\frac{i-4\sqrt{i}}{i-\sqrt{i}+3}\right)x_0 \quad (3.8)$$

Using (3.5) and (3.6), we have

$$\frac{x_n(IN)}{x_n(IM)} = \prod_{i=25}^{n}\left(\frac{i^{3/2}-4i}{i^{3/2}-i+6\sqrt{i}-9}\right)\left(\frac{\sqrt{i}-1}{\sqrt{i}-4}\right) = \prod_{i=25}^{n}\frac{i^2-5i^{3/2}+4i}{i^2-5i^{3/2}+10i-33\sqrt{i}+36}$$
$$= \prod_{i=25}^{n}\left[1-\frac{(6i-33\sqrt{i}+36)}{i^2-5i^{3/2}+10i-33\sqrt{i}+36}\right]$$

But

$$0 \le \lim_{n\to\infty}\prod_{i=25}^{n}\left(1-\frac{6i-33\sqrt{i}+36}{i^2-5i^{3/2}+10i-33\sqrt{i}+36}\right)$$
$$\le \lim_{n\to\infty}\prod_{i=25}^{n}\left(1-\frac{1}{i}\right) = \lim_{n\to\infty}\frac{24}{25}\times\frac{25}{26}\cdots\frac{n-1}{n} = \lim_{n\to\infty}\frac{24}{n} = 0.$$

Hence $\lim_{n\to\infty}\left|\frac{x_n(IN)-0}{x_n(IM)-0}\right| = 0$. Therefore, using definition 1.5, the new three-step implicit iteration (IN) converges faster than the implicit Mann iteration (IM) to the fixed-point $p = 0$.

Similarly, using (3.5) and (3.7), we arrive at

$$\frac{x_n(IM)}{x_n(M)} = \prod_{i=25}^{n}\left(\frac{\sqrt{i}-4}{\sqrt{i}-1}\right)\left(\frac{\sqrt{i}}{\sqrt{i}-3}\right) = \prod_{i=25}^{n}\left(\frac{i-4\sqrt{i}}{i-4\sqrt{i}+3}\right)$$

with

$$0 \le \lim_{n\to\infty}\prod_{i=25}^{n}\left(\frac{i-4\sqrt{i}}{i-4\sqrt{i}+3}\right) \le \lim_{n\to\infty}\prod_{i=25}^{n}\left(1-\frac{1}{i}\right) = \lim_{n\to\infty}\frac{24}{25}\times\frac{25}{26}\cdots\frac{n-1}{n} = \lim_{n\to\infty}\frac{24}{n} = 0.$$

Therefore $\lim_{n\to\infty}\left|\frac{x_n(IM)-0}{x_n(M)-0}\right| = 0$. That is implicit Mann iteration (IM) converges faster than the explicit Mann iteration (M) to the fixed-point $p = 0$.

Also, using (3.6) and (3.8), we get

$$\frac{x_n(IN)}{x_n(II)} = \prod_{i=25}^{n}\left(\frac{i^{3/2}-4i}{i^{3/2}-i+6\sqrt{i}-9}\right)\left(\frac{i-\sqrt{i}+3}{i-4\sqrt{i}}\right) = \prod_{i=25}^{n}\frac{i^{5/2}-i^2+3i^{3/2}-4i^2+4i^{3/2}-12i}{i^{5/2}-4i^2-i^2+4i^{3/2}+6i^{3/2}+36\sqrt{i}}$$

$$= \prod_{i=25}^{n}\frac{i^{5/2}-5i^2+7i^{3/2}-12i}{i^{5/2}-5i^2+10i^{3/2}} = \prod_{i=25}^{n}\left[1-\frac{(3i^{3/2}+12i)}{i^{5/2}-5i^2+10i^{3/2}}\right]$$

with

$$0 \le \lim_{n\to\infty}\prod_{i=25}^{n}\left[1-\frac{(3i^{3/2}+12i)}{(i^{5/2}-5i^2+10i^{3/2})}\right] \le \lim_{n\to\infty}\prod_{i=25}^{n}\left(1-\frac{1}{i}\right) \le 0$$

which implies

$$\left|\frac{x_n(IN)-0}{x_n(II)-0}\right| = 0$$

Therefore, the new three-step iteration converges fast as compared to two-step implicit Ishikawa-type iteration.

Using computer programming in C++, the convergence speed of various iterations is compared and observations are listed in the Table 1 by taking initial approximation $x_0 = 1$, $T(x) = \frac{x}{4}$ and $\alpha_n = \beta_n = \gamma_n = 1 - \frac{4}{\sqrt{n}}$, $n \ge 25$. The table reveals that newly introduced implicit iteration has better convergence rate as compared to implicit Ishikawa-type iteration, implicit Mann iteration as well as explicit Mann iteration and implicit Mann iteration converges faster than corresponding explicit Mann iteration to the fixed-point $p = 0$.

Table 1. Comparison of convergence rate of new iteration with other iterations

Number of iterations (n)	Mann iteration (M)	Implicit Mann iteration (IM)	Implicit Ishikawa type iteration (II)	Implicit new iteration (IN)
25	0.4	0.25	0.217391	0.206612
26	0.164661	0.0670294	0.0509705	0.0460629
27	0.0695938	0.0191074	0.0127723	0.0109812
28	0.0301378	0.00575025	0.0033952	0.00277867
29	0.0133485	0.00181636	0.000951573	0.000741746
30	0.00603721	0.000599294	0.000279742	0.000207812
31	0.00278426	0.000205692	8.58832e−005	6.08396e−005
32	0.00130768	7.31828e−005	2.74322e−005	1.85426e−005
33	0.000624769	2.69091e−005	9.08659e−006	5.8642e−006
34	0.000303328	1.01987e−005	3.11239e−006	1.91897e−006
35	0.000149513	3.97501e−006	1.09965e−006	6.48125e−007
36	7.47563e−005	1.59e−006	3.99872e−007	2.25435e−007
-	-	-	-	-

4. Data dependence of implicit iteration in hyperbolic spaces

THEOREM 4.1 *Let $T:K \to K$ be a mapping satisfying (1.7). Let T_1 be an approximate operator of T as in Definition 1.7, and $\{x_n\}_{n=0}^{\infty}$, $\{u_n\}_{n=0}^{\infty}$ be two implicit iterations associated to T, T_1 and defined by*

$$\begin{aligned} x_n &= W(x_{n-1}, Ty_n, \alpha_n) \\ y_n &= W(z_n, Tz_n, \beta_n) \\ z_n &= W(x_n, Tx_n, \gamma_n) \end{aligned} \tag{4.1}$$

and

$$\begin{aligned} u_n &= W(u_{n-1}, T_1 v_n, \alpha_n) \\ v_n &= W(w_n, T_1 w_n, \beta_n) \\ w_n &= W(u_n, T_1 u_n, \gamma_n) \end{aligned} \tag{4.2}$$

respectively, where $\alpha_{nn=0}^{\infty}$, $\beta_{nn=0}^{\infty}$ *and* $\gamma_{nn=0}^{\infty}$ are real sequences in [0, 1] satisfying $\sum_{n=0}^{\infty}(1-\alpha_n) = \infty$. Let $p = Tp$ *and* $q = T_1 q$,, then for $\varepsilon > 0$, we have the following estimate:

$$d(p,q) \leq \frac{\varepsilon}{(1-a)^2}.$$

Proof Using Definition 1.2, iterations (4.1) and (4.2) yield the following estimates:

$$\begin{aligned} d(x_n, u_n) = d(W(x_{n-1}, Ty_n, \alpha_n), W(u_{n-1}, T_1 v_n, \alpha_n)) &\leq \alpha_n d(x_{n-1}, u_{n-1}) + (1-\alpha_n) d(Ty_n, T_1 v_n) \\ &\leq \alpha_n d(x_{n-1}, u_{n-1}) + (1-\alpha_n)\{d(Ty_n, T_1 y_n) + d(T_1 y_n, T_1 v_n)\} \\ &\leq \alpha_n d(x_{n-1}, u_{n-1}) + (1-\alpha_n)\{\varepsilon + \varphi d(y_n, T_1 y_n) + a d(y_n, v_n)\} \\ &\leq \alpha_n d(x_{n-1}, u_{n-1}) + (1-\alpha_n)\varepsilon + (1-\alpha_n)\varphi d(y_n, T_1 y_n) + (1-\alpha_n) a d(y_n, v_n) \end{aligned} \tag{4.3}$$

$$d(y_n, v_n) = d(W(z_n, Tz_n, \beta_n), W(w_n, T_1 w_n, \beta_n)) \leq \beta_n d(z_n, w_n) + (1-\beta_n) d(Tz_n, T_1 w_n) \tag{4.4}$$

$$d(Tz_n, T_1 w_n) \leq d(Tz_n, T_1 z_n) + d(T_1 z_n, T_1 w_n) \leq \varepsilon + \varphi d(z_n, T_1 z_n) + a d(z_n, w_n) \tag{4.5}$$

$$d(z_n, w_n) \leq \gamma_n d(x_n, u_n) + (1-\gamma_n) d(Tx_n, T_1 u_n) \tag{4.6}$$

$$d(Tx_n, T_1 u_n) \leq d(Tx_n, T_1 x_n) + d(T_1 x_n, T_1 u_n) \leq \varepsilon + \varphi d(x_n, T_1 x_n) + a d(x_n, u_n) \tag{4.7}$$

Using (4.3)–(4.7), we arrive at

$$\begin{aligned} d(x_n, u_n) &\leq \alpha_n d(x_{n-1}, u_{n-1}) \\ &+\{a(1-\alpha_n)\beta_n\gamma_n - a^2(1-\alpha_n)\beta_n(1-\gamma_n) \\ &-a^2(1-\alpha_n)(1-\beta_n)\gamma_n - a^3(1-\alpha_n)(1-\beta_n)(1-\gamma_n) d(x_n, u_n) \\ &+(1-\alpha_n)a^2\{(1-\beta_n)(1-\gamma_n)\varphi d(x_n, T_1 x_n) + \varphi d(y_n, T_1 y_n) + a(1-\beta_n)\varphi d(z_n, T_1 z_n)\} \\ &+(1-\alpha_n)\varepsilon\{a(1-\beta_n) + a^2(1-\beta_n)(1-\gamma_n) + 1\} \end{aligned}$$

which further implies

$$\begin{aligned} d(x_n, u_n)&[1 - a(1-\alpha_n)\{\beta_n\gamma_n + a\beta_n(1-\gamma_n) + a(1-\beta_n)\gamma_n + a^2(1-\beta_n)(1-\gamma_n)\}] \\ &\leq \alpha_n d(x_{n-1}, u_{n-1}) \\ &+(1-\alpha_n)\{a^2(1-\beta_n)(1-\gamma_n)\varphi d(x_n, T_1 x_n) + \varphi d(y_n, T_1 y_n) + a(1-\beta_n)\varphi d(z_n, T_1 z_n)\} \\ &+(1-\alpha_n)\varepsilon\{a(1-\beta_n) + a^2(1-\beta_n)(1-\gamma_n) + 1\} \end{aligned} \tag{4.8}$$

and so

$$
\begin{aligned}
& d(x_n, u_n) \\
& \leq \frac{\alpha_n}{[1-a(1-\alpha_n)\{\beta_n\gamma_n+a\beta_n(1-\gamma_n)+a(1-\beta_n)\gamma_n+a^2(1-\beta_n)(1-\gamma_n)\}]} d(x_{n-1}, u_{n-1}) \\
& \quad + \frac{(1-\alpha_n)\{a^2(1-\beta_n)(1-\gamma_n)\varphi d(x_n,T_1x_n)+\varphi d(y_n,T_1y_n)+a(1-\beta_n)\varphi d(z_n,T_1z_n)\}}{[1-a(1-\alpha_n)\{\beta_n\gamma_n+a\beta_n(1-\gamma_n)+a(1-\beta_n)\gamma_n+a^2(1-\beta_n)(1-\gamma_n)\}]} \\
& \quad + \frac{(1-\alpha_n)\varepsilon\{a(1-\beta_n)+a^2(1-\beta_n)(1-\gamma_n)+1\}}{[1-a(1-\alpha_n)\{\beta_n\gamma_n+a\beta_n(1-\gamma_n)+a(1-\beta_n)\gamma_n+a^2(1-\beta_n)(1-\gamma_n)\}]}.
\end{aligned}
\tag{4.9}
$$

Let $\frac{C_n}{D_n} = \frac{\alpha_n}{[1-a(1-\alpha_n)\{\beta_n\gamma_n+a\beta_n(1-\gamma_n)+a(1-\beta_n)\gamma_n+a^2(1-\beta_n)(1-\gamma_n)\}]}$

then

$$
\begin{aligned}
1 - \frac{C_n}{D_n} &= \frac{1-a(1-\alpha_n)\{\beta_n\gamma_n+a\beta_n(1-\gamma_n)+a(1-\beta_n)\gamma_n+a^2(1-\beta_n)(1-\gamma_n)\}-\alpha_n}{[1-a(1-\alpha_n)\{\beta_n\gamma_n+a\beta_n(1-\gamma_n)+a(1-\beta_n)\gamma_n+a^2(1-\beta_n)(1-\gamma_n)\}]} \\
&\geq 1 - [a(1-\alpha_n) + \alpha_n]
\end{aligned}
$$

which further implies

$$
\frac{C_n}{D_n} \leq a(1-\alpha_n) + \alpha_n = 1 - (1-\alpha_n)(1-a) \tag{4.10}
$$

Using (4.10), (4.9) becomes

$$
d(x_n, u_n) \leq [1 - (1-\alpha_n)(1-a)]d(x_{n-1}, u_{n-1}) + \frac{(1-\alpha_n)(1-a)\left\{\begin{array}{l} a^2(1-\beta_n)(1-\gamma_n)\varphi d(x_n, T_1x_n) \\ +\varphi d(y_n, T_1y_n) + a(1-\beta_n)\varphi d(z_n, T_1z_n) + 3\varepsilon \end{array}\right\}}{(1-a)[1-a(1-\alpha_n)\{\beta_n\gamma_n+a\beta_n(1-\gamma_n)+a(1-\beta_n)\gamma_n+a^2(1-\beta_n)(1-\gamma_n)\}]} \tag{4.11}
$$

Now, it is easy to see that

$$
\begin{aligned}
&[1 - a(1-\alpha_n)\{\beta_n\gamma_n + a\beta_n(1-\gamma_n) + a(1-\beta_n)\gamma_n + a^2(1-\beta_n)(1-\gamma_n)\}] \\
&\quad = [1 - a(1-\alpha_n)\{1 - (1-\beta_n)(1-a)\}\{1 - (1-\gamma_n)(1-a)\}] \geq 1 - a
\end{aligned}
$$

and hence

$$
\frac{1}{[1 - a(1-\alpha_n)\{1 - (1-\beta_n)(1-a)\}\{1 - (1-\gamma_n)(1-a)\}]} \leq \frac{1}{1-a}
$$

So, (4.11) becomes

$$
d(x_n, u_n) \leq [1 - (1-\alpha_n)(1-a)]d(x_{n-1}, u_{n-1}) + \frac{(1-\alpha_n)(1-a)\left\{\begin{array}{l} a^2(1-\beta_n)(1-\gamma_n)\varphi d(x_n, T_1x_n) \\ +\varphi d(y_n, T_1y_n) + a(1-\beta_n)\varphi d(z_n, T_1z_n) + 3\varepsilon \end{array}\right\}}{(1-a)^2} \tag{4.12}
$$

or

$$
a_n \leq (1 - r_n)a_{n-1} + r_n t_n
$$

where

$$
a_n = d(x_n, u_n),\ r_n = (1-\alpha_n)(1-a)
$$

and

$$
t_n = \frac{\left\{\begin{array}{l} a^2(1-\beta_n)(1-\gamma_n)\varphi d(x_n, T_1x_n) \\ +\varphi d(y_n, T_1y_n) + a(1-\beta_n)\varphi d(z_n, T_1z_n) + 3\varepsilon \end{array}\right\}}{(1-a)^2}
$$

Now, from Theorem 2.1, we have $\lim_{n\to\infty} d(x_n, p) = 0$, $\lim_{n\to\infty} d(u_n, p) = 0$ and since ϕ is continuous, hence $\lim_{n\to\infty} \varphi d(x_n, Tx_n) = \lim_{n\to\infty} \varphi d(y_n, Ty_n) = \lim_{n\to\infty} \varphi d(z_n, Tz_n) = 0$.

Therefore, using Lemma (1.8), (4.12) yields

$$d(p, q) \leq \frac{3\varepsilon}{(1-a)^2}$$

Remark 4.2 Putting $\gamma_n = \beta_n = 1$ and $\gamma_n = 1$ in (4.1) and (4.2), respectively, data-dependence results for implicit Mann iteration and implicit Ishikawa-type iteration can be proved easily on the same lines as in Theorem 4.1.

Acknowledgements
The authors would like to thank the referee for his/her careful reading of manuscript and their valuable comments.

Funding
Preety Malik is supported by Council of Scientific & Industrial Research (CSIR), India, under Junior Research Fellowship with Ref. No. 17-06/2012(i)EU-V.

Author details
Renu Chugh[1]
E-mail: chughrenu@yahoo.com
Preety Malik[1]
E-mail: preety0709@gmail.com
Vivek Kumar[2]
E-mail: ratheevivek15@yahoo.com
[1] Department of Mathematics, Maharshi Dayanand University, Rohtak 124001, Haryana, India.
[2] Department of Mathematics, KLP College, Rewari, India.

References

Anh, P. Ky., & Binh, T. Q. (2004). Stability and convergence of implicit iteration processes. *Vietnam Journal of Mathematics, 32*, 467–473.

Berinde, V. (2004). Picard iteration converges faster than Mann iteration for a class of quasi-contractive operators. *Fixed Point Theory and Applications, 2*, 94–105.

Berinde, V. (2011). Stability of Picard iteration for contractive mappings satisfying an implicit relation. *Carpathian Journal of Mathematics, 27*, 13–23.

Chidume, C. E., & Shahzad, N. (2005). Strong convergence of an implicit iteration process for a finite family of nonexpansive mappings. *Nonlinear Analysis: Theory, Methods & Applications, 62*, 1149–1156. http://dx.doi.org/10.1016/j.na.2005.05.002

Chugh, R., & Kumar, V. (2013). Convergence of SP iterative scheme with mixed errors for accretive Lipschitzian and strongly accretive Lipschitzian operators in Banach spaces. *International Journal of Computer Mathematics, 90*, 1865–1880. http://dx.doi.org/10.1080/00207160.2013.765558

Ciric, L. B. (1971). Generalized contractions and fixed point theorems. *Publications de l'Institut Mathématique (Beograd) (N. S.), 12*, 19–26.

Ciric, Lj. B. (1974). A generalization of Banach's contraction principle. *Proceedings of the American Mathematical Society, 45*, 267–273. http://dx.doi.org/10.2307/2040075

Ciric, Lj. B. (1977). Quasi-contractions in Banach spaces. *Publications de l'Institut Mathématique, 21*, 41–48.

Ciric, Lj. B., Lee, B. S., & Rafiq, A. (2010). Faster Noor iterations. *Indian Journal of Mathematics, 52*, 429–436.

Ćirić, Lj. B., & Nikolić, N. T. (2008a). Convergence of the Ishikawa iterates for multi-valued mappings in convex metric spaces. *Georgian Mathematical Journal, 15*, 39–43.

Ćirić, Lj. B., & Nikolić, N. T. (2008b). Convergence of the Ishikawa iterates for multi-valued mappings in metric spaces of hyperbolic type. *Matematicki Vesnik, 60*, 149–154.

Ćirić, Lj. B., Rafiq, A., Cakić, N., & Ume, J. S. (2009). Implicit Mann fixed point iterations for pseudo-contractive mappings. *Applied Mathematics Letters, 22*, 581–584. http://dx.doi.org/10.1016/j.aml.2008.06.034

Ćirić, Lj. B., Rafiq, A., Radenović, S., Rajović, M., & Ume, J. S. (2008). On Mann implicit iterations for strongly accretive and strongly pseudo-contractive mappings. *Applied Mathematics and Computation, 198*, 128–137.

Ciric, Lj. B., Ume, J. S. M., & Khan, S. (2003). On the convergence of the Ishikawa iterates to a common fixed point of two mappings. *Archivum Mathematicum (Brno) Tomus, 39*, 123–127.

Glowinski, R., & Le Tallec, P. Le. (1989). *Augmented Lagrangian and operator-splitting methods in nonlinear mechanics*. Philadelphia, PA: SIAM. http://dx.doi.org/10.1137/1.9781611970838

Gursoy, F., Karakaya, V., & Rhoades, B. E. (2013). Data dependence results of new multi-step and S-iterative schemes for contractive- like operators. *Fixed Point Theory and Applications, 76*, 1–12.

Hussian, N., Chugh, R., Kumar, V., & Rafiq, A. (2012). On the rate of convergence of Kirk type iterative schemes. *Journal of Applied Mathematics, 2012*, 22p.

Khan, A. R., Fukhar-ud-din, H., & Khan, M. A. A. (2012). An implicit algorithm for two finite families of nonexpansive maps in hyperbolic spaces. *Fixed Point Theory and Applications, 54*, 1–12.

Khan, A. R., Kumar, V., & Hussain, N. (2014). Analytical and numerical treatment of Jungck-type iterative schemes. *Applied Mathematics and Computation, 231*, 521–535. http://dx.doi.org/10.1016/j.amc.2013.12.150

Kohlenbach, U. (2004). Some logical metatherems with applications in functional analysis. *Transactions of the American Mathematical Society, 357*, 89–128.

Kumar, V., Latif, A., Rafiq, A., & Hussain, N. (2013). S-iteration process for quasi-contractive mappings. *Journal of Inequalities and Applications, 2013*, 1–24.

Mann, W. R. (1953). Mean value methods in iteration. *Proceedings of the American Mathematical Society, 4*, 506–510. http://dx.doi.org/10.1090/S0002-9939-1953-0054846-3

Olatinwo, M. O. (2011). Stability results for some fixed point iterative processes in convex metric spaces. *International Journal of engineering, IX*, 103–106.

Olatinwo, M. O., & Imoru, C. O. (2008). Some convergence results for the Jungck-Mann and the Jungck Ishikawa iteration processes in the class of generalized Zamfirescu operators. *Acta Mathematica Universitatis Comenianae, LXXVII, 2*, 299–304.

Osilike, M. O. (1995). Some stability results for fixed point iteration procedures. *Journal of the Nigerian Mathematics Society, 14*, 17–29.

Ostrowski, A. M. (1967). The round-off stability of iterations. *ZAMM - Zeitschrift für Angewandte Mathematik und Mechanik, 47*, 77–81. http://dx.doi.org/10.1002/(ISSN)1521-4001

Rhoades, B. E. (1993). Fixed point theorems and stability results for fixed point iteration procedures. *Indian Journal of Pure and Applied Mathematics, 24*, 691–703.

Shahzad, N., & Zegeye, H. (2009). On Mann and Ishikawa iteration schemes for multi-valued maps in Banach spaces. *Nonlinear Analysis: Theory, Methods & Applications, 71*, 838–844. http://dx.doi.org/10.1016/j.na.2008.10.112

Takahashi, W. (1970). A convexity in metric space and nonexpansive mappings. *Kodai Mathematical Seminar Reports, 22*, 142–149. http://dx.doi.org/10.2996/kmj/1138846111

Timis, I. (2012). Stability of Jungck-type iterative procedure for some contractive type mappings via implicit relation. *Miskolc Mathematical Notes, 13*, 555–567.

Zamfirescu, T. (1972). Fix point theorems in metric spaces. *Archiv der Mathematik, 23*, 292–298. http://dx.doi.org/10.1007/BF01304884

3

Dissipativity and passivity analysis for discrete-time complex-valued neural networks with time-varying delay

G. Nagamani[1]* and S. Ramasamy[1]

*Corresponding author: G. Nagamani, Department of Mathematics, Gandhigram Rural Institute - Deemed University, Gandhigram, Tamil Nadu 624 302, India
E-mail: nagamanigru@gmail.com

Reviewing editor: Yong Hong Wu, Curtin University of Technology, Australia

Abstract: In this paper, we consider the problem of dissipativity and passivity analysis for complex-valued discrete-time neural networks with time-varying delays. The neural network under consideration is subject to time-varying. Based on an appropriate Lyapunov–Krasovskii functional and by using the latest free-weighting matrix method, a sufficient condition is established to ensure that the neural networks under consideration is strictly $(\mathcal{Q}, \mathcal{S}, \mathcal{R})$-dissipative. The derived conditions are presented in terms of linear matrix inequalities. A numerical example is presented to illustrate the effectiveness of the proposed results.

Subjects: Applied Mathematics; Mathematics & Statistics; Science

Keywords: complex-valued neural networks; dissipativity; Lyapunov–Krasovskii functional; Linear matrix inequalities (LMIs); time-varying delay

1. Introduction

In the past several decades, the neural networks are very important nonlinear circuit networks because of their wide applications in various fields such as associative memory, signal processing, data compression, system control (Hirose, 2003), optimization problem, and so on Liang, Wang, and Liu (2009), Wang, Ho, Liu, and Liu (2009), Liu, Wang, Liang, and Liu (2009), Bastinec, Diblik, and Smarda (2010), and

ABOUT THE AUTHOR

G. Nagamani served as a lecturer in Mathematics in Mahendra Arts and Science College, Namakkal, Tamilnadu, India, during 2001–2008. Currently she is working as an assistant professor in the Department of Mathematics, Gandhigram Rural University-Deemed University , Gandhigram, Tamilnadu, India, since June 2011. She has published more than 15 research papers in various SCI journals holding impact factors. She is also serving as a reviewer for few SCI journals. Her research interest is in the field of Modeling of Stochastic Differential Equations, Neural Networks, Dissipativity and Passivity Analysis.

The author research area is based on the passivity approach for dynamical systems and also for various types of neural networks such as Markovian jumping neural networks, Takagi–Sugeno fuzzy stochastic neural networks, and Cohen–Grossberg neural networks. The author has published 14 research articles in most reputed SCI journals in the thrust area of the project during the past six years.

PUBLIC INTEREST STATEMENT

The passivity approach for interconnection of passive systems provides a nice tool for controlling a large class of nonlinear systems and DNNs, and its potential applications have been found in the stability and stabilization schemes of electrical networks, and in the control of teleoperators.

Diblik, Schmeidel, and Ruzickova (2010). Recently, neural networks have been electronically implemented and they have been used in real-time applications. However in electronic implementation of neural networks, some essential parameters of neural networks such as release rate of neurons, connection weights between the neurons and transmission delays might be subject to some deviations due to the tolerances of electronic components employed in the design of neural networks (Aizenberg, Paliy, Zurada, & Astola, 2008; Hu & Wang, 2012; Mostafa, Teich, & Lindner, 2013; Wang, Xue, Fei, & Li, 2013; Wu, Shi, Su, & Chu, 2011). As we know, time delays commonly exist in the neural networks because of the network traffic congestions and the finite speed of information transmission in networks. So the study of dynamic properties with time delay is of great significance and importance. However, most of the studied networks are real number valued. Recently, in order to investigate the complex properties in complex-valued neural networks, some complex-valued network models are proposed.

The most notable feature of complex-valued neural networks (CVNNs) is the compatibility with wave phenomena and wave information related to, for example, electromagnetic wave, light wave, electron wave, and sonic wave (Hirose, 2011). Furthermore, CVNNs are widely applied in coherent electromagnetic wave signal processing. They are mainly used in adapting, processing of interferometric synthetic aperture radar (InSAR) images captured by satellite or airplane to observe land surface (Suksmono & Hirose, 2002; Yamaki & Hirose 2009). Another important application field is sonic and ultrasonic processing. Pioneering work has been done in various directions (Zhang & Ma, 1997). In communication systems, the CVNNS can be regarded as an extension of adaptive complex filters, i.e. modular multiple-stage and nonlinear version. From this view point, several groups worked on time sequential signal processing (Goh & Mandic, 2005, 2007). Furthermore, there are many ideas based on CVNNs in image processing. An example is the adaptive processing for blur compensation by identifying the point scatting function in the frequency domain (Aizenberg et al., 2008). Recently, many mathematicians and scientists have paid more attention to this field of research. Besides, CVNNs have different and more complicated properties than the real-valued ones. Therefore, it is necessary to study the dynamic behaviors of the systems deeply. Over the past decades, some work has been done to analyze the dynamic behavior of the equilibrium points of the various CVNNs. In Mostafa et al. (2013), local stability analysis of discrete-time, continuous-state, complex-valued recurrent neural networks with inner state feedback was presented. In Zhou and Song (2013), the authors studied boundedness and complete stability of complex-valued neural networks with time delay by using free weighting matrices.

It is well known that dissipativity theory gives a framework for the design and analysis of control systems using an input-output description on energy-related considerations (Jing, Yao, & Shen, 2014; Wu, Shi, Su, & Chu, 2013; Wu, Yang, & Lam, 2014) and it becomes a powerful tool in characterizing important system behaviors such as stability. The passivity theory, being an effective tool for analyzing the stability of systems, has been applied in complexity (Zhao, Song, & He, 2014), signal processing, especially for high-order systems and thus the passivity analysis approach has been used for a long time to deal with the control problems (Chua, 1999). However, to the best of our knowledge, there is no result addressed on the dissipativity and passivity analysis of discrete-time complex-valued neural networks with time-varying delay, which motivates the present study.

In this paper, we consider the problem of dissipativity and passivity analysis for discrete-time complex-valued neural networks with time-varying delay. Based on the lemma proposed in Zhou and Song (2013), a condition is derived for strict $(\mathcal{Q}, \mathcal{S}, \mathcal{R})$-dissipativity and passivity of the control neural networks, which depends only on the discrete delay. In established model, the delay-dependent dissipativity and passivity conditions are derived and the obtained linear matrix inequalities (LMIs) can be checked numerically using the effective LMI toolbox in MATLAB and accordingly the estimator gains are obtained. The effectiveness of the proposed design is finally demonstrated by a numerical example.

The rest of this paper is organized as follows: model description and preliminaries are given in Section 2. Dissipativity and passivity analysis for discrete-time complex-valued neural networks with time-varying delay are presented in Section 3. Illustrative example and its simulation results for dissipativity conditions have been given in Section 4.

Notations: $\mathbb{C}^n$ and $\mathbb{R}^n$ denote, respectively, the n-dimensional complex space and Euclidean space. $z(k) = x(k) + iy(k)$ denote the complex-valued function, where $x(k),\ y(k) \in \mathbb{R}^n$. $\mathbb{R}^{n\times m}$ is the set of real $n \times m$ matrices, I is the identity matrix of appropriate dimension. For any matrix P, $P > 0(P < 0)$ means P is a positive definite (negative definite) matrix. The superscript $*$ denotes the matrix complex conjugate transpose, $diag\{\cdots\}$ stands for a block-diagonal matrix. Let $C([-\tau_2, 0], \mathcal{D})$ be the Banach space of continuous functions mapping $[-\tau_2, 0]$ into $\mathcal{D} \subset \mathbb{C}^n$. For integers a and b with a < b, let N[a, b] = {a, a+1, ... , b−1, b}. X^T represents the transpose of matrix X, $\Delta V(k)$ denotes the difference of function $V(k)$ given by $\Delta V(k) = V(k+1) - V(k)$.

2. Model description and preliminaries

Consider the following discrete-time complex-valued neural networks with time-varying delays:

$$\begin{aligned} z(k+1) &= Af(z(k)) + Bf(z(k-\tau(k))) + u(k) \\ y(k) &= f(z(k)) \end{aligned} \tag{1}$$

where $z(k) = [z_1(k), z_2(k), \ldots z_n(k)]^T \in \mathbb{C}^n$ is the neuron state vector; $A \in \mathbb{C}^{n\times n}$, $B \in \mathbb{C}^{n\times n}$ are the connection weight matrix and the delayed connection weight matrix, respectively; $y(k)$ is the output of neural network (1). $u(k) = [u_1(k), u_2(k), \ldots u_n(k)]^T \in \mathbb{C}^n$ is the input vector; time delay $\tau(k)$ ranges from τ_1 to τ_2 as $\tau_1 \le \tau(k) \le \tau_2$; $f(z(k)) = [f_1(z_1(k)), \ldots, f_n(z_n(k))]^T \in \mathbb{C}^n$ and $f(z(k-\tau(k))) = [f_1(z_1(k-\tau_1(k))), \ldots, f_n(z(k-\tau_1(k)))]^T \in \mathbb{C}^n$ are the complex-valued neuron activation functions without and with time delays. The initial conditions of the CVNNs (1) are given by

$$z_i(s) = \phi_i(s),\ s \in [-\tau_2, 0], \quad i = 1, 2, \ldots, n$$

where $\phi_i \in C([-\tau_2, 0], \mathcal{D})$ are continuous. Complex-valued parameters in the neural network can be represented as $a_{ij} = a_{ij}^R + ia_{ij}^I$, $b_{ij} = b_{ij}^R + ib_{ij}^I$. Then (1) can be written as

$$\begin{aligned} z_i^R(k+1) &= \sum_{j=1}^{n} a_{ij}^R f(z_j^R(k)) - \sum_{j=1}^{n} a_{ij}^I f(z_j^I(k)) + \sum_{j=1}^{n} b_{ij}^R f(z_j^R(k-\tau(k))) - \sum_{j=1}^{n} b_{ij}^I f(z_j^I(k-\tau(k))) \\ &\quad + u_i^R(k) \end{aligned} \tag{2}$$

$$\begin{aligned} z_i^I(k+1) &= \sum_{j=1}^{n} a_{ij}^R f(z_j^I(k)) + \sum_{j=1}^{n} a_{ij}^I f(z_j^R(k)) + \sum_{j=1}^{n} b_{ij}^R f(z_j^I(k-\tau(k))) + \sum_{j=1}^{n} b_{ij}^I f(z_j^R(k-\tau(k))) \\ &\quad + u_i^I(k) \end{aligned} \tag{3}$$

where z_i^R and z_i^I are the real and imaginary parts of variable z_i, respectively. a_{ij}^R and a_{ij}^I are the real and imaginary parts of connection weight a_{ij}; b_{ij}^R and b_{ij}^I are the real and imaginary parts of delayed connection weight b_{ij}. $u_i^R(k)$ and $u_i^I(k)$ are the real and imaginary parts of $u(k)$. Connection weight matrices are represented as $A^R = (a_{ij}^R)_{n\times n} \in \mathbb{R}^{n\times n}$, $A^I = (a_{ij}^I)_{n\times n} \in \mathbb{R}^{n\times n}$, $B^R = (b_{ij}^R)_{n\times n} \in \mathbb{R}^{n\times n}$, and $B^I = (b_{ij}^I)_{n\times n} \in \mathbb{R}^{n\times n}$. Then, we have

$$\begin{aligned} z^R(k+1) &= A^R f^R(k) - A^I f^I(k) + B^R f^R(k-\tau(k))) - B^I f^I(k-\tau(k))) + u^R(k) \\ z^I(k+1) &= A^R f^I(k) + A^I f^R(k) + B^R f^I(k-\tau(k))) + B^I f^R(k-\tau(k))) + u^I(k) \end{aligned}$$

To derive the main results, we will introduce the following assumptions, definitions, and lemmas.

ASSUMPTION 2.1

The activation function $f_j(\cdot)$ can be separated into real and imaginary parts of the complex numbers z. It follows that $f_j(z)$ is expressed by

$$f_j(z) = f_j^R(Re(z)) + if_j^I(Im(z))$$

where $f_i^R(\cdot), f_j^I(\cdot)$: $\mathbb{R} \to \mathbb{R}$ for all $j = 1, 2, \ldots, n$. Then,

$$l_j^{R-} \leq \frac{f_j^R(\alpha_1) - f_j^R(\alpha_2)}{\alpha_1 - \alpha_2} \leq l_j^{R+}$$

$$l_j^{I-} \leq \frac{f_j^I(\alpha_1) - f_j^I(\alpha_2)}{\alpha_1 - \alpha_2} \leq l_j^{I+}, \ \forall \alpha_1, \alpha_2 \in \mathbb{R}, \ \alpha_1 \neq \alpha_2$$

Definition 2.1 Zhou and Song (2013): The neural network (1) is said to be $(\mathcal{Q}, \mathcal{S}, \mathcal{R})$-dissipative, if the following dissipation inequality

$$\sum_{k=0}^{k_p} r(u(k), y(k)) \geq 0, \quad \forall k_p \geq 0 \tag{4}$$

holds under zero initial condition for any nonzero input $u \in l_2[0, +\infty)$. Furthermore, if for some scalar $\gamma > 0$, the dissipation inequality

$$\sum_{k=0}^{k_p} r(u(k), y(k)) \geq \gamma \sum_{k=0}^{k_p} u^T(k)u(k) \quad \forall k_p \geq 0 \tag{5}$$

holds under zero initial condition for any nonzero input $u \in l_2[0, +\infty)$, then the neural network (1) is said to be strictly $(\mathcal{Q}, \mathcal{S}, \mathcal{R})$-$\gamma$-dissipative. In this paper, we define a quadratic supply rate $r(u, y)$ associated with neural network (1) as follows:

$$r(u, y) = y^T \mathcal{Q} y + 2y^T \mathcal{S} u + u^T \mathcal{R} u \tag{6}$$

where $\mathcal{Q} \leq 0$, $\mathcal{S}$, and $\mathcal{R}$ are real symmetric matrices of appropriate dimensions.

Definition 2.2 Wu et al. (2011): The neural network (1) is said to be passive if there exists a scalar $\gamma > 0$ such that, for all $k_0 \geq 0$

$$2\sum_{j=0}^{k_0} y^T(j)u(j) \geq -\gamma \sum_{j=0}^{k_0} u^T(j)u(j)$$

under the zero initial condition.

LEMMA 2.1 *Liu et al. (2009): Let X and Y be any n-dimensional real vectors, and let P be an $n \times n$ positive semidefinite matrix. Then, the following inequality holds:*

$$2X^T PY \leq X^T PX + Y^T PY$$

LEMMA 2.2 *For any constant matrix $M \in \mathbb{R}^{n \times n}$, $M = M^T > 0$, integers r_1 and r_2 satisfying $r_2 > r_1$, and vector function ω: $\mathbb{N}[r_1, r_2] \to \mathbb{R}^n$, such that the sums concerned are well defined, then*

$$-(r_2 - r_1 + 1)\sum_{j=r_1}^{r_2} \omega^T(j)M\omega(j) \leq -\sum_{j=r_1}^{r_2} \omega^T(j)M \sum_{j=r_1}^{r_2} \omega(j) \tag{7}$$

Proof

$$
\begin{aligned}
-(r_2 - r_1 + 1)\sum_{j=r_1}^{r_2}\omega^T(j)M\omega(j) &= -\frac{1}{2}(r_2 - r_1 + 1)\sum_{j=r_1}^{r_2}\omega^T(j)M\omega(j) - \frac{1}{2}(r_2 - r_1 + 1)\sum_{j=r_1}^{r_2}\omega^T(j)M\omega(j) \\
&= -\frac{1}{2}(r_2 - r_1 + 1)\sum_{i=r_1}^{r_2}\omega^T(i)M\omega(i) - \frac{1}{2}(r_2 - r_1 + 1)\sum_{j=r_1}^{r_2}\omega^T(j)M\omega(j) \\
&= -\frac{1}{2}\Big[\sum_{i=r_1}^{r_2}(r_2 - r_1 + 1)\omega^T(i)M\omega(i) + \sum_{i=r_1}^{r_2}\sum_{j=r_1}^{r_2}\omega^T(j)M\omega(j)\Big] \\
&= -\frac{1}{2}\sum_{j=r_1}^{r_2}\Big[(r_2 - r_1 + 1)\omega^T(i)M\omega(i) + \sum_{j=r_1}^{r_2}\omega^T(j)M\omega(j)\Big] \\
&= -\frac{1}{2}\sum_{j=r_1}^{r_2}\sum_{j=r_1}^{r_2}[\omega^T(j)M\omega(j) + \omega^T(i)M\omega(i)] \\
&\leq -\sum_{j=r_1}^{r_2}\sum_{i=r_1}^{r_2}\omega^T(j)M\omega(i) \qquad \text{[Using Lemma 2.1]} \\
&= -\sum_{j=r_1}^{r_2}\omega^T(j)M\sum_{i=r_1}^{r_2}\omega(i)
\end{aligned}
$$

LEMMA 2.3 *Let $M \in \mathbb{R}^{n\times n}$ be a positive semidefinite matrix, $\xi_j \in \mathbb{R}^n$, and scalar constant $a_j \geq 0$ $(j = 1, 2, \ldots)$. If the series concerned is convergent, then the following inequality holds:*

$$
\Big(\sum_{j=1}^{\infty} a_j\xi_j\Big)^T M\Big(\sum_{j=1}^{\infty} a_j\xi_j\Big) \leq \Big(\sum_{j=1}^{\infty} a_j\Big)\sum_{j=1}^{\infty} a_j\xi_j^T M\xi_j \tag{8}
$$

Proof Letting m be a positive integer, we have

$$
\begin{aligned}
\Big(\sum_{j=1}^{m} a_j\xi_j\Big)^T M\Big(\sum_{j=1}^{m} a_j\xi_j\Big) &= \Big(\sum_{j=1}^{m} a_j\xi_j\Big)^T M\Big(\sum_{i=1}^{m} a_i\xi_i\Big) \\
&= \sum_{j=1}^{m}\sum_{i=1}^{m} a_j a_i \xi_j^T M\xi_i \\
&\leq \sum_{j=1}^{m}\sum_{i=1}^{m} \frac{1}{2}a_j a_i(\xi_j^T M\xi_j + \xi_i^T M\xi_i) \quad \text{(by the Lemma 2.1)} \\
&= \Big(\sum_{j=1}^{m} a_j\Big)\sum_{j=1}^{m} a_j\xi_j^T M\xi_j
\end{aligned}
$$

and then (8) follows directly by letting $m \to \infty$, which completes the proof.

LEMMA 2.4 *Given a Hermitian matrix Q, The inequality $Q < 0$ is equivalent to*

Table 1. Dimensions of matrices concerned in Theorem 3.1

Matrices (Nonzero Ones)	Dimensions (Row by Column)
$P_i, Q_i, R_i, S_i, T_i, U_i, V_i, W_i, G_i, H_i, \quad i = 1, 2$	$n \times n$
$F_1 = diag\{s_1, s_2, \ldots, s_n\}$	$n \times n$
$F_2 = diag\{s_1', s_2', \ldots, s_n'\}$	$n \times n$
$\Gamma = diag\{g_1^2, g_2^2, \ldots, g_n^2\}$	$n \times n$

$$\begin{pmatrix} \mathbb{Q}^R & -\mathbb{Q}^I \\ \mathbb{Q}^I & \mathbb{Q}^R \end{pmatrix} < 0$$

where $\mathbb{Q}^R = Re(\mathbb{Q})$ *and* $\mathbb{Q}^I = Im(\mathbb{Q})$.

3. Main results

In this section, we derive the dissipativity criterion for discrete-time complex-valued neural networks (1) with time-varying delays using the Lyapunov functional method combining with LMI approach. For convenience, we use the following notations: $\psi(k) = \frac{1}{\alpha(k)}\left[\sum_{i=k-\tau(k)}^{k-\tau_1-1} z(i)\right]$, $\phi(k) = \frac{1}{\beta(k)}\left[\sum_{i=k-\tau_2}^{k-\tau(k)-1} z(i)\right]$, $M = \frac{\tau_1^4}{4}G + \frac{(\tau_2-\tau_1)^2}{4}H$, $\alpha(k) = \tau(k) - \tau_1$, $\beta(k) = \tau_2 - \tau(k)$. Table 1 describes the matrices along with the dimensions that are used in the following Theorem 3.1.

THEOREM 3.1 *Assume that Assumption 2.1 holds, then the complex-valued neural networks (1) are dissipative if there exist positive Hermitian matrices* $P = P_1 + iP_2$, $Q = Q_1 + iQ_2$, $R = R_1 + iR_2$, $S = S_1 + iS_2$, $T = T_1 + iT_2$, $U = U_1 + iU_2$, $V = V_1 + iV_2$, $W = W_1 + iW_2$, $G = G_1 + iG_2$, $H = H_1 + iH_2$, *two positive diagonal matrices* $F_1 > 0$, $F_2 > 0$, *and a scalar* $\gamma > 0$ *such that the following LMI holds.*

$$\Theta = \begin{pmatrix} \Theta^R & -\Theta^I \\ \Theta^I & \Theta^R \end{pmatrix} < 0 \tag{9}$$

where

$$\Theta^R = \begin{pmatrix}
\Theta^R_{1,1} & 0 & 0 & 0 & \Theta^R_{1,5} & \Theta^R_{1,6} & \Theta^R_{1,7} & 0 & 0 & 0 & \Theta^R_{1,11} \\
0 & \Theta^R_{2,2} & 0 & 0 & 0 & 0 & 0 & 0 & \Theta^R_{2,9} & 0 & 0 \\
0 & 0 & \Theta^R_{3,3} & 0 & 0 & 0 & 0 & 0 & \Theta^R_{3,9} & 0 & 0 \\
0 & 0 & 0 & \Theta^R_{4,4} & 0 & 0 & 0 & 0 & 0 & 0 & 0 \\
0 & 0 & 0 & 0 & \Theta^R_{5,5} & \Theta^R_{5,6} & 0 & 0 & 0 & 0 & \Theta^R_{5,11} \\
0 & 0 & 0 & 0 & 0 & \Theta^R_{6,6} & 0 & 0 & 0 & 0 & \Theta^R_{6,11} \\
0 & 0 & 0 & 0 & 0 & 0 & \Theta^R_{7,7} & 0 & 0 & 0 & 0 \\
0 & 0 & 0 & 0 & 0 & 0 & 0 & \Theta^R_{8,8} & 0 & 0 & 0 \\
0 & 0 & 0 & 0 & 0 & 0 & 0 & 0 & \Theta^R_{9,9} & 0 & 0 \\
0 & 0 & 0 & 0 & 0 & 0 & 0 & 0 & 0 & \Theta^R_{10,10} & 0 \\
0 & 0 & 0 & 0 & 0 & 0 & 0 & 0 & 0 & 0 & \Theta^R_{11,11}
\end{pmatrix} \tag{10}$$

and

$$\Theta^I = \begin{pmatrix}
\Theta^I_{1,1} & 0 & 0 & 0 & \Theta^I_{1,5} & \Theta^I_{1,6} & \Theta^I_{1,7} & 0 & 0 & 0 & \Theta^I_{1,11} \\
0 & \Theta^I_{2,2} & 0 & 0 & 0 & 0 & 0 & 0 & \Theta^I_{2,9} & 0 & 0 \\
0 & 0 & \Theta^I_{3,3} & 0 & 0 & 0 & 0 & 0 & \Theta^I_{3,9} & 0 & 0 \\
0 & 0 & 0 & \Theta^I_{4,4} & 0 & 0 & 0 & 0 & 0 & 0 & 0 \\
0 & 0 & 0 & 0 & \Theta^I_{5,5} & \Theta^I_{5,6} & 0 & 0 & 0 & 0 & \Theta^I_{5,11} \\
0 & 0 & 0 & 0 & 0 & \Theta^I_{6,6} & 0 & 0 & 0 & 0 & \Theta^I_{6,11} \\
0 & 0 & 0 & 0 & 0 & 0 & \Theta^I_{7,7} & 0 & 0 & 0 & 0 \\
0 & 0 & 0 & 0 & 0 & 0 & 0 & \Theta^I_{8,8} & 0 & 0 & 0 \\
0 & 0 & 0 & 0 & 0 & 0 & 0 & 0 & \Theta^I_{9,9} & 0 & 0 \\
0 & 0 & 0 & 0 & 0 & 0 & 0 & 0 & 0 & \Theta^I_{10,10} & 0 \\
0 & 0 & 0 & 0 & 0 & 0 & 0 & 0 & 0 & 0 & \Theta^I_{11,11}
\end{pmatrix} \tag{11}$$

with

$$\Theta^R_{1,1} = -P_1 + Q_1 + R_1 + T_1 + \tau_1^2 U_1 + \tau_2^2 V_1 + M_1 - \tau_1^2 G_1 + F_1\Gamma$$
$$\Theta^R_{1,5} = -A_1^T M_1 - A_2^T M_2, \Theta^R_{1,6} = -B_1^T M_1 - B_2^T M_2, \Theta^R_{1,7} = -\tau_1 G_1, \Theta^R_{1,11} = -M_1$$
$$\Theta^R_{2,2} = -Q_1 + S_1 - H_1, \Theta^R_{2,9} = H_1, \Theta^R_{3,3} = -H_1 - R_1 - S_1, \Theta^R_{3,9} = H_1$$
$$\Theta^R_{4,4} = -T_1 + B_1^T P_1 B_1 - B_1^T P_2 B_2 + B_2^T P_2 B_1 + B_2^T P_1 B_2 - F_2\Gamma$$
$$\Theta^R_{5,5} = A_1^T P_1 A_1 - A_1^T P_2 A_2 + A_2^T P_2 A_1 + A_2^T P_1 A_2 + A_1^T M_1 A_1 - A_1^T M_2 A_2 + A_2^T M_2 A_1 + A_2^T M_1 A_2 - \mathcal{Q}_1 - F_1, \Theta^R_{5,6} = A_1^T P_1 B_1 - A_1^T P_2 B_2 + A_2^T P_2 B_1 + A_2^T P_1 B_2$$
$$\Theta^R_{5,11} = P_1 A_1 - P_2 A_2 + M_1 A_1 - M_2 A_2 - I - \mathcal{S}_1$$
$$\Theta^R_{6,6} = B_1^T M_1 B_1 - B_1^T M_2 B_2 + B_2^T M_2 B_1 + B_2^T P_1 B_2 - F_2, \Theta^R_{6,11} = B_1^T P_1 + B_2^T P_2$$
$$\Theta^R_{7,7} = -U_1 + G_1, \Theta^R_{8,8} = -V_1, \Theta^R_{9,9} = -H_1, \Theta^R_{10,10} = -H_1, \Theta^R_{11,11} = -\mathcal{R}_1 + \gamma I + P_1 M_1$$
$$\Theta^I_{1,1} = -P_1 + Q_1 + R_1 + T_1 + \tau_1^2 U_1 + \tau_2^2 V_1 + M_1 - \tau_1^2 G_1$$
$$\Theta^I_{1,5} = -A_1^T M_2 - A_2^T M_1, \Theta^I_{1,6} = -B_1^T M_2 - B_2^T M_1, \Theta^I_{1,7} = -\tau_1 G_2, \Theta^I_{1,11} = -M_2$$
$$\Theta^I_{2,2} = -Q_2 + S_2 - H_2, \Theta^I_{2,9} = H_2, \Theta^I_{3,3} = -H_2 - R_2 - S_2, \Theta^I_{3,9} = H_2$$
$$\Theta^I_{4,4} = -T_2 + B_1^T P_2 A_1 + B_1^T P_1 B_2 - B_2^T P_1 B_1 + B_2^T P_2 B_2$$
$$\Theta^I_{5,5} = A_1^T P_2 A_1 + A_1^T P_1 A_2 - A_2^T P_1 A_1 + A_2^T P_2 A_2 + A_1^T M_2 A_1 + A_1^T M_1 A_2 - A_2^T M_1 A_1 + A_2^T M_2 A_2 - \mathcal{Q}_2, \Theta^I_{5,6} = A_1^T P_2 B_1 + A_1^T P_1 B_2 - A_2^T P_1 B_1 + A_2^T P_2 B_2$$
$$\Theta^I_{5,11} = P_1 A_2 - P_2 A_1 + M_1 A_2 - M_2 A_1 - \mathcal{S}_2$$
$$\Theta^I_{6,6} = B_1^T M_2 B_1 + B_1^T M_1 B_2 - B_2^T M_1 B_1 + B_2^T P_2 B_2, \Theta^I_{6,11} = B_1^T P_2 - B_2^T P_1$$
$$\Theta^I_{7,7} = -U_2 + G_2, \Theta^I_{8,8} = -V_2, \Theta^I_{9,9} = -H_2, \Theta^I_{10,10} = -H_2, \Theta^I_{11,11} = -\mathcal{R}_2 + P_2 M_2$$

Proof Defining $\eta(k) = z(k+1) - z(k)$, we consider the following Lyapunov-Krasovskii functional for neural network in (1):

$$V(k) = \sum_{i=1}^{6} V_i(k)$$

where

$$V_1(k) = z^*(k)Pz(k)$$
$$V_2(k) = \sum_{i=k-\tau_1}^{k-1} z^*(i)Qz(i) + \sum_{i=k-\tau_2}^{k-1} z^*(i)Rz(i) + \sum_{i=k-\tau_2}^{k-\tau_1-1} z^*(i)Sz(i) + \sum_{i=k-\tau(k)}^{k-1} z^*(i)Tz(i)$$
$$V_3(k) = \tau_1 \sum_{j=-\tau_1}^{-1} \sum_{i=k+j}^{k-1} z^*(i)Uz(i) + \tau_2 \sum_{j=-\tau_2}^{-1} \sum_{i=k+j}^{k-1} z^*(i)Vz(i)$$
$$V_4(k) = (\tau_2 - \tau_1) \sum_{j=-\tau_2}^{-1-\tau_1} \sum_{i=k+j}^{k-1} z^*(i)Wz(i) + \sum_{j=\tau_1}^{-1+\tau_2} \sum_{i=k-j}^{k-1} z^*(i)Tz(i)$$
$$V_5(k) = \frac{\tau_1^2}{2} \sum_{i=-\tau_1}^{-1} \sum_{j=i}^{0} \sum_{l=k+j}^{k-1} \eta^*(l)G\eta(l) + \sum_{i=k-\tau_1}^{k-\tau_1-1} \sum_{j=i}^{0} \sum_{l=k+j}^{k-1} \eta^*(l)H\eta(l)$$

Letting $\Delta V(k) = V(k+1) - V(k)$, along the solution of the neural network (1), we have

$$\begin{aligned}\Delta V_1(k) &= z^*(k+1)Pz(k+1) - z^*(k)Pz(k)\\ &= f(z(k))^*A^*PA + f(z(k-\tau(k)))^*B^TPBf(z(k-\tau(k))) + u(k)^*Pu(k)\\ &\quad + 2f(z(k))^*A^*PBf(z(k-\tau(k))) + 2f(z(k-\tau(k)))^*B^*Pu(k) + 2u(k)^*PAf(z(k))\\ &\quad - z^*(k)Pz(k)\end{aligned} \tag{12}$$

$$\begin{aligned}\Delta V_2(k) &= z^*(k)(Q+R+T)z(k) + z^*(k-\tau_1)(-Q+S)z(k-\tau_1) + z^*(k-\tau_2)(-R-S)z(k-\tau_2)\\ &\quad - z^*(k-\tau(k))Tz(k-\tau(k)) + \sum_{i=k+1-\tau(k+1)}^{k-\tau(k)} z^*(i)Tz(i)\end{aligned} \tag{13}$$

$$\begin{aligned}\Delta V_3(k) &= z^*(k)(\tau_1^2U + \tau_2^2)Vz(k) - \Big[\sum_{i=k-\tau_1}^{k-1} z^*(i)\Big]U\Big[\sum_{i=k-\tau_1}^{k-1} z^*(i)\Big]\\ &\quad - \Big[\sum_{i=k-\tau_2}^{k-1} z^*(i)\Big]V\Big[\sum_{i=k-\tau_2}^{k-1} z^*(i)\Big]\end{aligned} \tag{14}$$

$$\begin{aligned}\Delta V_4(k) &= z^*(k)\Big[(\tau_2-\tau_1)^2W + (\tau_2-\tau_1)T\Big]z(k) - \Big[\sum_{i=k-\tau_2}^{k-\tau_1-1} z^*(i)\Big]W\Big[\sum_{i=i=k-\tau_2}^{k-\tau_1-1} z(i)\Big]\\ &\quad - \sum_{i=k+1-\tau_2}^{k-\tau_1}\Big[(z^*(i))Tz(i)\Big]\end{aligned} \tag{15}$$

$$\begin{aligned}\Delta V_5(k) &= z^*(k+1)\Big[\frac{\tau_1^4}{4}G + \frac{(\tau_2-\tau_1)^2}{4}H\Big]z(k+1) + z^*(k)\Big[\frac{\tau_1^4}{4}G + \frac{(\tau_2-\tau_1)^2}{4}H\Big]z(k)\\ &\quad - 2z^*(k+1)\Big[\frac{\tau_1^4}{4}G + \frac{(\tau_2-\tau_1)^2}{4}H\Big]z(k)\\ &\quad - \Big[\tau_1 z(k) - \sum_{i=k-\tau_1}^{k-1} z(k)\Big]^T G\Big[\tau_1 z(k) - \sum_{i=k-\tau_1}^{k-1} z(k)\Big]\\ &\quad - (z(k-\tau_1) - \psi(k))^*H(z(k-\tau_1) - \psi(k)) - (z(k-\tau_1) - \phi(k))^*H(z(k-\tau_1) - \phi(k))\\ \Delta V_5(k) &= f(z(k))^*A^*M\,Af(z(k)) + f(z(k-\tau(k))^*B^*M\,Bf(z(k-\tau(k)) + u(k)^*M\,u(k)\\ &\quad + 2f(z(k))^*A^*M\,Bf(z(k-\tau(k))) + 2f(z(k-\tau(k)))B^*Mu(k) + 2u(k)^*M\,Af(z(k))\\ &\quad + z(k)^*Mz(k) - 2f(z(k))^*A^*M\,z(k) - f(k-2\tau(k))^*B^*M\,z(k) - 2u(k)^*M\,z(k)\\ &\quad - \Big[\sum_{i=k-\tau_1}^{k-1} z(i)\Big]^* G\Big[\sum_{i=k-\tau_1}^{k-1} z(i)\Big] - 2\tau_1 z^*(k)G\Big[\sum_{i=k-\tau_1}^{k-1} z(i)\Big] - z^*(k-\tau_1))H\,z(k-\tau_1)\\ &\quad - z^*(k-\tau_2)H\,z(k-\tau_2) + 2z^*(k-\tau_1)H\,\psi(k) + 2z^*(k-\tau_2)H\,\phi(k)\\ &\quad - \psi^*(k)H\,\psi(k) - \phi^*(k))H\,\phi(k)\end{aligned} \tag{16}$$

Furthermore, from the Assumption 2.1, the activation function $f_j(\cdot)$ of (1) can be written as $l_j^{R-} \le \frac{f_j^R(x_j(k))}{x_j(k)} \le l_j^{R+}, l_j^{I-} \le \frac{f_j^I(y_j(k))}{y_j(k)} \le l_j^{I+}$ for all $j = 1, 2, \ldots, n$. Hence, we have

$$|f_j^R(x_j(k))| \le |g_j^R(x_j(k))|, \quad |f_j^I(x_j(k))| \le |g_j^I(x_j(k))| \tag{17}$$

where $g_j^R = max\{|l_j^{R-}|, |l_j^{R+}|\}, g_j^R = max\{|l_j^{I-}|, |l_j^{I+}|\}$ for all $j = 1, 2, \ldots, n$.

From (17), we get

$$s_j f(z_j(k))^* f(z_j(k)) \le s_j g_j^2 z_j(k)^* z_j(k) \tag{18}$$

where $g_j = max\{g_j^R, g_j^I\}$ and s_j is a positive constant for all $j = 1, 2, \ldots, n$. Therefore, we can write the vector form of (18) as follows:

$$f(z(k))^* F_1\, f(z(k)) \leq z(k)^* F_1 \Gamma\, z(k)$$
$$f(z(k))^* F_1\, f(z(k)) - z(k)^* F_1 \Gamma\, z(k) \leq 0 \tag{19}$$

where $F_1 = diag\{s_1, s_2, \ldots, s_n\}$.

Similarly,

$$f(z(k-\tau(k)))^* F_2\, f(z(k-\tau(k))) \leq z(k-\tau(k))^* F_2 \Gamma\, z(k-\tau(k))$$
$$f(z(k-\tau(k)))^* F_2\, f(z(k-\tau(k))) - z(k-\tau(k))^* F_2 \Gamma\, z(k-\tau(k)) \leq 0 \tag{20}$$

where $F_2 = diag\{s_1^{'}, s_2^{'}, \ldots, s_n^{'}\}$ and $\Gamma = diag\{g_1^2, g_2^2, \ldots, g_n^2\}$.

Now, $\Delta V(k) = \Delta V_1(k) + \Delta V_1(k) + \Delta V_1(k) + \Delta V_1(k) + \Delta V_1(k) + \Delta V_1(k) + 0 + 0$.

Substituting equations from (12) to (16) in $\Delta V(k)$ and using the inequalities (19) and (20) in the RHS of $\Delta V(k)$, we get

$$\Delta V(k) - y^*(k)\mathcal{Q}y(k) - 2y^*(k)\mathcal{S}u(k) - u^*(k)(\mathcal{R} - \gamma I)u(k) \leq \xi^*(k)\Theta\xi(k) \tag{21}$$

where

$$\xi^*(k) = \left[z(k)\ z(k-\tau_1)\ z(k-\tau_2)\ z(k-\tau(k))\ f(z(k))\ f(z(k-\tau(k)))\ \sum_{k-\tau_1}^{k-1} z(k)\ \sum_{k-\tau_2}^{k-1} z(k)\ \phi(k)\ u(k)\right]^*.$$

Thus,

$$\sum_{k=0}^{k_p} \Delta V(k) - \sum_{k=0}^{k_p} y^*(k)\mathcal{Q}y(k) - 2u^*(k)\mathcal{S}y(k) - u^*(k)(\mathcal{R} - \gamma I)u(k) \leq \sum_{k=0}^{k_p} \xi^*(k)\Theta\xi(k) \tag{22}$$

for all $k_p \in \mathbb{N}$.

Suppose $\Theta < 0$, then (22) yields

$$\sum_{k=0}^{k_p} \Delta V(k) \leq \sum_{k=0}^{k_p} [y^*(k)\mathcal{Q}y(k) + 2u^*(k)\mathcal{S}y(k) + u^*(k)(\mathcal{R} - \gamma I)u(k)]$$

$$V(x(k+1)) - V(x(0)) \leq \sum_{k=0}^{k_p} y^*(k)\mathcal{Q}y(k) + 2u^*(k)\mathcal{S}y(k) + u^*(k)(\mathcal{R} - \gamma I)u(k)]$$

for all $k_p \in \mathbb{N}$.

Thus (5) holds under the zero initial condition. Therefore, according to Definition 2.1, neural network (1) is strictly $(\mathcal{Q}, S, R)$-γ-dissipative. This completes the proof. □

The LMIs obtained in Theorem 3.1 ensures the $(\mathcal{Q}, S, R) - \gamma$-dissipativity of discrete-time complex-valued neural network (1). Further, we specialize Theorem 3.1 to obtain the passivity conditions for the system (1), by assuming $\mathcal{Q} = 0, \mathcal{S} = I$, and $\mathcal{R} = 2\gamma I$. The derived passivity conditions are presented in the following corollary.

COROLLARY 3.2 *Assume that Assumption 2.1 holds, then the complex-valued neural networks (1) are passive if there exist positive Hermitian matrices* $P = P_1 + iP_2$, $Q = Q_1 + iQ_2$, $R = R_1 + iR_2$, $S = S_1 + iS_2$, $T = T_1 + iT_2$, $U = U_1 + iU_2$, $V = V_1 + iV_2$, $W = W_1 + iW_2$, $G = G_1 + iG_2$, $H = H_1 + iH_2$, *two positive diagonal matrices* $F_1 > 0$, $F_2 > 0$, *and a scalar* $\gamma > 0$ *such that the following LMI holds.*

$$\Sigma = \begin{pmatrix} \Sigma^R & -\Sigma^I \\ \Sigma^I & \Sigma^R \end{pmatrix} < 0 \tag{23}$$

where

$$\Sigma^R = \begin{pmatrix} \Sigma^R_{1,1} & 0 & 0 & 0 & \Sigma^R_{1,5} & \Sigma^R_{1,6} & \Sigma^R_{1,7} & 0 & 0 & 0 & \Sigma^R_{1,11} \\ 0 & \Sigma^R_{2,2} & 0 & 0 & 0 & 0 & 0 & 0 & \Sigma^R_{2,9} & 0 & 0 \\ 0 & 0 & \Sigma^R_{3,3} & 0 & 0 & 0 & 0 & 0 & \Sigma^R_{3,9} & 0 & 0 \\ 0 & 0 & 0 & \Sigma^R_{4,4} & 0 & 0 & 0 & 0 & 0 & 0 & 0 \\ 0 & 0 & 0 & 0 & \Sigma^R_{5,5} & \Sigma^R_{5,6} & 0 & 0 & 0 & 0 & \Sigma^R_{5,11} \\ 0 & 0 & 0 & 0 & 0 & \Sigma^R_{6,6} & 0 & 0 & 0 & 0 & \Sigma^R_{6,11} \\ 0 & 0 & 0 & 0 & 0 & 0 & \Sigma^R_{7,7} & 0 & 0 & 0 & 0 \\ 0 & 0 & 0 & 0 & 0 & 0 & 0 & \Sigma^R_{8,8} & 0 & 0 & 0 \\ 0 & 0 & 0 & 0 & 0 & 0 & 0 & 0 & \Sigma^R_{9,9} & 0 & 0 \\ 0 & 0 & 0 & 0 & 0 & 0 & 0 & 0 & 0 & \Sigma^R_{10,10} & 0 \\ 0 & 0 & 0 & 0 & 0 & 0 & 0 & 0 & 0 & 0 & \Sigma^R_{11,11} \end{pmatrix} \tag{24}$$

and

$$\Sigma^I = \begin{pmatrix} \Sigma^I_{1,1} & 0 & 0 & 0 & \Sigma^I_{1,5} & \Sigma^I_{1,6} & \Sigma^I_{1,7} & 0 & 0 & 0 & \Sigma^I_{1,11} \\ 0 & \Sigma^I_{2,2} & 0 & 0 & 0 & 0 & 0 & 0 & \Sigma^I_{2,9} & 0 & 0 \\ 0 & 0 & \Sigma^I_{3,3} & 0 & 0 & 0 & 0 & 0 & \Sigma^I_{3,9} & 0 & 0 \\ 0 & 0 & 0 & \Sigma^I_{4,4} & 0 & 0 & 0 & 0 & 0 & 0 & 0 \\ 0 & 0 & 0 & 0 & \Sigma^I_{5,5} & \Sigma^I_{5,6} & 0 & 0 & 0 & 0 & \Sigma^I_{5,11} \\ 0 & 0 & 0 & 0 & 0 & \Sigma^I_{6,6} & 0 & 0 & 0 & 0 & \Sigma^I_{6,11} \\ 0 & 0 & 0 & 0 & 0 & 0 & \Sigma^I_{7,7} & 0 & 0 & 0 & 0 \\ 0 & 0 & 0 & 0 & 0 & 0 & 0 & \Sigma^I_{8,8} & 0 & 0 & 0 \\ 0 & 0 & 0 & 0 & 0 & 0 & 0 & 0 & \Sigma^I_{9,9} & 0 & 0 \\ 0 & 0 & 0 & 0 & 0 & 0 & 0 & 0 & 0 & \Sigma^I_{10,10} & 0 \\ 0 & 0 & 0 & 0 & 0 & 0 & 0 & 0 & 0 & 0 & \Sigma^I_{11,11} \end{pmatrix} \tag{25}$$

with

$$\begin{aligned}
\Sigma^R_{1,1} &= \Theta^R_{1,1}, \Sigma^R_{1,5} = \Theta^R_{1,5}, \Sigma^R_{1,6} = \Theta^R_{1,6}, \Sigma^R_{1,7} = \Theta^R_{1,7}, \Sigma^R_{1,11} = \Theta^R_{1,11}, \Sigma^R_{2,2} = \Theta^R_{2,2}, \Sigma^R_{2,9} = \Theta^R_{2,9} \\
\Sigma^R_{3,3} &= \Theta^R_{3,3}, \Sigma^R_{3,9} = \Theta^R_{3,9}, \Sigma^R_{4,4} = \Theta^R_{4,4}, \Sigma^R_{5,6} = \Theta^R_{5,6}, \Sigma^R_{6,6} = \Theta^R_{6,6}, \Sigma^R_{6,11} = \Theta^R_{6,11}, \Sigma^R_{7,7} = \Theta^R_{7,7} \\
\Sigma^R_{8,8} &= \Theta^R_{8,8}, \Sigma^R_{9,9} = \Theta^R_{9,9}, \Sigma^R_{10,10} = \Theta^R_{10,10}, \Sigma^I_{1,1} = \Theta^I_{1,1}, \Sigma^I_{1,5} = \Theta^I_{1,5}, \Sigma^I_{1,6} = \Theta^I_{1,6}, \Sigma^I_{1,7} = \Theta^I_{1,7} \\
\Sigma^I_{1,11} &= \Theta^I_{1,11}, \Sigma^I_{2,2} = \Theta^I_{2,2}, \Sigma^I_{2,9} = \Theta^I_{2,9} \Sigma^I_{3,3} = \Theta^I_{3,3}, \Sigma^I_{3,9} = \Theta^I_{3,9}, \Sigma^I_{4,4} = \Theta^I_{4,4}, \Sigma^I_{6,6} = \Theta^I_{6,6} \\
\Sigma^I_{6,11} &= \Theta^I_{6,11}, \Sigma^I_{7,7} = \Theta^I_{7,7}, \Sigma^I_{8,8} = \Theta^I_{8,8}, \Sigma^I_{9,9} = \Theta^I_{9,9}, \Sigma^I_{10,10} = \Theta^I_{10,10} \\
\Sigma^R_{5,5} &= A_1^T P_1 A_1 - A_1^T P_2 A_2 + A_2^T P_2 A_1 + A_2^T P_1 A_2 + A_1^T M_1 A_1 - A_1^T M_2 A_2 + A_2^T M_2 A_1 + A_2^T M_1 A_2 - F_1 \\
\Sigma^R_{5,11} &= P_1 A_1 - P_2 A_2 + M_1 A_1 - M_2 A_2 - I, \Sigma^R_{11,11} = -\gamma I + P_1 M_1 \\
\Sigma^I_{5,5} &= A_1^T P_2 A_1 + A_1^T P_1 A_2 - A_2^T P_1 A_1 + A_2^T P_2 A_2 + A_1^T M_2 A_1 + A_1^T M_1 A_2 - A_2^T M_1 A_1 + A_2^T M_2 A_2 \\
\Sigma^I_{5,6} &= A_1^T P_2 B_1 + A_1^T P_1 B_2 - A_2^T P_1 B_1 + A_2^T P_2 B_2, \Sigma^I_{5,11} = P_1 A_2 - P_2 A_1 + M_1 A_2 - M_2 A_1 - I \\
\Sigma^I_{11,11} &= P_2 M_2
\end{aligned}$$

Proof The proof is same as that of Theorem 3.1 and hence it is omitted.

4. Numerical examples

In this section, we will give an example showing the effectiveness of established theories.

Example 4.1 Consider the discrete-time complex-valued neural networks (1), where the interconnected matrices are, respectively,

$$A = \begin{bmatrix} 0.2 + 0.1j & -0.2 + 0.2j \\ -0.1 + 0.1j & 0.2 \end{bmatrix}, \quad B = \begin{bmatrix} -0.2 - 0.2j & -0.1 \\ 0.1 + 0.3j & -0.1 + 0.2j \end{bmatrix}$$

$$\mathcal{Q} = \begin{bmatrix} 5 & 2.2 + 1.4i \\ 2.2 - 1.4i & 3 \end{bmatrix}, \mathcal{R} = \begin{bmatrix} 4.5 & -0.5 - i \\ -0.5 + i & 2.5 \end{bmatrix}, \mathcal{S} = \begin{bmatrix} 0.3 + 0.5i & -0.6 - 0.2i \\ 0.4 - 0.6i & 0.5 + 0.3i \end{bmatrix}$$

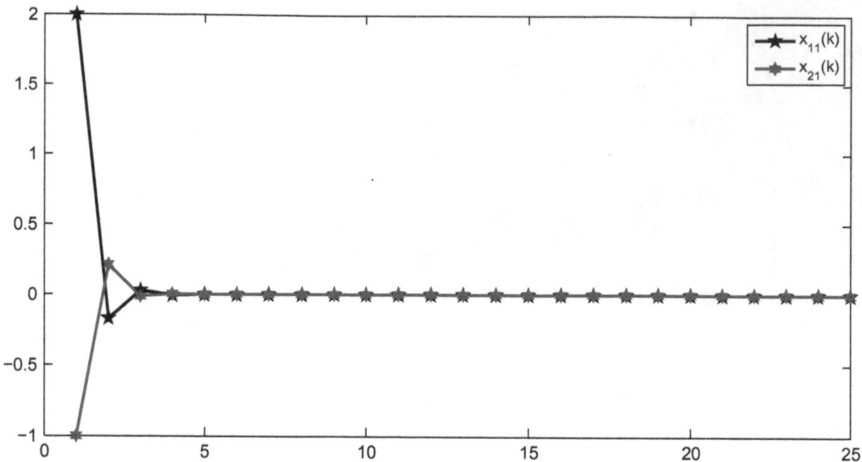

Figure 1. State trajectories of real part of two-neuron complex-valued neural networks for $\tau_1(k) = 2.5 + 0.5\sin(0.5k\pi)$ **and** $\tau_2(k) = 4.5 + 0.5\sin(0.5k\pi)$ **with initial states** $x_{11} = 2 + 2j$.

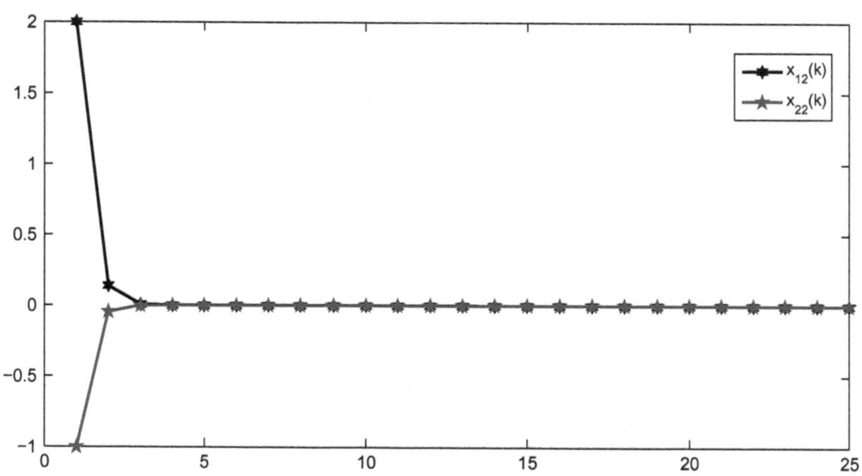

Figure 2. State trajectories of imaginary part of two-neuron complex-valued neural networks for $\tau_1(k) = 2.5 + 0.5\sin(0.5k\pi)$ **and** $\tau_2(k) = 4.5 + 0.5\sin(0.5k\pi)$ **with initial states** $x_{12} = -1 - j$.

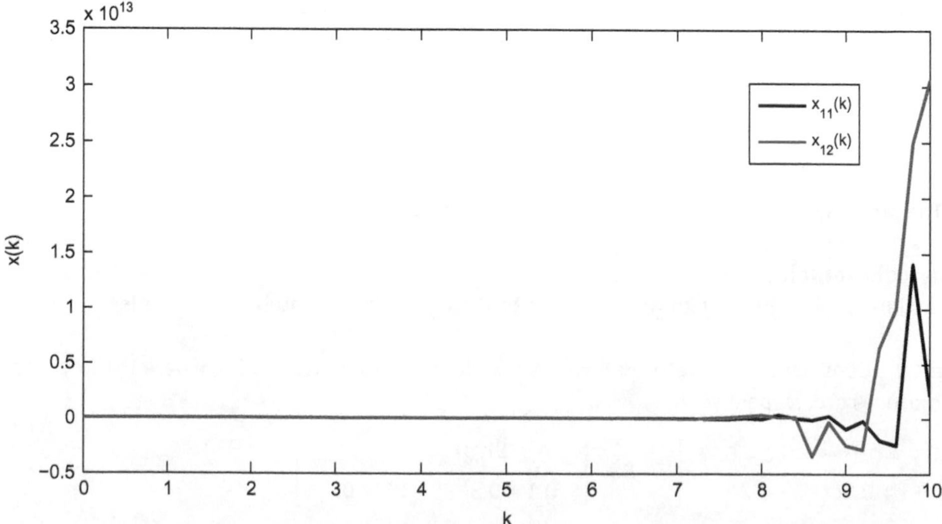

Figure 3. State trajectories of real part of two-neuron complex-valued neural networks for $\tau_1(k) = 3.5 + 0.5\sin(0.5k\pi)$ **and** $\tau_2(k) = 5.5 + 0.5\sin(0.5k\pi)$ **with initial states** $x_{11} = 2 + 2j$.

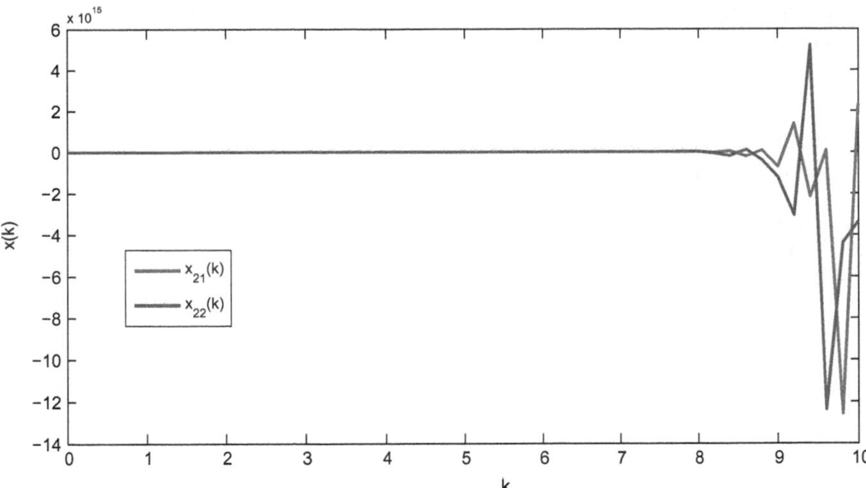

Figure 4. State trajectories of imaginary part of two-neuron complex-valued neural networks for $\tau_1(k) = 3.5 + 0.5\sin(0.5k\pi)$ **and** $\tau_2(k) = 5.5 + 0.5\sin(0.5k\pi)$ **with initial states** $x_{12} = -1 - j$.

Here, the activation functions are assumed to be $f(x) = \frac{1-e^{-x}}{1+e^{x}}$, $f(y) = \frac{1}{1+e^{x}}$ with $F_1 = diag\{0.01, 0.01\}$, $F_2 = diag\{0.1, 0.1\}$, and $\Gamma = diag\{0.1, 0.1\}$. Taking $\tau_1(k) = 2.5 + 0.5\sin(0.5k\pi)$, $\tau_2(k) = 4.5 + 0.5\sin(0.5k\pi)$, using the Matlab LMI control toolbox for LMI (9), the feasible matrices are sought as

$$P_1 = \begin{bmatrix} 17.1077 & 3.3520 \\ * & 20.3736 \end{bmatrix}, \quad P_2 = \begin{bmatrix} 3.4822 & 6.9148 \\ * & 18.1144 \end{bmatrix},$$

$$Q_1 = \begin{bmatrix} 2.2241 & 0.8264 \\ * & 3.1205 \end{bmatrix}, \quad Q_2 = \begin{bmatrix} 1.7735 & 0.6018 \\ * & 3.0145 \end{bmatrix}$$

$$R_1 = \begin{bmatrix} 0.9142 & 0.3672 \\ * & 1.3126 \end{bmatrix}, \quad R_2 = \begin{bmatrix} 0.7175 & 0.2633 \\ * & 1.2604 \end{bmatrix},$$

$$S_1 = \begin{bmatrix} 1.3086 & 0.4592 \\ * & 1.8068 \end{bmatrix}, \quad S_2 = \begin{bmatrix} 1.0548 & 0.3385 \\ * & 1.7529 \end{bmatrix}$$

$$T_1 = \begin{bmatrix} 1.9012 & 0.6074 \\ * & 2.5871 \end{bmatrix}, \quad T_2 = \begin{bmatrix} 2.3327 & 1.1674 \\ * & 3.9043 \end{bmatrix},$$

$$U_1 = \begin{bmatrix} 0.3527 & 0.1549 \\ * & 0.5209 \end{bmatrix}, \quad U_2 = \begin{bmatrix} 0.2547 & 0.1052 \\ * & 0.4923 \end{bmatrix}$$

$$V_1 = \begin{bmatrix} 0.0626 & 0.0309 \\ * & 0.0985 \end{bmatrix}, \quad V_2 = \begin{bmatrix} 0.0587 & 0.0240 \\ * & 0.0951 \end{bmatrix},$$

$$W_1 = \begin{bmatrix} 0.1692 & 0.0778 \\ * & 0.2535 \end{bmatrix}, \quad W_2 = \begin{bmatrix} 0.1288 & 0.0539 \\ * & 0.2400 \end{bmatrix}$$

$$G_1 = \begin{bmatrix} 0.0338 & 0.0231 \\ * & 0.0585 \end{bmatrix}, \quad G_2 = \begin{bmatrix} 0.0615 & 0.0236 \\ * & 0.0608 \end{bmatrix},$$

$$H_1 = \begin{bmatrix} 0.1549 & 0.1186 \\ * & 0.2980 \end{bmatrix}, \quad H_2 = \begin{bmatrix} 0.4757 & 0.1626 \\ * & 0.2913 \end{bmatrix}$$

Setting the initial states as $x_{11} = 2 + 2j$ and $x_{12} = -1 - j$, Figures 1 and 2 show that the model (1) with above given parameters is dissipative in the sense of Definition 2.1 with $\gamma = 58.9167$. Further, the state curves for the real and imaginary parts of the discrete-time complex-valued neural networks (1) have been given in Figures 1 and 2. When $\tau_1(k) = 3.5 + 0.5\sin(0.5k\pi)$ and

$\tau_2(k) = 5.5 + 0.5\sin(0.5k\pi)$, the LMI (9) in Theorem 3.1 is not feasible and hence the CVNNs (1) is not $(\mathcal{Q}, \mathcal{S}, \mathcal{R})$-dissipative. In this case, Figures 3 and 4 describe the unstable behavior of the trajectories of the CVNNs (1).

Remark 4.1 Different from the Lyapunov functional $V(k)$ given in Zhang, Wang, Lin, and Liu (2014), in our paper, we have constructed the appropriate Lyapunov functional involving the terms

$$V_4(k) = (\tau_2 - \tau_1) \sum_{j=-\tau_2}^{-1-\tau_1} \sum_{i=k+j}^{k-1} \left[Z^*(l)HZ(l)\right]$$

$$+ \sum_{j=\tau_1}^{-1+\tau_2} \sum_{i=k-j}^{k-1} \left[(z^*(i)Tz(i)\right]$$

$$V_5(k) = \frac{\tau_1^2}{2} \sum_{i=-\tau_1}^{-1} \sum_{j=i}^{0} \sum_{l=k+j}^{k-1} \left[\eta^*(l)H\eta(l)\right]$$

$$+ \sum_{i=k-\tau_1}^{k-\tau_1-1} \sum_{j=i}^{0} \sum_{l=k+j}^{k-1} \left[(\eta^*(l)H\eta(l)\right]$$

Further, Lemma 2.3 is used to reduce the triple summation terms in $\Delta V_5(k)$. In Zhang et al. (2014), the maximum values of upper bounds are obtained as $\tau_1 = 1$ and $\tau_2 = 2$ whereas the proposed results in our paper yield $\tau_1 = 3$ and $\tau_2 = 5$. Hence, the results proposed in Theorem 3.1 are less conservative than those obtained in Zhang et al. (2014).

5. Conclusions

In this paper, dissipativity and passivity analysis for discrete-time complex-valued neural networks with time-varying delays was studied. A delay-dependent condition has been provided to ensure the considered neural network to be strictly $(\mathcal{Q}, S, R)$-γ dissipative. An effective LMI approach has been proposed to derive the dissipativity criterion. Based on the new bounding technique and appropriate type of Lyapunov functional, a sufficient condition for the solvability of this problem is established for the dissipativity criterion. One numerical example is given to show the effectiveness of the established results. We would like to point out that it is possible to generalize our main results to more complex systems, such as neural networks with parameter uncertainties, stochastic perturbations, and Markovian jumping parameters.

Funding
The work of authors was supported by UGC-BSR Research Start-Up-Grant, New Delhi, India, under the sanctioned No. F. 20-1/2012 (BSR)/20-5(13)/2012(BSR).

Author details
G. Nagamani[1]
E-mail: nagamanigru@gmail.com
S. Ramasamy[1]
E-mail: ramasamygru@gmail.com
1 Department of Mathematics, Gandhigram Rural Institute - Deemed University, Gandhigram, Tamil Nadu, 624 302 India.

References

Aizenberg, I., Paliy, D. V., Zurada, J. M., & Astola, J. T. (2008). Blur identification by multilayer neural network based on multivalued neurons. *IEEE Transaction on Neural Networks, 19*, 883–898.

Bastinec, J., Diblik, J., & Smarda, Z. (2010). Existence of positive solutions of discrete linear equations with a single delay. *Journal of Difference Equations and Applications, 16*, 1165–1177.

Chua, L. O. (1999). Passivity and complexity. *IEEE Transactions on Circuit and Systems, 46*, 71–82.

Diblik, J., Schmeidel, E., & Ruzickova, M. (2010). Asymptotically periodic solutions of Volterra system of difference equations. *Computers and Mathematics with Applications, 59*, 2854–2867.

Goh, S. L., & Mandic, D. P. (2007). An augmented extended Kalman filter algorithm for complex valued recurrent neural networks. *Neural Computation, 19*(4), 1–17.

Goh, S. L., & Mandic, D. P. (2005). Nonlinear adaptive prediction of complex valued nonstationary signals. *IEEE Transactions on Signal Processing, 53*, 1827–1836.

Hirose, A. (2003). *Complex-valued neural networks: Theory and applications*. Vol. 5, *Series on innovative intelligence*. River Edge, NJ: World Scientific.

Hirose, A. (2011). Nature of complex number and complex valued neural networks. *Frontiers of Electrical and Electronic Engineering in China, 6*, 171–180.

Hu, J., & Wang, J. (2012). Global stability of complex-valued recurrent neural networks with time-delays. *IEEE Transaction on Neural Networks and Learning Systems, 23*, 853–865.

Jing, W., Yao, F., & Shen, H. (2014). Dissipativity-based state estimation for Markov jump discrete-time neural networks with unreliable communication links. *Neurocomputing, 139*, 107–113.

Liang, J., Wang, Z., & Liu, X. (2009). State estimation for coupled uncertain stochastic networks with missing measurements and time-varying delays: The discrete-time case. *IEEE Transactions on Neural Networks, 20*, 781–793.

Liu, Y., Wang, Z., Liang, J., & Liu, X. (2009). Stability and synchronization of discrete-time Markovian jumping neural networks with mixed mode-dependent time delays. *IEEE Transactions on Neural Networks, 20*, 1102–1116.

Mostafa, M., Teich, W. G., & Lindner, J. (2013). Local stability analysis of discrete-time, continuous-state, complex-valued recurrent neural networks with inner state feedback. *IEEE Transactions on Neural Networks and Learning Systems, 25*, 830–836. doi:10.1109/TNNLS.2013.2281217

Suksmono, A. B., & Hirose, A. (2002). Adaptive noise reduction of insar image based on complex-valued MRF model and its application to phase unwrapping problem. *IEEE Transactions on Geoscience and Remote Sensing, 40*, 699–709.

Wang, T., Xue, M., Fei, S., & Li, T. (2013). Triple Lyapunov functional technique on delay-dependent stability for discrete-time dynamical networks. *Neurocomputing, 122*, 221–228.

Wang, Z., Ho, D. W. C., Liu, Y., & Liu, X. (2009). Robust H8 control for a class of nonlinear discrete time delay stochastic systems with missing measurements. *Automatica, 45*, 684–691.

Wu, L., Yang, X., & Lam, H. K. (2014). Dissipativity analysis and synthesis for discrete-time T-S fuzzy stochastic systems with time-varying delay. *IEEE Transactions on Fuzzy Systems, 22*, 380–394.

Wu, Z. G., Shi, P., Su, H., & Chu, J. (2011). Passivity analysis for discrete-time stochastic Markovian jump neural networks with mixed time delays. *IEEE Transaction on Neural Networks, 22*, 1566–1575.

Wu, Z. G., Shi, P., Su, H., & Chu, J. (2013). Dissipativity analysis for discrete-time stochastic neural networks with time-varying delays. *IEEE Transactions on Neural networks, 22*, 345–355.

Yamaki, R., & Hirose, A. (2009). Singular unit restoration in interferograms based on complex valued Markov random field model for phase unwrapping. *IEEE Geoscience and Remote Sensing Letters, 6*, 18–22.

Zhang, H., Wang, X. Y., Lin, X. H., & Liu, C. X. (2014). Stability and synchronization for discrete-time complex-valued neural networks with time-varying delays. *PLoS ONE, 9*, e93838. doi:10.1371/journal.pone.0093838

Zhang, Y., & Ma, Y. (1997). CGHA for principal component extraction in the complex domain. *IEEE Transactions on Neural Networks, 8*, 1031–1036.

Zhao, Z., Song, Q., & He, S. (2014). Passivity analysis of stochastic neural networks with time-varying delays and leakage delay. *Neurocomputing, 125*, 22–27.

Zhou, B., & Song, Q. (2013). Boundedness and complete stability of complex-valued neural networks with time delay. *IEEE Transactions on Neural Networks and Learning Systems, 24*, 1227–1238.

4

Fuzzy bi-criteria scheduling on parallel machines involving weighted flow time and maximum tardiness

Kewal Krishan Nailwal[1], Deepak Gupta[2] and Sameer Sharma[3*]

*Corresponding author: Sameer Sharma, Department of Mathematics, D.A.V. College, Jalandhar, Punjab 144008, India

E-mail: samsharma31@gmail.com

Reviewing editor: Yong Hong Wu, Curtin University of Technology, Australia

Abstract: This paper considers a bi-criteria scheduling on parallel machines in fuzzy environment which optimizes the weighted flow time and maximum tardiness. The fuzziness, vagueness or uncertainty in processing time of jobs is represented by triangular fuzzy membership function. The bi-criteria problem with weighted flow time and maximum tardiness as primary and secondary criteria, respectively, for any number of parallel machines is NP-hard. So, a heuristic algorithm is proposed to find the optimal sequence of jobs processing on parallel machines so as to minimize the secondary criterion of maximum tardiness without violating the primary criterion of weighted flow time. A numerical illustration is also given in support of the algorithm proposed.

Subjects: Mathematics & Statistics; Statistics & Probability; Operations Research; Optimization

Keywords: fuzzy processing time; average high ranking; maximum tardiness; weighted flow time; due date; weighted shortest processing time

1. Introduction

Scheduling is concerned with the optimal allocation of scare resources with objective of optimizing one or several criteria's (Baker, 1974). Scheduling of jobs has been a fruitful area of research for many decades, in which order of processing of jobs is decided. If the jobs are scheduled properly, it

ABOUT THE AUTHORS

The group of authors involved in this critically researched paper consists in Deepak Gupta, Sameer Sharma, and Kewal Krishan Nailwal. The team under the competent supervision of Deepak Gupta works in the area of scheduling problems. Some research work on optimizing scheduling problems in industry with various parameters along with the linkage models has been contributed. This paper particularly optimizes a fuzzy bi-criteria problem involving weighted flow time and tardiness. This paper will go in a long way to solve the problems faced in production management where a quality service means product is to be processed with minimum weighted flow time and meeting customer's required due date or satisfaction of the customer. Besides this, a fuzzy environment of the problem brings the problem closer to real-life application area as fuzzy theory tackles well with vagueness and uncertainty in processing times of jobs.

PUBLIC INTEREST STATEMENT

In this bi-objective parallel machine scheduling problem a system of n-independent jobs are to be processed on any of the available m-identical parallel machines. The processing times of jobs are considered to be uncertain or vague in nature due to the presence of some external sources. Fuzzy set theory is used to handle the uncertainty involved. The delay in processing of jobs is called tardiness or lateness and the delayed job is called tardy job. The flow time of a job is the total time it spends in the system i.e. sum of waiting times and processing times. In case when the jobs have different degree of importance, indicated by the weight of the job, the total weighted flow time is the simplest measure of the quality of service received by the jobs. Therefore, the paper seeks to highlight optimization of maximum tardiness and minimum weighted flow time of jobs.

not only saves time but also increases the efficiency of the system. The classical parallel machine scheduling problem includes the scheduling of n- independent jobs say {1, 2, ..., n}, on a set of m-identical parallel machines say {1, 2, ..., m}. Every job is considered with a deterministic processing time p_i, a release date r_i, and a due date d_i while optimizing the objective function as number of tardy jobs, maximum tardiness, weighted flow time, maximum completion time, etc. However, due to the sources of uncertainty and complex nature of the problems in real-time situations, it is quite erratic to set them as precise values. Some researchers work on the concept of probability distribution to model parallel machine problem that is usually predicted from chronological data. But, whenever production data are unreliable or presented in vague way, stochastic models may not be the best choice. Fuzzy set theory may provide an alternative approach for dealing with the vagueness. Zadeh (1965) was first to introduce fuzzy sets as a mathematical way of representing impreciseness or vagueness in everyday life. Fuzzy set theory has numerous applications in various fields such as engineering, medicine, manufacturing, and others. In real-life situations, the processing times of jobs are not always exact due to incomplete knowledge or uncertain environment which implies the existence of various external sources and types of uncertainty. Fuzzy set theory can be used to handle uncertainty inherent in actual scheduling problems.

On single machine bi-criteria, Smith (1956) was the first to consider minimization of weighted completion time subject to minimal value of T_{max}. The other literature on single machine bi-criteria includes (Chand & Schneeberger, 1986; Gupta, Hariri, & Potts, 1999). Su, Chen, and Chen (2012) dealt with scheduling on parallel machines to minimize the maximum tardiness. For parallel-machine scheduling problems concerning bi-criterion, Parkash (1997) studied the bi-criteria scheduling problems on parallel machines and analyzed that the bi-criteria function is NP-hard. Sarin and Hariharan (2000) proposed a heuristic for a two-machine case with objective as minimizing number of tardy jobs under the constraint of minimizing maximum tardiness. A review of scheduling on parallel machines was given by Mokotoff (2001). Gupta, Ruiztorres, and Webster (2003) considered the scheduling of jobs on identical parallel machines for minimizing maximum tardiness with optimal flow time. In his paper, the type of scheduling problems is presented as: Identical parallel machines; Uniform parallel machines and unrelated parallel machines. Optimizing dual performance measures on parallel machines in fuzzy environment is fairly an open area of research. A survey of literature has revealed little work reported on the bi-criteria scheduling problems on parallel machine in fuzzy environment. Jha, Indumati, Singh, and Gupta (2011) formulated a fuzzy bi-criterion release time problem related to cost minimization and reliability maximization objectives under budgetary constraint. Sunita and Singh (2010a, 2010b) studied the optimization of various characteristics on parallel machines in fuzzy environment. Gupta, Gupta, Sharma, and Aggarwal (2012) developed an algorithm for the bi-objective problem which optimizes the number of tardy jobs without violating the value of T_{max} with uncertain processing time. Sharma, Gupta, and Seema (2013) developed an algorithm to schedule jobs on parallel identical machines so as to minimize the secondary criteria of weighted flow time without violating the primary criteria of maximum tardiness in fuzzy environment. Sharma, Gupta, and Seema (2012) studied the bi-objective problem with total tardiness and number of tardy jobs as primary and secondary criteria, respectively, for any number of parallel machines in fuzzy environment.

The two approaches can be involved for bi-criteria problems: both the criteria's are optimized simultaneously by using suitable weights for the criteria's. Second, the criteria's are optimized sequentially by first optimizing the primary criterion and then the secondary criterion subject to the value obtained for the primary criterion. In the present paper, we study the bi-objective scheduling on parallel machines in which the problem is divided into two steps: in the primary step, the weighted flow time of jobs is calculated and in the secondary step, the maximum tardiness is minimized with bi-objective function as T_{max}/WFT. The concept of fuzziness in processing time of jobs is introduced as in real-life situations the processing time of jobs may vary due to the lack of complete knowledge and the decision-making process gets more complicated with increment in the number of constraints.

The rest of the paper is organized as follows: In Section 2, problem is formulated. Section 3 describes the basics of fuzzy set theory. Section 4 deals with various theorems for optimizing the bi-criteria problem. Section 5 describes the algorithm proposed to find the optimal sequence for bi-criteria problem T_{max}/WFT. In Section 6, numerical illustrations are given to support the proposed algorithm. The paper is concluded in Section 7 followed by the references.

2. Problem formulation

The following notations have been used throughout the present paper.

i	designate the ith job, $i = 1, 2, 3, ..., n$
d_i	due date of the ith job
C_i	completion time of the ith job
w_i	weightage of the ith job
T_i	tardiness of the ith job = max $(C_i - d_i, 0)$
T_{max}	maximum tardiness
n	total number of jobs to be scheduled
m	total number of parallel machines
WFT(S)	weighted flow time of jobs for sequence S

The following assumptions are taken into consideration for developing bi-criteria algorithm on parallel machines.

(i) The jobs are available at time zero.

(ii) No pre-emption of jobs is allowed.

(iii) The machines are identical in all respects.

(iv) The machines are available all the time.

(v) The jobs are independent of each other.

(vi) Each machine can process at most one job at a time.

Before formulating the bi-criteria problem, the mathematical formulation for the single criterion's are represented first. These are:

2.1. Criterion: weighted flow time

The weighted flow time is the weighted version of flow time. The formulation to minimize the WFT is as follows:

$$\text{Minimize WFT}(S) = \sum_{i=1}^{n} w_i C_i$$

Subject to the condition that all weights for jobs are non-negative,

Here, S is the given sequence and WFT(S) is the corresponding objective function of weighted flow time.

2.2. Criterion: maximum tardiness

Tardiness is given by $T_i = \max(C_i - d_i, 0)$, where C_i and d_i are the completion time and due date of the ith job, respectively. Maximum tardiness is given by $T_{max} = \max(T_i)$.

The formulation is as follows:

$$\text{Minimize } Z = T_{max},$$
$$\text{Subject to constraint: } T_{max} \geq C_i - d_i \quad \forall i$$

along with the non-negativity constraints.

The formulation of the bi-criteria problems is similar to that of single criteria problems but with some additional constraints requiring that the optimal value of the primary objective function is not violated. Here, the problem is divided into two steps: one is primary in which weighted flow time of jobs is minimized and in secondary step, the maximum tardiness of jobs is minimized under the objective function value of primary criterion.

3. Basic fuzzy set theory

A fuzzy set $\tilde{A}$ defined on the universal set of real numbers R, is said to be a fuzzy number if its membership function has the following characteristics:

(i) $\mu_{\tilde{A}}: R \to [0, 1]$ is continuous.

(ii) $\mu_{\tilde{A}} = 0$ for all $x \in (-\infty, a_1) \cup (a_3, \infty)$.

(iii) $\mu_{\tilde{A}}$ is strictly increasing on $[a_1, a_2]$ and strictly decreasing on $[a_2, a_3]$.

(iv) $\mu_{\tilde{A}} = 1$ for $x = a_2$.

3.1. Triangular fuzzy number

Triangular fuzzy number (TFN) is a fuzzy number represented with three points as $\tilde{A} = (a_1, a_2, a_3)$, where a_1 and a_3 denote the lower and upper limits of support of a fuzzy set $\tilde{A}$. The membership value of the x denoted by $\mu_{\tilde{A}}(x), x \in R^+$, can be calculated according to the following formula (Figure 1).

$$\mu_{\tilde{A}}(x) = \begin{cases} 0; & x \leq a_1 \\ \frac{x-a_1}{a_2-a_1}; & a_1 < x < a_2 \\ \frac{a_3-x}{a_3-a_2}; & a_2 < x < a_3 \\ 0; & x \geq a_3 \end{cases}$$

3.2. Average high ranking

The system characteristics are described by membership function; it preserves the fuzziness of input information. However, the designer would prefer one crisp value for one of the system characteristics rather than fuzzy set. In order to overcome this problem, we defuzzify the fuzzy values of system characteristic by using the Yager's (1981) formula. If $\tilde{A} = (a_1, a_2, a_3)$, then its crisp value is defined as:

$$\text{crisp}\,(\tilde{A}) = h(\tilde{A}) = \frac{3a_2 + a_3 - a_1}{3} \tag{1}$$

3.3. Fuzzy arithmetic operations

If $\tilde{A}_1 = (m_{A_1}, \alpha_{A_1}, \beta_{A_1})$ and $\tilde{A}_2 = (m_{A_2}, \alpha_{A_2}, \beta_{A_2})$ be the two TFNs, then

(i) $\tilde{A}_1 + \tilde{A}_2 = (m_{A_1}, \alpha_{A_1}, \beta_{A_1}) + (m_{A_2}, \alpha_{A_2}, \beta_{A_2}) = (m_{A_1} + m_{A_2}, \alpha_{A_1} + \alpha_{A_2}, \beta_{A_1} + \beta_{A_1})$.

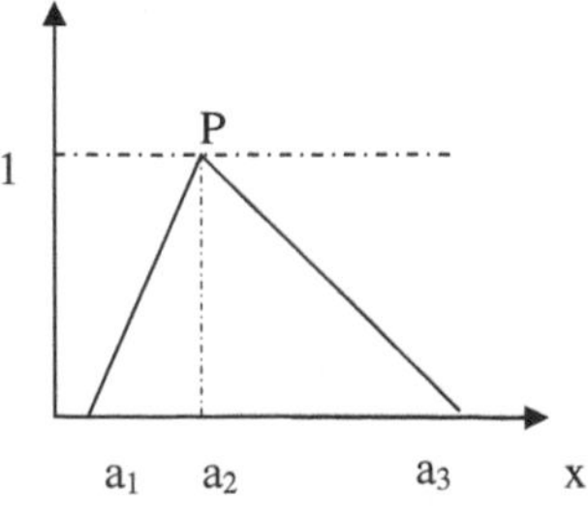

Figure 1. Triangular fuzzy number.

(ii) $\tilde{A}_1 - \tilde{A}_2 = (m_{A_1}, \alpha_{A_1}, \beta_{A_1}) - (m_{A_2}, \alpha_{A_2}, \beta_{A_2}) = (m_{A_1} - m_{A_2}, \alpha_{A_1} - \alpha_{A_2}, \beta_{A_1} - \beta_{A_2})$, if the following condition is satisfied $DP\left(\tilde{A}_1\right) \geq DP\left(\tilde{A}_2\right)$, where $DP(\tilde{A}_1) = \frac{\beta_{A_1} - m_{A_1}}{2}$ and $DP(\tilde{A}_2) = \frac{\beta_{A_2} - m_{A_2}}{2}$. Here, *DP* denotes difference point of a TFN.

Otherwise; $\tilde{A}_1 - \tilde{A}_2 = (m_{A_1}, \alpha_{A_1}, \beta_{A_1}) - (m_{A_2}, \alpha_{A_2}, \beta_{A_2}) = (m_{A_1} - \beta_{A_2}, \alpha_{A_1} - \alpha_{A_2}, \beta_{A_1} - m_{A_2})$.

(iii) $k\tilde{A}_1 = k(m_{A_1}, \alpha_{A_1}, \beta_{A_1}) = (km_{A_1}, k\alpha_{A_1}, k\beta_{A_1})$; if $k > 0$.

(iv) $k\tilde{A}_1 = k(m_{A_1}, \alpha_{A_1}, \beta_{A_1}) = (k\beta_{A_1}, k\alpha_{A_1}, km_{A_1})$; if $k < 0$.

4. Theorems

The following theorems have been established to optimize the bi-criteria scheduling on parallel machines involving weighted flow time and maximum tardiness.

4.1. Theorem

A sequence *S* of jobs following weighted smallest processing time (WSPT) rule, in which the jobs are placed to the earliest available location on the machines in WSPT order, minimizes the weighted flow time.

Proof Let if possible sequence *S* obtained by using the WSPT rule, i.e. by arranging the jobs in the decreasing order of their weights; break the ties (if any) arbitrary is not optimal. Let there exist a better sequence of jobs S′ (say) in which ith and jth adjacent jobs are interchanged.

Let $C_i(S)$ and $C_j(S)$ be the completion times of *i*th and *j*th job, respectively, in schedule *S*, respectively. Similarly, let $C_i(S')$ and $C_j(S')$ be the completion times of *i*th and *j*th job, respectively, in schedule *S*′.

For sequence *S*: We have

$$C_i(S) = t_a + 1, \; C_j(S) = t_a + 2$$

For sequence *S*′: We have

$$C_i(S') = t_a + 2, \; C_j(S') = t_a + 1$$

Next, the weighted flow time for the sequence *S* of jobs is

$$\begin{aligned} \text{WFT}(S) &= w_i \times C_i(S) + w_j \times C_j(S) \\ &= w_i \times (t_a + 1) + w_j \times (t_a + 2) = w_i t_a + w_i + w_j t_a + 2w_j = (w_i + w_j)t_a + w_i + 2w_j \end{aligned} \quad (2)$$

Similarly, the weighted flow time for the sequence *S*′ jobs is

$$\begin{aligned} \text{WFT}(S') &= w_i \times C_i(S') + w_j \times C_j(S') \\ &= w_i \times (t_a + 2) + w_j \times (t_a + 1) = w_i t_a + 2w_i + w_j t_a + w_j = (w_i + w_j)t_a + 2w_i + w_j \end{aligned} \quad (3)$$

Since, the *i*th and *j*th jobs are placed by WSPT rule. Therefore, we have $w_i \geq w_j$.

Hence, from results (2) and (3), we have

WFT(*S*′) ≥ WFT(*S*), i.e. weighted flow time for the sequence *S*′ is more as compared to the sequence *S*.

Hence, the sequence *S* following the WSPT rule minimizes the weighted flow time.

4.2. Theorem

A sequence S of jobs following early due date (EDD) order is an optimal sequence with maximum tardiness (T_{max}).

Proof Let if possible sequence S following EDD is not an optimal sequence. Let there exist a better sequence of jobs S′ (say) i.e. a late job in S is early in S′. Let S′ be the schedule in which ith and jth adjacent jobs are interchanged, i.e.

$$S = 1 - 2 - 3 - 4 - \ldots - i - j - \ldots - n$$

$$S' = 1 - 2 - 3 - 4 - \ldots - j - i - \ldots - n$$

The *i*th and *j*th jobs are adjacent on parallel machines if they are placed on adjacent positions as shown in Figure 2. Further, an adjacent pair of jobs may be located on the same machine, say *k*, or on different machines, say *k* and *l*. The jobs appearing on the machines before an adjacent pair are designated by *A* and those appearing after the pair are designated by *B* with an appropriate subscript for the corresponding machine. Let t_a be the completion time of the job before *i*th job on a machine.

In schedule *S*: Irrespective of the fact that the *i*th and *j*th adjacent jobs are on same machine or on different machines, we have

$$T_i = \max\{t_a + 1 - d_i, 0\}, \quad T_j = \max\{t_a + 2 - d_j, 0\}$$

In schedule *S′*: We have

$$T_i' = \max\{t_a + 2 - d_i, 0\}, \quad T_j' = \max\{t_a + 1 - d_j, 0\}$$

Since, $d_i \le d_j$, we have $T_i \le T_i'$ and $T_j \le T_j'$.

Therefore, $\max\{T_i', T_j'\} \ge \max\{T_i, T_j\}$.

This implies that the sequence *S* is better.

Hence, sequence *S* of jobs following EDD order is the optimal sequence for T_{max}.

4.3. Theorem

If all the jobs have distinct weights, then no improvement can be made to the values of any secondary criterion of maximum tardiness.

Proof Let *S* be the schedule obtained by arranging the jobs by WSPT order. Let C_i be the completion time of *i*th job and C_j be the completion time of *j*th job. Let S_i and S_j be the corresponding secondary criterion values. When *i*th and *j*th jobs are switched, let *S′* be the schedule so obtained.

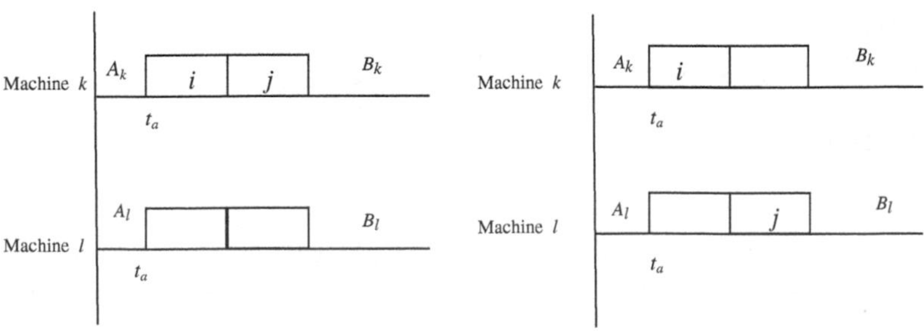

Figure 2. Adjacent jobs on parallel machines.

If $C_i < C_j$ then, $C_i' = C_j$ and $C_i = C_j'$. Hence $w_iC_i + w_jC_j < w_iC_i' + w_jC_j'$, which is a violation to primary criterion as all the other jobs contribute the same. Therefore, it implies that no improvement to the value of secondary criterion can be made with any possible exchange of jobs. The only possibility left is to rearrange the jobs having the same weights so that the value of a secondary criterion can be improved.

4.4. Theorem

If the problem of single criteria, maximum tardiness is NP-hard, the scheduling problem on parallel machines optimizing the bi-objective function maximum tardiness/WFT will also be NP-hard.

Proof We shall prove the result by the method of contradiction:

Let if possible the bi-objective function maximum tardiness/WFT is not NP-hard. Therefore, there must exist a polynomial algorithm which can solve the problem of optimizing the bi-objective function maximum tardiness/WFT on parallel processing machines.

This implies that single criteria of maximum tardiness can be optimized in polynomial time, i.e. maximum tardiness is not NP-hard. This is a contradiction, as maximum tardiness is NP-hard.

Hence, the scheduling problem optimizing the bi-objective function maximum tardiness/WFT on parallel processing machines will also be NP-hard.

5. Algorithm

The following algorithm is proposed to find the optimal sequence for bi-criteria problem T_{max}/WFT:

Step 1: Find average high ranking (AHR) of the fuzzy processing time of various jobs on different machines.

Step 2: Arrange all the jobs in weighted shortest processing time (WSPT) order.

Step 3: If all the jobs have distinct weights then go to Step 5.

Step 4: If all the jobs do not have distinct weights then arrange all the jobs with same weights in EDD order. Break the tie if any arbitrarily. Note the improvement in T_{max} subject to the primary set of constraints.

Step 5: Assign the jobs one by one to the earliest available machine. This optimizes the objective T_{max}/WFT.

6. Numerical illustration

The following numerical illustrations are given to test the efficiency of algorithm proposed to optimize the bi-criteria T_{max}/WFT on parallel machines in fuzzy environment.

Consider an example of jobs with processing time in fuzzy environment, due date and with jobs weightage (being distinct) on three parallel machines in a flow shop. The objective is to obtain a sequence of the jobs processing optimizing the bi-criteria taken as T_{max}/WFT (Table 1).

Solution: The AHR of processing time of given jobs as per the Step 1 using Equation 1 is as follows: (Table 2)

On arranging the jobs as per Step 2 in WSPT order on the parallel available machines M_1, M_2, and M_3, the order of jobs becomes 4–2–1 –3–5. The in-out i.e. the flow of jobs on machines for WSPT order is as in Table 3.

Therefore, T_{max} = 25/3 units and weighted flow time, $WFT = \sum_{i=1}^{n} w_i C_i = 5 \times \frac{23}{3} + 4 \times \frac{53}{3} + 3 \times \frac{32}{3} + 2 \times \frac{55}{3} + 1 \times \frac{70}{3} = 201$ units.

As per the Step 3, all the jobs have distinct weights; therefore, assigning the jobs to the available machines optimizes secondary criteria of T_{max}. Hence, the optimal sequence which optimizes the objective T_{max}/WFT is 4-2-1-3-5 with WFT = 201 units and minimum T_{max} = 25/3.

Consider another example of jobs with processing time in fuzzy environment, due date and with jobs weightage (being not distinct) on three parallel machines in a flow shop is given. The objective is to obtain sequence of the jobs processing optimizing the bi-criteria taken as T_{max}/WFT (Table 4).

Solution: The AHR of processing time of given jobs as per Step 1 using Equation 1 is as follows: (Table 5)

On arranging the jobs as per Step 2 in WSPT order on the parallel available machines M_1, M_2, and M_3, the order of jobs becomes1-6-2-3-4-7-5. The in-out flow of jobs on machines is as in Table 6.

Therefore, maximum tardiness, T_{max} = 11 units and $WFT = \sum_{i=1}^{n} w_i C_i = 5 \times \frac{26}{3} + 4 \times \frac{38}{3} + 3 \times \frac{23}{3} + 3 \times \frac{58}{3} + 2 \times \frac{55}{3} + 2 \times \frac{67}{3} + 1 \times \frac{78}{3} = 282.3$ units.

Table 1. Jobs with fuzzy processing time

Jobs	1	2	3	4	5
Processing time (in fuzzy environment)	(9, 10, 11)	(16, 17, 18)	(9, 10, 11)	(6, 7, 8)	(11, 12, 13)
Due date (d_i)	13	19	18	11	15
Weightage (w_i)	3	4	2	5	1

Table 2. Jobs with AHR of processing time

Jobs	1	2	3	4	5
AHR of processing time	32/3	53/3	32/3	23/3	38/3
Due date (d_i)	13	19	18	11	15
Weightage (w_i)	3	4	2	5	1

Table 3. In-out of jobs following WSPT order

Jobs	4	2	1	3	5
M_1	0–23/3			23/3–55/3	
M_2		0–53/3			
M_3			0–32/3		32/3–70/3
Due date (d_i)	11	19	13	18	15
Weightage (w_i)	5	4	3	2	1
Tardiness (T_i)				1/3	25/3

Table 4. Jobs with fuzzy processing time

Jobs	1	2	3	4	5	6	7
Processing time (in fuzzy environment)	(7, 8, 9)	(6, 7, 8)	(10, 11, 12)	(8, 9, 10)	(6, 7, 8)	(11, 12, 13)	(8, 9, 10)
Due date (d_i)	12	18	19	16	10	13	15
Weightage (w_i)	5	3	3	2	1	4	2

Table 5. Jobs with AHR of processing time

Jobs	1	2	3	4	5	6	7
AHR of processing time	26/3	23/3	35/3	29/3	23/3	38/3	29/3
Due date (d_i)	12	18	19	16	10	13	15
Weightage (w_i)	5	3	3	2	1	4	2

Table 6. In-out of jobs in WSPT order

Jobs	1	6	2	3	4	7	5
M_1	0–26/3				26/3–55/3		55/3–78/3
M_2		0–38/3				38/3–67/3	
M_3			0–23/3	23/3–58/3			
Due date (d_i)	12	18	19	16	10	13	15
Weightage (w_i)	5	4	3	3	2	2	1
Tardiness (T_i)				10/3	25/3	28/3	11

Table 7. In-out of jobs following EDD order

Jobs	1	6	3	2	4	7	5
M_1	0–26/3			26/3–49/3			49/3–72/3
M_2		0–38/3				38/3–67/3	
M_3			0–35/3		35/3–64/3		
Weightage (w_i)	5	4	3	3	2	2	1
Due date (d_i)	12	18	16	19	10	13	15
Tardiness (T_i)	-	-	-	-	34/3	28/3	27/3

Since the weights corresponding to jobs are not distinct, therefore, as per the Step 3 of algorithm we optimize T_{max} with given WFT. Following this we arrange the jobs 2 and 3 with same weightage as 3 in EDD order. Also, arrange the jobs 4 and 7 with same weightage as 2 in EDD order. Since the job 3 has EDD than the job 2, therefore, the job 3 will come before the job 2 in EDD order. The jobs 4 and 7 are already in EDD order. Following this we have the in-out of jobs as in Table 7.

Therefore, maximum tardiness, T_{max} = 34/3 units and WFT = $\sum_{i=1}^{n} w_i C = 5 \times \frac{26}{3} + 4 \times \frac{38}{3} + 3 \times \frac{35}{3} + 3 \times \frac{49}{3} + 2 \times \frac{64}{3} + 2 \times \frac{67}{3} + 1 \times \frac{72}{3} = 288$ units which violates the primary criterion of WFT having value 288 units. Therefore, reject the sequence 1-6-3-2-4-7-5. Hence, the optimal sequence of jobs processing is 1-6-2-3-4-7-5 with minimum WFT as 282.3 units and minimum T_{max} as 11 units.

7. Conclusion

In this paper, an algorithm to optimize the bi-criteria scheduling problem involving maximum tardiness and weighted flow time on parallel machines in fuzzy environment is considered. The concept of fuzzy processing time is introduced in processing of jobs, as in many real-life situations the processing time of jobs may vary due to the lack of complete knowledge, uncertainty, and vagueness. The study may further be extended by introducing the concepts of non-availability constraints, machines processing the jobs with different speeds, and also by considering trapezoidal fuzzy membership functions to represent the fuzziness in processing time.

Funding

The authors received no direct funding for this research.

Author details

Kewal Krishan Nailwal[1]

E-mail: kk_nailwal@yahoo.co.in

Deepak Gupta[2]

E-mail: guptadeepak2003@yahoo.co.in

Sameer Sharma[3]

E-mail: samsharma31@gmail.com

ORCID ID: http://orcid.org/0000-0002-7198-0371

[1] Department of Mathematics, A.P.J. College of Fine Arts, Jalandhar, Punjab 144001, India.

[2] Department of Mathematics, M. M. University, Mullana, Ambala, Haryana, India.

[3] Department of Mathematics, D.A.V. College, Jalandhar, Punjab 144008, India.

References

Baker, K. R. (1974). *Introduction to sequencing and scheduling.* New York, NY: Wiley.

Chand, S., & Schneeberger, H. (1986). A note on the single-machine scheduling problem with minimum weighted completion time and maximum allowable tardiness. *Naval Research Logistics Quarterly, 33*, 551–557. http://dx.doi.org/10.1002/(ISSN)1931-9193

Gupta, D., Sharma, S., & Aggarwal, S. (2012). Bi-objective scheduling on parallel machines with uncertain processing time. *Advances in Applied Science Research, 3*, 1020–1026.

Gupta, J. N. D., Hariri, A. M. A., & Potts, C. N. (1999). Single-machine scheduling to minimize maximum tardiness with minimum number of tardy jobs. *Annals of Operations Research, 92*, 107–123. http://dx.doi.org/10.1023/A:1018974428912

Gupta, J. N. D., Ruiz-torres, A. J., & Webster, S. (2003). Minimizing maximum tardiness and number of tardy jobs on parallel machines subject to minimum flow-time. *Journal of the Operational Research Society, 54*, 1263–1274. http://dx.doi.org/10.1057/palgrave.jors.2601638

Jha, P. C., Indumati, N. A., Singh, O., & Gupta, D. (2011). Bi-criterion release time problem for a discrete SRGM under fuzzy environment. *International Journal of Mathematics in Operational Research, 3*, 680–696. http://dx.doi.org/10.1504/IJMOR.2011.043016

Mokotoff, E. (2001). Parallel machine scheduling problems: A survey. *Asia-Pacific Journal of Operational Research, 18*, 193–242.

Parkash, D. (1997). *Bi-criteria scheduling problems on parallel machines* (PhD thesis). University of Birekshurg, Virginia, VA.

Sarin, S. C., & Hariharan, R. (2000). A two machine bicriteria scheduling problem. *International Journal of Production Economics, 65*, 125–139. http://dx.doi.org/10.1016/S0925-5273(99)00050-X

Sharma, S., Gupta, D., & Seema., S. (2012). Bi-objective scheduling on parallel machines in fuzzy environment. *Proceedings of the Second International Conference on Soft Computing for Problem Solving (SocProS 2012), 1*, 365–372.

Sharma, S., Gupta, D., & Seema, S. (2013). Bicriteria scheduling on parallel machines: Total tardiness and weighted flowtime in fuzzy environment. *International Journal of Mathematics in Operational Research, 5*, 492–507. http://dx.doi.org/10.1504/IJMOR.2013.054733

Smith, W. E. (1956). Various optimizers for single stage production. *Naval Research Logistics Quarterly, 3*, 59–66. http://dx.doi.org/10.1002/(ISSN)1931-9193

Su, L.-H., Chen, P.-S., & Chen, S.-Y. (2012). Scheduling on parallel machines to minimize maximum tardiness for the customer order problem. *International Journal of System Sciences, 44*, 926–936. doi:10.1080/00207721.2011.649366

Sunita, & Singh, T. P. (2010a). Bi-objective in fuzzy scheduling on parallel machines. *Arya Bhatta Journal of Mathematics and Informatics, 2*, 149–152.

Sunita, & Singh, T. P. (2010b). Single criteria scheduling with fuzzy processing time on parallel machines. In *Proceedings of International Conference in IISN 2010, ISTK* (pp. 444–446). Yamuna Nagar, Haryana.

Yager, R. R. (1981). A procedure for ordering fuzzy subsets of the unit interval. *Information Sciences, 24*, 143–161. http://dx.doi.org/10.1016/0020-0255(81)90017-7

Zadeh, L. A. (1965). Fuzzy sets. *Information and Control, 8*, 338–353. http://dx.doi.org/10.1016/S0019-9958(65)90241-X

Numerical solution of singularly perturbed problems using Haar wavelet collocation method

Firdous A. Shah[1*], R. Abass[2] and J. Iqbal[2]
*Corresponding author: Firdous A. Shah, Department of Mathematics, University of Kashmir, South Campus, Anantnag 192 101, Jammu and Kashmir, India
E-mail: fashah79@gmail.com

Reviewing editor: Yong Hong Wu, Curtin University of Technology, Australia

Abstract: In this paper, a collocation method based on Haar wavelets is proposed for the numerical solutions of singularly perturbed boundary value problems. The properties of the Haar wavelet expansions together with operational matrix of integration are utilized to convert the problems into systems of algebraic equations with unknown coefficients. To demonstrate the effectiveness and efficiency of the method various benchmark problems are implemented and the comparisons are given with other methods existing in the recent literature. The demonstrated results confirm that the proposed method is considerably efficient, accurate, simple, and computationally attractive.

Subjects: Advanced Mathematics; Applied Mathematics; Astrophysics; Mathematics & Statistics; Physical Sciences; Physics; Plasmas & Fluids; Science; Statistical Physics; Thermodynamics & Kinetic Theory

Keywords: Haar wavelet; operational matrix; singularly perturbed; boundary layer

AMS Subject Classifications: 42C40; 65L10; 65L12; 65M70; 65N35

1. Introduction

Singularly perturbed problems (SPPs) arise in various branches of applied mathematics and physics such as fluid mechanics, quantum mechanics, elasticity, plasticity, semi-conductor device physics,

ABOUT THE AUTHORS

Firdous A. Shah is a senior assistant professor in the Department of Mathematics at the University of Kashmir, India. His primary research interests include basic theory of wavelets and their applications in differential and integral equations, Economics and Finance, and Computer Networking. He has authored/co-authored over 60 research papers in international journals of high repute. He has recently co-authored a book on wavelets entitled Wavelet Transforms and Their Applications, Springer, New York, 2015.

R. Abass received his MSc in Applied Mathematics from BGSB University Rajouri, India. His research interests are focused on different wavelet methods for the numerical treatment of integral and differential equations.

J. Iqbal received his MPhil and PhD degrees in Mathematics from Aligarh Muslim University, Aligarh, India. Currently, he is a senior assistant professor in the Department of Mathematical Sciences, BGSB University, Rajouri.

PUBLIC INTEREST STATEMENT

Singular perturbation problems (SPPs) have applications in various disciplines of knowledge, for instance, neurobiology, fluid mechanics, elasticity, quantum mechanics, geophysics, aerodynamics, oceanography, chemical reactor theory, convection–diffusion processes, and optimal control. It is a well-known fact that the solution of these problems exhibit multi-scale character so the usual numerical treatment of SPPs gives major computational difficulties due to the presence of boundary and interior layers and, in recent years, a large number of numerical methods have been developed to provide accurate numerical solutions. In this work, a new collocation method based on Haar wavelets is proposed for the numerical solution of singularly perturbed two-point boundary value problems. Accuracy and efficiency of the suggested method is established through comparison with the existing methods available in the open literature.

geophysics, optimal control theory, aerodynamics, oceanography, and mathematical models of chemical reactions. Mathematically, self-adjoint SPPs are defined as

$$\left.\begin{array}{lll} Ly(x) & \equiv -\epsilon y''(x) + f(x)y(x) = g(x), & f(x) \geq 0, x \in [0,1], \\ y(0) & = \alpha_0, & y(1) = \alpha_1, \end{array}\right\} \tag{1.1}$$

where α_0, α_1 are given constants, ϵ is a small positive parameter such that $0 < \epsilon \ll 1$ and $f(x)$, $g(x)$ are sufficiently smooth functions. It is known that Problem (1.1) has a unique solution y, which in general displays boundary layers at $x = 0$ and $x = 1$. These type of problems are characterized by the presence of a small parameter ϵ that multiplies the highest order derivative, and they are stiff and there exists a boundary or interior layer where the solutions change rapidly. That is, there are thin transition layers where the solution varies rapidly or jumps abruptly, while away from the layers the solution behaves regularly and varies slowly. For more details on singular perturbation problems, we refer to the monographs (Miller, O'Riordan, & Shishkin, 1996; Roos, Stynes, & Tobiska, 1996).

In recent years, the studies of SPPs problems have been tackled by many researchers but the majority of these problems cannot be solved analytically, so one would like ideally to use the numerical methods available in the open literature such as homotopy perturbation method (Chun & Sakthivel, 2010), Adomian decomposition method (Wazwaz, 2002), sinc approximation solution (Mohsen & EL-Gamel, 2008), spline collocation method (Aziz & Khan 2002; Kadalbajoo & Arora, 2010; Kadalbajoo & Gupta, 2009; Kadalbajoo, Gupta, & Awasthi, 2008; Khan, Khan, & Aziz, 2006; Khan & Khandelwala, 2014; Kumar & Mehra, 2007; Rashidinia, Ghasemi, & Mahmoodi, 2007; Surla & Stojanovic, 1988), reproducing kernel method (Geng & Cui, 2011), finite element method (Lenferink, 2002), the finite-volume element method (Phongthanapanich & Dechaumphai, 2009), and wavelet method (Pandit & Kumar, 2014). An alternative solution is proposed in the present paper in the form of operational matrix method, which is based on Haar wavelets for the numerical solution of singularly perturbed reaction–diffusion problems.

Wavelets became an active field of research in the 1980s, with the works of researchers such as Morlet, Grossman, and Daubechies (1992) on signal processing. Starting as an alternative to Fourier analysis, their popularity soon expanded, owing mainly to the localized nature of wavelet basis in frequency and time, as well as their hierarchical structure. Wavelets have numerous applications in approximation theory and have been extensively used in the context of numerical approximation in the relevant literature during the last two decades. Different types of wavelets and approximating functions have been used in numerical solution of boundary value problems such as Daubechies, Battle–Lemarie, *B*-spline, Chebyshev, Legendre, and Haar wavelets. Among all the wavelet families, the Haar wavelets have gained popularity among researchers due to their useful properties such as simple applicability, orthogonality, and compact support. Compact support of the Haar wavelet basis permits straight inclusion of the different types of boundary conditions in the numeric algorithms. The basic idea of Haar wavelet method is to convert the differential equations to a system of algebraic equations by the operational matrices of integral or derivative (Chen & Hsiao, 1997; Lepik, 2008). Recently, many authors have used Haar wavelet method for solving ordinary and partial differential equations. For a historical background and an overview of wavelets in general and Haar wavelets in particular, the reader can refer to (Lepik, 2014).

The objective of this research is to construct a simple collocation method based on Haar wavelets for the numerical solution of singularly perturbed reaction–diffusion problems of the type (1.1) which arise in mathematical modeling of different engineering applications. The proposed method has the following advantages in comparison to the existing methods available in the open literature:

(1) Haar wavelets collocation method (HWCM) uses simple box functions and consequently the formulation of numerical method based on these functions involves lesser manual labor.
(2) HWCM does not require to calculate the inverse of Haar wavelet matrix.
(3) Contrary to reproducing kernel method (RKM), HWCM performs very well for a boundary value problem defined on a very long interval.

(4) Unlike RKM, HWCM does not require conversion of a boundary value problem into initial value problem using a procedure like shooting and hence this method eliminates the possibility of unstable solution due to missing initial condition in the case of RKM.

(5) Contrary to RKM, the boundary value problem needs not to be reduced into a system of first-order ODE's.

(6) A variety of boundary conditions can be handled with equal ease.

Finally, the obtained numerical approximate results of this method are then compared with the exact solutions as well as solutions available in open literature. The numerical outcomes indicate that the proposed method yields highly accurate results.

The organization of this paper is as follows. In Section 2, Haar wavelets and their integral are introduced. In Section 3, the Haar wavelet collocation method is presented and described for the numerical solution of the class of singularly perturbed reaction–diffusion equations. In Section 4, our method has been tested by several problems and the obtained results are compared with results of the existing methods. Finally, in Section 5, the conclusion of the study is given.

2. Haar wavelets and operational matrix of integration

Haar wavelets have been used from 1910 when they were introduced by the Hungarian mathematician Alfred Haar (Lepik, 2014). The Haar wavelet, being an odd rectangular pulse pair, is the simplest and the oldest orthonormal wavelet with compact support. The Haar wavelet family for $x \in [0, 1]$ is defined as follows:

$$h_i(x) = \begin{cases} 1, & \text{for } x \in [\alpha, \beta) \\ -1, & \text{for } x \in [\beta, \gamma) \\ 0, & \text{elsewhere.} \end{cases} \tag{2.1}$$

where

$$\alpha = \frac{k}{m}, \beta = \frac{k+0.5}{m}, \gamma = \frac{k+1}{m}.$$

Here, m and k have integer values as $m = 2^j, j = 0, 1, \dots, J$ and J show the resolution of the wavelet and $k = 0, 1, \dots, m-1$ is the translation parameter. Maximal level of resolution is J. The index of h_i in Equation (2.1) is calculated by $i = m + k + 1$. In the case with minimal values $m = 1, k = 0$, we have $i = 2$, the maximal value of i is $2M = 2^{J+1}$. We also have $i = 1$ corresponding to the scaling function of Haar wavelet family, i.e. $h_1(x) = 1$ in [0, 1]. For more about Haar wavelets and their applications, we refer to the monographs (Debnath & Shah, 2015; Lepik, 2014).

Next, we shall establish an operational matrix for integration by means of Haar wavelets for which we follow the same notations as used in (Lepik, 2008) for Haar function and their integrals as follows:

$$P_{i,1}(x) = \int_0^x h_i(x)\,dx, P_{i,n+1}(x) = \int_0^x P_{i,n}(x)\,dx, \text{ and } C_{i,n}(x) = \int_0^1 P_{i,n}(x)\,dx, \; n = 1, 2, \dots. \tag{2.2}$$

These integrals can be calculated analytically with the help of Equation (2.1); by doing so we get the following equations

$$P_{i,n}(x) = \begin{cases} 0 & \text{for } x \in [0, \alpha) \\ \frac{1}{n!}(x-\alpha)^n & \text{for } x \in [\alpha, \beta) \\ \frac{1}{n!}[(x-\alpha)^n - 2(x-\beta)^n] & \text{for } x \in [\beta, \gamma) \\ \frac{1}{n!}[(t-\alpha)^n - 2(t-\beta)^n + (t-\gamma)^n], & \text{for } x \in [\gamma, 1), \end{cases} \tag{2.3}$$

where $i = 2, 3, \ldots$ and $n = 1, 2, \ldots$. Note that

$$P_{1,n}(x) = \frac{x^n}{n!}, \quad C_{1,n}(x) = \frac{1}{(n+1)!}, n = 1, 2, \ldots.$$

Any square integrable function $y(x)$ defined on [0, 1] can be expressed in term of the Haar basis as follows:

$$y(x) = c_1 h_1(x) + c_2 h_2(x) + \cdots = \sum_{i=1}^{\infty} c_i h_i(x), \tag{2.4}$$

where the Haar coefficients c_i, $i = 1, 2, \ldots$, are determined by

$$c_i = \langle y, h_i \rangle = 2^j \int_0^1 y(x) h_i(x)\, dx. \tag{2.5}$$

Even though the series expansion of $y(x)$ involves infinite terms, if $y(x)$ is a piecewise constant or it may be approximated as a piecewise constant for each sub-interval, then $y(x)$ can be terminated at finite terms. That means $y(x)$ can be expressed as follows:

$$y(x) = \sum_{i=1}^{2M} c_i h_i(x). \tag{2.6}$$

Equivalently, above relation can be written in the matrix form as follows:

$$\mathbf{Y} = \mathbf{C}_m^T \mathbf{H}_m, \tag{2.7}$$

where $\mathbf{Y}$ is the discrete form of the continuous function $y(x)$ and $\mathbf{C}_m^T = [c_1, c_2, \ldots, c_m]$ is the m-dimensional row vector. Moreover, $\mathbf{H}_m$ is the Haar wavelet matrix of order m and is defined by $\mathbf{H}_m = [\mathbf{h}_1, \mathbf{h}_2, \ldots, \mathbf{h}_m]^T$; that is,

$$\mathbf{H}_m = \begin{pmatrix} \mathbf{h}_1 \\ \mathbf{h}_2 \\ \vdots \\ \mathbf{h}_m \end{pmatrix} = \begin{pmatrix} h_{1,1} & h_{1,2} & \cdots & h_{1,m} \\ h_{2,1} & h_{2,2} & \cdots & h_{2,m} \\ \vdots & \vdots & \vdots & \vdots \\ h_{m,1} & h_{m,2} & \cdots & h_{m,m} \end{pmatrix}, \tag{2.8}$$

where $\mathbf{h}_1, \mathbf{h}_2, \ldots, \mathbf{h}_m$ are the discrete form of the Haar wavelet basis. For Haar wavelet approximations, the following collocation points are considered:

$$x_\ell = \frac{\ell - 0.5}{m}, \ell = 1, 2, \ldots, m. \tag{2.9}$$

For example, if $j = 2 \Rightarrow 2M = 8$, so that the Haar matrix can be expressed as follows:

$$\mathbf{H}_8 = \begin{pmatrix} 1 & 1 & 1 & 1 & 1 & 1 & 1 & 1 \\ 1 & 1 & 1 & 1 & -1 & -1 & -1 & -1 \\ 1 & 1 & -1 & -1 & 0 & 0 & 0 & 0 \\ 0 & 0 & 0 & 0 & 1 & 1 & -1 & -1 \\ 1 & -1 & 0 & 0 & 0 & 0 & 0 & 0 \\ 0 & 0 & 1 & -1 & 0 & 0 & 0 & 0 \\ 0 & 0 & 0 & 0 & 1 & -1 & 0 & 0 \\ 0 & 0 & 0 & 0 & 0 & 0 & 1 & -1 \end{pmatrix}.$$

At the collocation points as defined by (2.9), equation (2.6) becomes

$$y(x_\ell) = \sum_{i=1}^{2M} c_i h_i(x_\ell). \tag{2.10}$$

The Haar approximation y_m of y is given by

$$y_m(x) = \sum_{i=1}^{N} c_i h_i(x), N = 2M = 2^{j+1}, j = 0, 1, \ldots, J. \tag{2.11}$$

Therefore, the corresponding error at the mth level may be defined as follows:

$$\|y(x) - y_m(x)\|_2 = \left\| y(x) - \sum_{i=1}^{N} c_i h_i(x) \right\|_2 = \left\| \sum_{i=2^{j+1}}^{\infty} c_i h_i(x) \right\|_2 .$$

3. Method of solution

With the aid of Haar operational matrices as defined in Section 2, we solve the following singularly perturbed reaction–diffusion problem of the form

$$Ly(x) \equiv -\epsilon y''(x) + f(x)y(x) = g(x), y(0) = \alpha_0, y(1) = \alpha_1, \tag{3.1}$$

where $f(x)$, $g(x)$ are real-valued sufficiently smooth functions in [0, 1]. Let us assume that

$$y''(x) = \sum_{i=1}^{N} c_i h_i(x), \tag{3.2}$$

where c_i's, $i = 1, 2, \ldots, N$ are Haar coefficients to be determined. Integrating Equation (3.2) from 0 to x together with the given initial condition, we obtain

$$y'(x) = \sum_{i=1}^{N} c_i P_{i,1}(x) + y'(0). \tag{3.3}$$

Again integrating Equation (3.3) with the initial condition, then we have

$$y(x) = \sum_{i=1}^{N} c_i P_{i,2}(x) + xy'(0) + y(0). \tag{3.4}$$

To calculate $y'(0)$, we integrate Equation (3.3) between the limits 0 to 1 to get

$$y'(0) = y(1) - y(0) - \sum_{i=1}^{N} c_i C_{i,1}. \tag{3.5}$$

Hence, Equation (3.4) becomes

$$y(x) = \sum_{i=1}^{N} c_i \left[P_{i,2}(x) - xC_{i,1}\right] + x\left[y(1) - y(0)\right] + y(0). \tag{3.6}$$

Substituting the values of $y''(x)$ and $y(x)$ in Equation (3.1) and the discretization is applied using the collocation points given by (2.9) resulting into a system of algebraic equations which contains unknowns vectors c_i's. Solving this system of algebraic equations using the classical Newton's method, we obtain the Haar wavelet coefficients c_i's and then substituting these values in (3.4), we obtain the Haar wavelet collocation method for the numerical solution of singularly perturbed boundary value problem of the type (1.1).

4. Numerical experiments and discussion

In this section, we are going to study numerically the SPPs (3.1) with the known boundary condition. The main aim here is to show the accuracy and applicability of the present method, described in Section 3, for solving the SPPs. Performance of the proposed method is compared with the existing methods in literature (Khan et al., 2014; Kumar, Singh, & Mishra, 2007; Lubuma & Patidar, 2006; Natesan, Kumar, & Vigo-Aguiar, 2003; O'Riordan & Stynes, 1986; Rashidinia et al., 2007; Schatz & Wahlbin, 1983). The numerical results infer that the proposed method is very effective and superior in comparison with other existing methods. Numerical computations have been done with the software package MATLAB 7.0 and graphical outputs were generated by MAPLE 14 package.

Example 4.1 Consider the following singularly perturbed boundary value problem (Khan et al., 2014; Rashidinia et al., 2007):

$$\left.\begin{aligned} -\epsilon y''(x)+y(x) &= x, \quad x\in[0,1],\\ y(0) &= 1, \quad y(1)=1+\exp\left(\frac{-1}{\sqrt{\epsilon}}\right) \end{aligned}\right\}. \tag{4.1}$$

The exact solution of this problem is

$$y(t)=t+\exp\left(\frac{-t}{\sqrt{\epsilon}}\right). \tag{4.2}$$

The obtained maximum absolute errors of (4.1) is presented in comparison with the existing methods and exact solution (4.2) in Table 1 for different values of ϵ and N. From Table 1, it is clear that HWCM performs much better than existing methods (Khan et al., 2014; Rashidinia et al., 2007). Table 1 also shows improved convergence of HWCM, as with the increase in number of collocation points the maximum absolute errors decrease for the solution. The numerical result for $N=32$ and different values of ϵ is shown in Figure 1.

Table 1. Maximum absolute errors for different values of N and ϵ

$\epsilon \downarrow$	**N = 16**	**N = 32**	**N = 64**	**N = 128**
Our method				
1/16	5.0862E-05	2.4680E-05	6.1100E-07	6.1100E-09
1/32	2.4680E-05	1.2244E-06	2.6471E-07	1.3139E-09
1/64	1.2244E-05	6.1100E-06	2.7082E-07	1.3161E-08
1/128	6.1100E-06	2.7082E-07	1.1038E-08	4.9550E-09
1/256	3.8149E-06	6.2121E-08	1.4415E-09	5.1021E-010
Rashidinia et al. (2007)				
1/16	2.96E-06	1.18E-05d	1.15E-08	7.24E-10
1/32	1.18E-05	7.54E-07	4.67E-08	2.96E-09
1/64	4.74E-05	2.96E-06	1.86E-07	1.16E-08
1/128	1.78E-04	1.18E-05	7.46E-07	4.67E-08
1/256	7.41E-04	4.74E-05	2.98E-08	1.86E-10
Khan et al. (2014)				
1/16	7.376E-05	4.938E-06	3.147E-07	1.977E-08
1/32	2.771E-04	1.947E-05	1.260E-06	7.959E-08
1/64	9.787E-04	7.448E-05	4.982E-06	3.174E-07
1/128	3.645E-03	2.773E-04	1.948E-05	1.260E-06
1/256	1.292E-02	9.787E-04	7.448E-05	4.982E-06

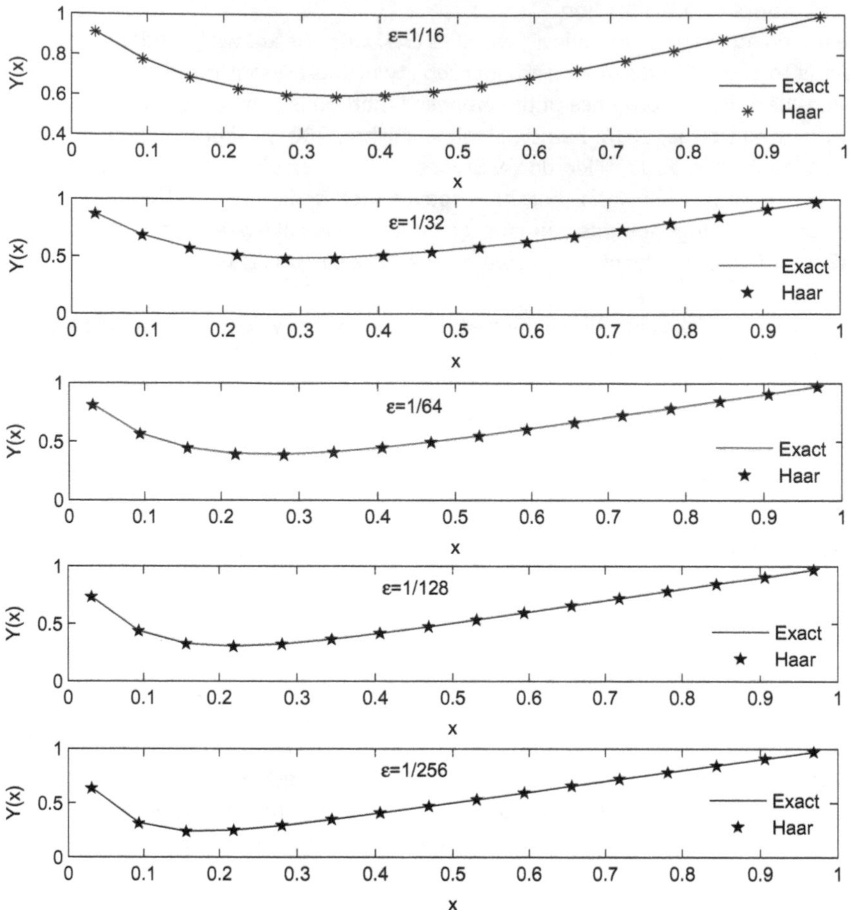

Figure 1. Comparison of numerical and exact solution at $N = 32$ and different values of ϵ.

Example 4.2 Let us consider the following singular perturbation problem (Lubuma et al., 2006; O'Riordan et al., 1986):

$$\left.\begin{aligned} &-\epsilon y''(x) + \frac{4}{(x+1)^4}\left[1+\sqrt{\epsilon}(x+1)\right]y(x) = f(x), \quad x \in [0,1], \\ &y(0) = 2, \qquad\qquad\qquad\qquad y(1) = -1, \end{aligned}\right\} \tag{4.3}$$

where

$$f(x) = -\frac{4}{(x+1)^4}\left[\left(1+\sqrt{\epsilon}(x+1)+4\pi^2\epsilon\right)\cos\left(\frac{4\pi x}{x+1}\right)\right.$$
$$\left. -2\pi\epsilon(x+1)\sin\left(\frac{4\pi x}{x+1}\right) + \frac{3(1+\sqrt{\epsilon}(x+1))}{1-\exp(-1/\sqrt{\epsilon})}\right].$$

The exact solution of the above problem is

$$y(x) = -\cos\left(\frac{4\pi x}{x+1}\right) + \frac{3\left[\exp(-2x/\sqrt{\epsilon}(x+1)) - \exp(-1/\sqrt{\epsilon})\right]}{1-\exp(-1/\sqrt{\epsilon})}.$$

Table 2 gives the maximum absolute error for test Example 4.2. It is obvious that the error bound is inversely proportional to the level of resolution J of Haar wavelet. Hence, the accuracy in the proposed method (HWCM) improves as we increase the level of resolution J. The solution produced through HWCM for the singularly perturbed boundary value problem (4.3) is shown in Figure 2 for different values of ϵ and $N = 32$.

Example 4.3 We next consider the singularly perturbed problem (Schatz et al., 1983)

Table 2. Maximum absolute errors for different values of N and ϵ

$\epsilon \downarrow$	$N = 16$	$N = 32$	$N = 64$	$N = 128$
Our method				
1	0.32E-02	0.78E-03	0.27E-03	0.34E-04
$(1/16)^{0.25}$	0.75E-03	0.31E-03	0.89E-04	0.18E-04
$(1/32)^{0.5}$	0.95E-03	0.99E-04	0.31E-04	0.01E-04
$(1/64)^{0.75}$	0.87E-03	0.22E-03	0.91E-04	0.11E-04
(1 / 128)	0.71E-03	0.44E-03	0.58E-04	0.21E-04
O'Riordan et al. (1986)				
1	0.11E+00	0.27E-01	0.69E-02	0.17E-02
$(1/16)^{0.25}$	0.95E-01	0.23E-01	0.56E-02	0.13E-02
$(1/32)^{0.5}$	0.78E-01	0.18E-01	0.42E-02	0.10E-02
$(1/64)^{0.75}$	0.66E-01	0.16E-01	0.40E-02	0.10E-02
(1 / 128)	0.64E-01	0.17E-01	0.42E-02	0.13E-02
Lubuma et al. (2006)				
1	0.51E-01	0.13E-01	0.32E-02	0.79E-03
$(1/16)^{0.25}$	0.38E-01	0.96E-02	0.24E-02	0.60E-03
$(1/32)^{0.5}$	0.25E-01	0.63E-02	0.16E-02	0.39E-03
$(1/64)^{0.75}$	0.16E-01	0.43E-02	0.11E-02	0.27E-03
(1 / 128)	0.14E-01	0.79E-02	0.24E-02	0.62E-03

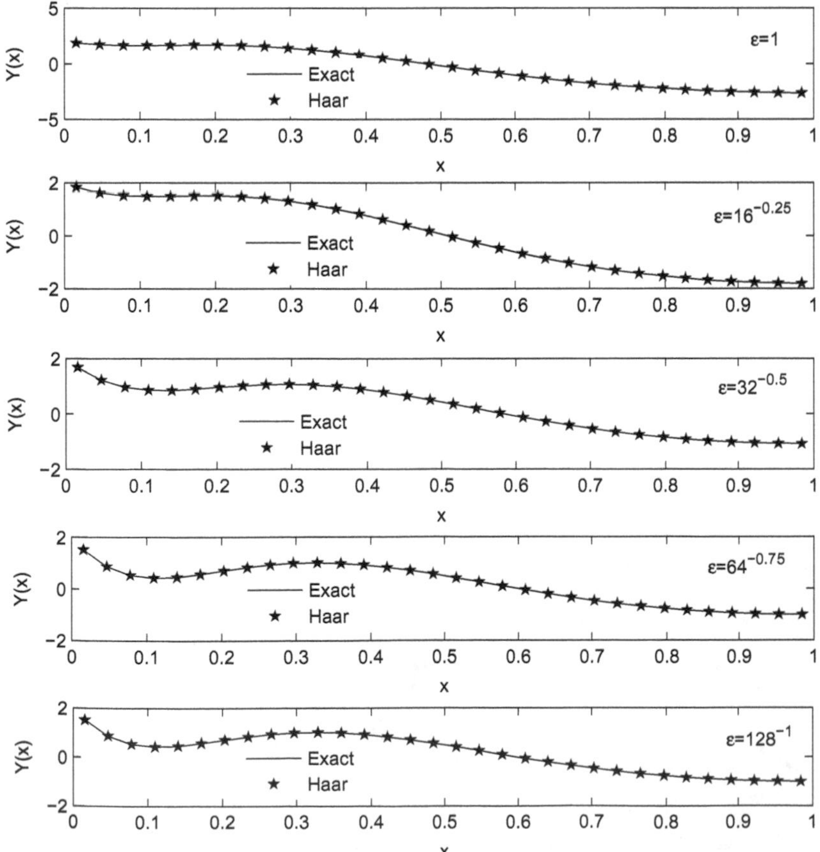

Figure 2. Comparison of numerical and exact solution at $N = 32$ and different values of ϵ.

$$\left.\begin{aligned} -\epsilon y''(x)+y(x) &= (x-1)-x\exp(-1/\epsilon), \quad x\in[0,1], \\ y(0)=0, &\quad y(1)=0 \end{aligned}\right\}, \tag{4.4}$$

whose exact solution is given by

$$y(x)=(x-1)-x\exp(-1/\epsilon)+\exp(-x/\epsilon).$$

We solve this problem by HWCM method which developed in Section 2 and compare our numerical results with finite element method (Schatz et al., 1983). In this example, the computed maximum absolute errors are compared with the exact solution at specified points in reference (Schatz et al., 1983) and the results are shown in Table 3 and graphically shown in Figure 3. This table shows that our method is considerably accurate in comparison with method in (Schatz et al., 1983).

Table 3. Maximum absolute errors for different values of N and ϵ

$\epsilon \downarrow$	$N=16$	$N=32$	$N=64$	$N=128$
Our method				
1	1.1420E-06	2.8699E-07	7.1820E-08	1.7962E-08
1 / 5	7.2799E-05	1.8568E-05	4.6624E-06	1.1673E-06
$1/5^2$	1.0533E-03	4.0689E-04	1.1226E-04	2.8906E-05
$1/5^3$	7.0483E-03	1.3723E-04	1.2529E-05	5.8909E-06
$1/5^4$	3.6848E-04	1.4255E-04	4.9873E-05	1.2027E-05
$1/5^5$	1.4897E-05	5.9510E-06	2.3677E-06	9.2707E-07
$1/5^6$	5.9615E-05	2.3845E-05	9.5358E-06	3.8111E-06
Schatz et al. (1983)				
1	2.2123E-04	1.9925E-04	5.2043E-05	2.7654E-05
1 / 5	3.8988E-03	1.1005E-03	5.8736E-04	2.8643E-05
$1/5^2$	3.8710E-02	1.7634E-03	6.8954E-04	1.2733E-04
$1/5^3$	5.9764E-03	5.9534E-04	2.8733E-04	7.8823E-05
$1/5^4$	4.7681E-05	1.1276E-05	7.8754E-06	2.9087E-06
$1/5^5$	2.7612E-03	3.8751E-05	1.7643E-05	4.8965E-06
$1/5^6$	7.7896E-06	4.8964E-06	3.6754E-07	1.9971E-07

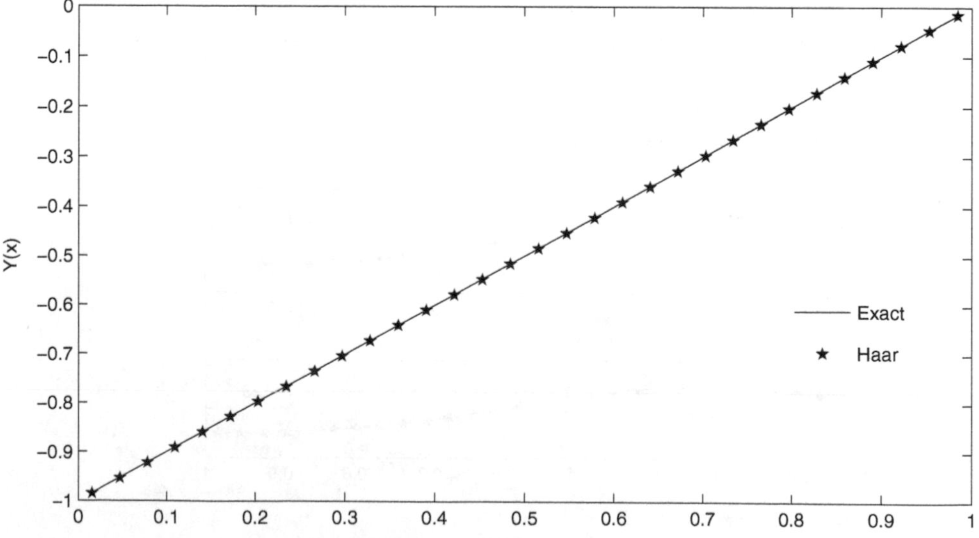

Figure 3. Comparison of numerical and exact solution at $N=32$ and $\epsilon=1/5^6$.

Example 4.4 Consider the following singular perturbed problem

$$\left.\begin{aligned} -\epsilon^2 y''(x) + y(x) &= (-\epsilon^2 + 1)e^x - x(e^1 + e^{-1/\epsilon}) - 2(1-x), \quad x \in [0,1], \\ y(0) &= 0, \qquad\qquad\qquad\qquad\qquad\qquad y(1) = 0 \end{aligned}\right\}. \tag{4.5}$$

The exact solution of the above problem is

$$y(x) = e^{-x/\epsilon} + e^x - x(e^1 + e^{-1/\epsilon}) - 2(1-x).$$

Next, we consider a nonlinear singular perturbation problem with left-end boundary layer. First, we convert the underlying problem as a sequence of linear singular perturbation problem using Newton's quasi-linearization technique, i.e. replacing the nonlinear problem by a sequence of linear problems. Then, the outer solution (the solution of the given problem by putting $\epsilon = 0$) is taken to be the initial approximation (Table 4 and graphically shown in Figure 4).

Example 4.5 We consider the following nonlinear singular perturbation problem (Kadalbajoo et al., 2010; Kumar, Singh, & Mishra, 2007)

$$\epsilon y''(x) + y(x)y'(x) - y(x) = 0, x \in [0,1] \tag{4.6}$$

with $y(0) = -1$ and $y(1) = 3.9995$. The exact solution of the problem is

Table 4. Maximum absolute errors for different values of N and ϵ

$\epsilon \downarrow$	$N = 16$	$N = 32$	$N = 64$	$N = 128$
1	4.2215E-06	1.0603E-06	2.6540E-07	6.6370E-08
1 / 5	7.3380E-05	1.8707E-05	4.6962E-06	1.1759E-06
$1/5^2$	1.0533E-03	4.0690E-04	1.1226E-04	2.8906E-05
$1/5^3$	7.0483E-04	1.3723E-04	1.2529E-05	5.8909E-06
$1/5^4$	3.6848E-05	1.4255E-04	4.9873E-05	1.2027E-05
$1/5^5$	5.4897E-06	1.9510E-06	6.3677E-07	1.2707E-07
$1/5^6$	5.9615E-06	2.3845E-06	9.5358E-07	3.8111E-07

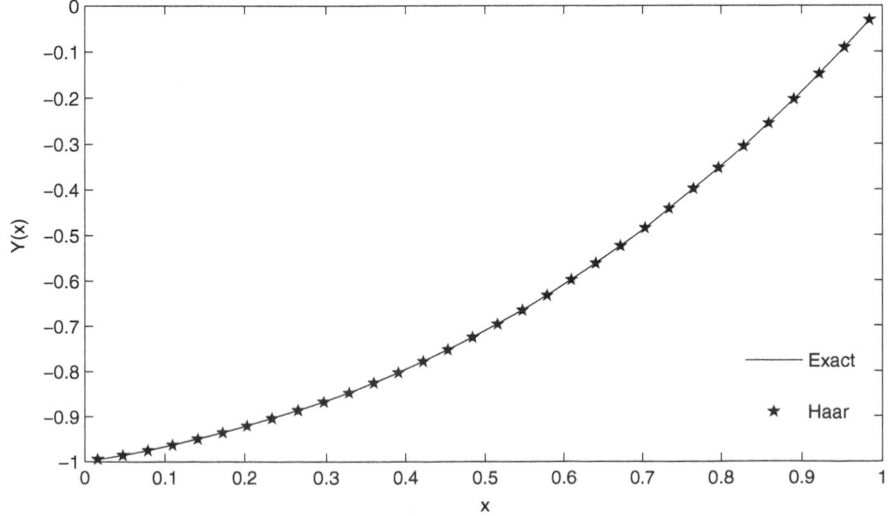

Figure 4. Comparison of numerical and exact solution at $N = 32$ and $\epsilon = 1/5^6$.

$$y(x) = x + c_1 \tanh\left(c_1(x/\varepsilon + c_2)/2\right), \tag{4.7}$$

where $c_1 = 2.9995$ and $c_2 = (1/c_1)\log_e[(c_1 - 1)/(c_1 + 1)]$.

For convenience, we convert the nonlinear Equation (4.6) into linear by taking the initial approximation from the problem $y(x)y'(x) + y(x) = 0$. For $x = 0$, we assume that $y'(x) = 0$, and $y(x) = C$. Further, in order to satisfy the condition at $x = 1$, we take $y(x) = x + 2.9995$ so as a result, we obtain the linear version of (4.6) as follows:

$$\varepsilon y''(x) - (x + 2.9995)y'(x) - (x + 2.9995) = 0, x \in [0, 1] \tag{4.8}$$

with $y(0) = -1$ and $y(1) = 3.9995$. The numerical results of Example 4.5 are presented in Table 5 and graphically shown in Figure 5 for $\varepsilon = 10^{-3}$ and $\varepsilon = 10^{-4}$.

Table 5. Numerical result of Example 4.5 at $\varepsilon = 10^{-3}$ and $\varepsilon = 10^{-4}$

	$\varepsilon = 10^{-3}$			$\varepsilon = 10^{-4}$		
x	**Our method**	**Kumar et al. (2007)**	**Exact sol.**	**Our method**	**Kumar et al. (2007)**	**Exact sol.**
0.0	−1.0000000	−1.0000000	−1.0000000	−1.0000000	−1.0000000	−1.0000000
0.1	3.0995421	3.0995652	3.0995000	3.1004189	3.1004235	3.0995000
0.2	3.1995369	-	3.1995000	3.1995153	-	3.1995000
0.3	3.2995311	3.2995507	3.2995000	3.3002127	3.3002183	3.2995000
0.4	3.3995295	-	3.3995000	3.3995104	-	3.3995000
0.5	3.4995268	3.4995362	3.4995000	3.5000081	3.5000131	3.4995000
0.6	3.5995116	-	3.5995000	3.5995059	-	3.5995000
0.7	3.6995171	3.6995217	3.6995000	3.6998043	0.1899999	3.6995000
0.8	3.7995101	-	3.7995000	3.7995031	-	3.7995000
0.9	3.8995049	3.8995072	3.8995000	3.8996009	3.8996026	3.8995000
1.0	3.9995000	3.9995000	3.9995000	3.9995000	3.9995000	3.9995000

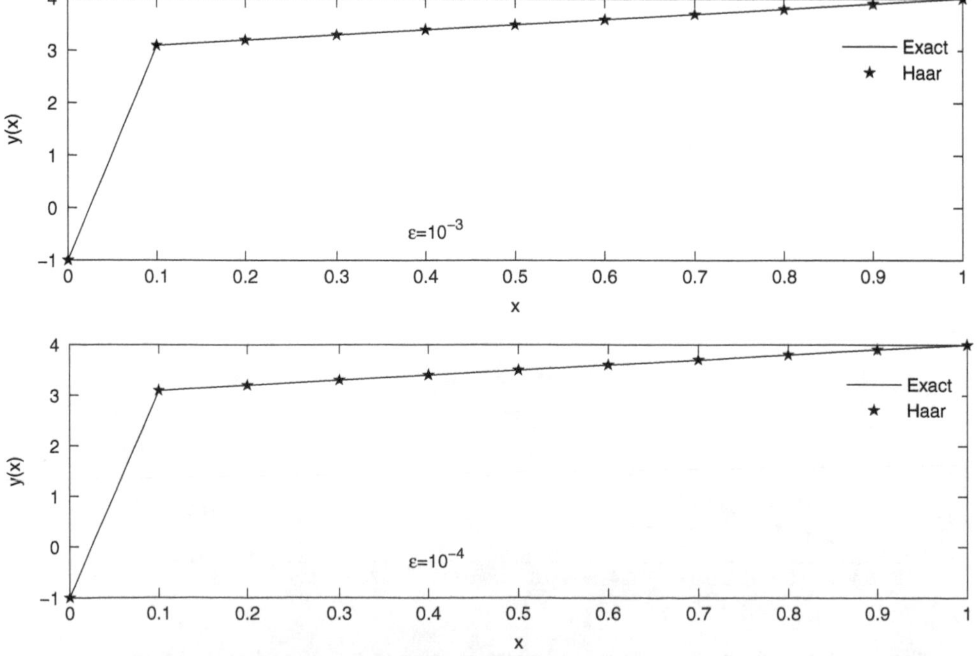

Figure 5. Comparison of numerical and exact solution at $\varepsilon = 10^{-3}$ and 10^{-4}.

Table 6. Maximum absolute errors of Example 4.6 for different values of ε and N

$\varepsilon \downarrow$	$N = 16$		$N = 32$		$N = 64$		$N = 128$		$N = 25$	
	HWCM	**Natesan et al. (2003)**	**HWCM**	**Natesan et al. (2003)**	**HWCM**	**Natesan et al. (2003)**	**HWCM**	**Natesan et al. (2003)**	**HWCM**	**Natesan et al. (2003)**
1.0e-00	0.0048	0.0079	0.0029	0.0038	0.0011	0.0019	0.0003	0.0009	0.0000	0.0005
1.0e-01	0.1288	0.1354	0.0604	0.0785	0.0379	0.0432	0.0113	0.0229	0.0047	0.0118
1.0e-02	0.1621	0.1753	0.1009	0.1156	0.0613	0.0786	0.0305	0.0487	0.0138	0.0293
1.0e-03	0.1619	0.1792	0.1013	0.1176	0.0621	0.0798	0.0311	0.0494	0.0141	0.0298
1.0e-04	0.1624	0.1796	0.1014	0.1177	0.0631	0.0800	0.0309	0.0495	0.0141	0.0298
1.0e-05	0.1624	0.1796	0.1016	0.1178	0.0631	0.0800	0.0311	0.0495	0.0139	0.0298
1.0e-06	0.1623	0.1796	0.1015	0.1178	0.0629	0.0800	0.0311	0.0495	0.0136	0.0298
1.0e-07	0.1622	0.1796	0.1015	0.1178	0.0632	0.0800	0.0311	0.0495	0.0136	0.0298
1.0e-08	0.1626	0.1796	0.1011	0.1178	0.0635	0.0800	0.0309	0.0495	0.0141	0.0298
1.0e-09	0.1623	0.1796	0.1015	0.1178	0.0635	0.0800	0.0309	0.0495	0.0141	0.0298

Example 4.6 Finally, consider the following singularly perturbed turning point problem (Natesan et al., 2003):

$$\varepsilon y''(x) - 2(2x-1)y'(x) - 4y(x) = 0, x \in [0,1] \tag{4.9}$$

with $y(0) = 1$ and $y(1) = 1$. The exact solution of the above problem is

$$y(x) = e^{-2x(1-x)/\varepsilon}. \tag{4.10}$$

Table 6 gives the maximum absolute error for test Example 4.6. It is obvious that the error bound is inversely proportional to the level of resolution J of Haar wavelet. Hence, the accuracy in the proposed method (HWCM) improves as we increase the level of resolution J. It further strengthens the claim that the proposed method gives excellent results even for very small ε.

5. Conclusion

The Haar wavelet collocation method is applied in order to find the numerical solution of one-dimensional singularly perturbed boundary value problems. The method is tested on several benchmark problems from the literature. The numerical results are compared with a few existing methods reported recently in the literature. The numerical evidence shows superiority of the new method in terms of fast convergence and better accuracy. The proposed method can safely and quickly be used for the solution of a wide range of similar problems.

Funding
The authors received no direct funding for this research.

Author details
Firdous A. Shah[1]
E-mail: fashah79@gmail.com
ORCID ID: http://orcid.org/0000-0001-8461-869X
R. Abass[2]
E-mail: rustamabass13@gmail.com
J. Iqbal[2]
E-mail: javid2iqbal@yahoo.co.in
[1] Department of Mathematics, University of Kashmir, South Campus, Anantnag 192 101, Jammu and Kashmir, India.
[2] Department of Mathematical Sciences, BGSB University, Rajouri 185234, Jammu and Kashmir, India.

AQ7

References
Aziz, T., & Khan, A. (2002). A spline method for second-order singularly perturbed boundary-value problems. *The Journal of Computational and Applied Mathematics, 147*, 445–452.
Chen, C. F., & Hsiao, C. H. (1997). Haar wavelet method for solving lumped and distributed-parameter systems. *IEE Proceedings - Control Theory and Applications, 144*, 87–94.
Chun, C., & Sakthivel, R. (2010). Homotopy perturbation technique for solving two-point boundary value problems-comparison with other methods. *Computer Physics Communications, 181*, 1021–1024.

Daubechies, I. (1992). *Ten Lectures on Wavelets*. Philadelphia: SIAM.

Debnath, L., & Shah, F. A. (2015). *Wavelet transforms and their applications*. New York, NY: Birkhäuser.

Geng, F. Z., & Cui, M. G. (2011). A novel method for solving a class of singularly perturbed boundary value problems based on reproducing kernel method. *Applied Mathematics and Computation, 218*, 4211–4215.

Kadalbajoo, M. K., & Arora, P. (2010). *B*-splines with artificial viscosity for solving singularly perturbed boundary value problems. *Mathematical and Computer Modelling, 52*, 654–666.

Kadalbajoo, M. K., & Gupta, V. (2009). Numerical solution of singularly perturbed convection-diffusion problem using parameter uniform *B*-spline collocation method. *The Journal of Mathematical Analysis and Applications, 355*, 439–452.

Kadalbajoo, M. K., Gupta, V., & Awasthi, A. (2008). A uniformly convergent *B*-spline collocation method on a nonuniform mesh for singularly perturbed one-dimensional time-dependent linear convection-diffusion problem. *The Journal of Computational and Applied Mathematics, 220*, 271–289.

Khan, A., Khan, I., & Aziz, T. (2006). Sextic spline solution of singularly perturbed boundary-value problems. *Applied Mathematics and Computation, 181*, 432–439.

Khan, A., & Khandelwala, P. (2014). Non-polynomial sextic spline solution of singularly perturbed boundary-value problems. *International Journal of Computer Mathematics, 91*, 1122–1135.

Kumar, V., & Mehra, M. (2007). Cubic spline adaptive wavelet scheme to solve singularly perturbed reaction diffusion problems. *International Journal of Wavelets, Multiresolution and Information Processing, 5*, 317–331.

Kumar, M., Singh, P., & Mishra, H. K. (2007). An initial-value technique for singularly perturbed boundary value problems via cubic spline. *International Journal for Computational Methods in Engineering Science and Mechanics, 8*, 419–427.

Lenferink, W. (2002). A second order scheme for a time-dependent, singularly perturbed convection-diffusion equation. *The Journal of Computational and Applied Mathematics, 143*, 49–68.

Lepik, U. (2008). Solving integral and differential equations by the aid of nonuniform Haar wavelets. *Applied Mathematics and Computation, 198*, 326–332.

Lepik, U., & Hein, H. (2014). *Haar wavelets with applications*. New York, NY: Springer.

Lubuma, J. S., & Patidar, K. C. (2006). Uniformly convergent non-standard finite difference methods for self-adjoint singular perturbation problems. *The Journal of Computational and Applied Mathematics, 191*, 228–238.

Miller, J. H., O'Riordan, E., & Shishkin, G. I. (1996). *Fitted numerical methods for singular perturbation problems*. Singapore: World Scientific.

Mohsen, A., & EL-Gamel, M. (2008). On the Galerkin and collocation methods for two-point boundary value problem using Sinc bases. *International Journal of Computer Mathematics, 56*, 930–941.

Natesan, S., Kumar, J. J., & Vigo-Aguiar, J. (2003). Parameter uniform numerical method for singularly perturbed turning point problems exhibiting boundary layers. *The Journal of Computational and Applied Mathematics, 158*, 121–134.

O'Riordan, E., & Stynes, M. (1986). A uniformly accurate finite-element method for a singularly perturbed one-dimensional reaction-diffusion problem. *Mathematics of Computation, 47*, 555–570.

Pandit, S., & Kumar, M. (2014). Haar wavelet approach for numerical solution of two parameters singularly perturbed boundary value problems. *Applied Mathematics & Information Sciences, 8*, 2965–2974.

Phongthanapanich, S., & Dechaumphai, P. (2009). Combined finite volume element method for singularly perturbed reaction-diffusion problems. *Applied Mathematics and Computation, 209*, 177–185.

Rashidinia, J., Ghasemi, M., & Mahmoodi, Z. (2007). Spline approach to the solution of a singularly-perturbed boundary-value problems. *Applied Mathematics and Computation, 189*, 72–78.

Roos, H. G., Stynes, M., & Tobiska, L. (1996). *Numerical Methods for singularly perturbed differential equations*. New York, NY: Springer.

Schatz, A. H., & Wahlbin, L. B. (1983). On the finite element method for singularly perturbed reaction-diffusion problems in two and one dimensions. *Mathematics of Computation, 40*, 47–89.

Surla, K., & Stojanovic, M. (1988). Solving singularly-perturbed boundary-value problems by splines in tension. *The Journal of Computational and Applied Mathematics, 24*, 355–363.

Wazwaz, A. M. (2002). A new method for solving singular initial value problems in the second-order ordinary differential equations. *Applied Mathematics and Computation, 128*, 45–57.

6

An algorithm for variational inequalities with equilibrium and fixed point constraints

Bui Van Dinh[1]*

*Corresponding author: Bui Van Dinh, Faculty of Information Technology, Department of Mathematics, Le Quy Don Technical University, No. 236 Hoang Quoc Viet Road, Hanoi, Vietnam
E-mail: vandinhb@gmail.com

Reviewing editor: Yong Hong Wu, Curtin University of Technology, Australia

Abstract: In this paper, we propose a new hybrid extragradient-viscosity algorithm for solving variational inequality problems, where the constraint set is the common elements of the set of solutions of a pseudomonotone equilibrium problem and the set of fixed points of a demicontractive mapping. Using the hybrid extragradient-viscosity method and combining with hybrid plane cutting techniques, we obtain the algorithm for this problem. Under certain conditions on parameters, the convergence of the iteration sequences generated by the algorithms is obtained.

Subjects: Applied Mathematics; Computer Mathematics; Financial Mathematics; Inverse Problems; Linear Programming; Mathematics & Statistics; Parallel Algorithms; Science

Keywords: variational inequalities; equilibrium problems; KyFan inequality; auxiliary subproblem principle; demicontractive mapping; projection method; Armijo linesearch; pseudomonotonicity

AMS 2010 Mathematics subject classifications: 65 K10; 65 K15; 90 C25; 90 C33

1. Introduction

Let $\mathbb{R}^n$ be a n-dimentional Euclidean space with the inner product $\langle\cdot,\cdot\rangle$ and associated norm $\|\cdot\|$. Let C be a nonempty closed convex subset in $\mathbb{R}^n$ and $G: C \to \mathbb{R}^n, T: C \to C$ be operators, and $f: C \times C \to \mathbb{R}$ be a bifunction satisfying $f(x,x) = 0$ for every $x \in C$. We consider the following variational inequality problem over the set is the common elements of the set of a pseudomonotone equilibrium problem and the set of fixed points of a demicontractive mapping (shortly VIEFP(C, f, T, G):

$$\text{Find } x^* \in S \text{ such that } \langle G(x^*), y - x^*\rangle \geq 0, \quad \forall y \in S, \tag{1}$$

ABOUT THE AUTHOR

Dr Bui Van Dinh is a lecturer at the Department of Mathematics, Faculty of Information Technology, Le Quy Don Technical University, Hanoi, Vietnam. His interest includes solution methods for optimization problems, fixed point problems, variational inequality, and equilibrium problems.

PUBLIC INTEREST STATEMENT

Variational inequality and equilibrium problems as well as fixed point problems are very useful and efficient tools in mathematics. They provided a unified framework for studying many problems arising in engineering, economics, and other fields. In this paper, we propose a new algorithm for solving variational inequality problems where the constraint set is the common elements of the set of solutions of a pseudomonotone equilibrium problem and the set of fixed points of a demicontractive mapping. One difficulty of this problem is that the constraint set is not given explicitly. The proposed method allows us to solve this problem by solving a sequence of convex programs in which they are much easier to solve.

where $S = S_f \cap Fix(T)$, $S_f = \{u \in C: f(u,y) \geq 0, \forall y \in C\}$, i.e. S_f is the solution set of the following equilibrium problem (EP(C, f) for short)

$$\text{Find } u \in C \text{ such that } f(u,y) \geq 0, \forall y \in C, \tag{2}$$

and $Fix(T)$ is the fixed points of the mapping T, i.e. $Fix(T) = \{v \in C \text{ such that } T(v) = v\}$.

We call problem (1) the upper problem and (2) the lower one. Problem (1) is a special case of mathematical programs with equilibrium constraints. Sources for such problems can be found in Luo, Pang, and Ralph (1996), Migdalas, Pardalos, and Varbrand (1988), Muu and Oettli (1992). Bilevel variational inequalities were considered in Anh, Kim, and Muu (2012), Moudafi (2010) and Yao, Liou, and Kang (2010) suggested the use of the proximal point method for monotone bilevel equilibrium problems, which contain monotone variational inequalites as a special case. Recently, Ding (2010) used the auxiliary problem principle to monotone bilevel equilibrium problems. In those papers, the lower problem is required to be monotone. In this case, the subproblems to be solved are monotone.

It should be noticed that the solution set S_f of the lower problem (2) is convex whenever f is pseudomonotone on C. However, the main difficulty is that, even the constrained set S_f is convex, it is not given explicitly as in a standard mathematical programming problem, and therefore the available methods of convex optimization and variational inequality cannot be applied directly to problem (1).

In this paper, we extend the hybrid extragradient-viscosity methods introduced by Maingé (2008b) for solving bilevel problem (1) when the lower problem is pseudomonotone with respect to its solution set equilibrium problems rather than monotone variational inequalities as in Maingé (2008b), the later pseudomonotonicity is somewhat more general than pseudomonotone. We show that the sequence of iterates generated by the proposed algorithm converges to the unique solution of the bilevel problem (1).

The paper is organized as follows. Section 2 contains some preliminaries on the Euclidean projection and equilibrium problems. Section 3 is devoted to presentation of the algorithm and its convergence. In Section 4, we describe a special case of variational inequalities with variational inequalities and fixed points constraints, where the lower variational inequality is pseudomonotone with respect to its solution set.

2. Preliminaries

In the rest of the paper, by P_C we denote as the projection operator on C, that is

$$P_C(x) \in C: \|x - P_C(x)\| \leq \|y - x\|, \quad \forall y \in C.$$

The following well-known results on the projection operator onto a closed convex set will be used in the sequel.

LEMMA 2.1 *Suppose that C is a nonempty closed convex set in $\mathbb{R}^n$. Then,*

(1) *$P_C(x)$ is singleton and well defined for every x;*
(2) $\pi = P_C(x)$ if and only if $\langle x - \pi, y - \pi \rangle \leq 0, \forall y \in C$; and
(3) $\|P_C(x) - P_C(y)\|^2 \leq \|x - y\|^2 - \|P_C(x) - x + y - P_C(y)\|^2, \forall x, y \in C$.

We recall some well-known definitions which will be useful in the sequel (see e.g. Blum & Oettli, 1994; Facchinei & Pang, 2003; Konnov, 2001; Muu & Oettli, 1992; Solodov & Svaiter, 1999).

Definition 2.1 A mapping $T: C \to C$ is called

(1) nonexpansive if $\|T(x) - T(y)\| \leq \|x - y\|$ for all $x, y \in C$;

(2) quasi-nonexpansive if $Fix(T) \neq \emptyset$ and $\|T(x) - x^*\| \leq \|x - x^*\|$ for all $x \in C$ and $x^* \in Fix(T)$;

(3) θ-strict pseudocontractive if there exists $\theta \in [0;1)$, such that $\|T(x) - T(y)\|^2 \leq \|x - y\|^2 + \theta\|(x - T(x)) - (y - T(y))\|^2$ for all $x, y \in C$; and

(4) μ-demicontractive if $Fix(T)$ is not empty, and there exists $\mu \in [0;1)$, such that $\|T(x) - T(x^*)\|^2 \leq \|x - x^*\|^2 + \mu\|x - T(x)\|^2$ for all $x \in C$ and $x^* \in Fix(T)$

Definition 2.2 A bifunction $\varphi: C \times C \to \mathbb{R}$ is said to be

(1) strongly monotone on C with modulus $\beta > 0$, if

$$\varphi(x, y) + \varphi(y, x) \leq -\beta\|x - y\|^2, \ \forall x, y \in C;$$

(2) monotone on C if

$$\varphi(x, y) + \varphi(y, x) \leq 0, \ \forall x, y \in C;$$

(3) pseudomonotone on C if

$$\varphi(x, y) \geq 0 \Longrightarrow \varphi(y, x) \leq 0, \ \forall x, y \in C; \text{ and}$$

(4) pseudomonotone on C with respect to x^* if

$$\varphi(x^*, y) \geq 0 \Longrightarrow \varphi(y, x^*) \leq 0, \ \forall y \in C.$$

We say that φ is pseudomonotone on C with respect to a set S if it is pseudomonotone on C with respect to every point $x^* \in S$.

From Definition 2.2, it follows that (1) $\Rightarrow$ (2) $\Rightarrow$ (3) $\Rightarrow$ (4) $\forall x^* \in C$.

When $\varphi(x, y) = \langle \phi(x), y - x \rangle$, where $\phi: C \to \mathbb{R}^n$ is an operator, then the definition (1) becomes:

$$\langle \varphi(x) - \varphi(y), x - y \rangle \geq \beta\|x - y\|^2, \ \forall x, y \in C$$

that is ϕ is β-strongly monotone on C. Similarly, if ϕ satisfies (2) ((3), (4) resp.) on C, then φ becomes monotone, (pseudomonotone, pseudomonotone with respect to x^* resp.) on C.

In the sequel, we need the following blanket assumptions

(A1) $f(., y)$ is continuous on Ω for every $y \in C$;

(A2) $f(x, .)$ is convex on Ω for every $x \in C$;

(A3) f is pseudomonotone on C with respect to the solution set S_f of EP(C, f);

(A4) T is μ-demicontractive and closed mapping;

(A5) G is L-Lipschitz and β-strongly monotone on C;

(B1) $h(.)$ is δ-strongly convex, continuously differentiable on Ω;

(B2) $\{\lambda_k\}$ is a positive sequence, such that $\sum_{k=0}^{\infty} \lambda_k = \infty$ and $\sum_{k=0}^{\infty} \lambda_k^2 < \infty$; and

(B3) $\{\mu_k\}$ is a positive sequences, such that $0 < \mu \leq \mu_k \leq \frac{1-\mu}{2}$.

LEMMA 2.2 *Suppose Problem EP(C, f) has a solution. Then, under assumptions (A1), (A2), and (A3), the solution set S_f is closed, convex, and*

$f(x^*,y) \geq 0 \quad \forall y \in C$ *if and only if* $f(y,x^*) \leq 0 \quad \forall y \in C.$

The following lemmas are well known from the auxiliary problem principle for equilibrium problems.

LEMMA 2.3 (Mastroeni, 2003) *Suppose that h is a continuously differentiable and strongly convex function on C with modulus $\delta > 0$. Then, under assumptions (A1) and (A2), a point $x^* \in C$ is a solution of EP(C, f) if and only if it is a solution to the equilibrium problem:*

Find $x^* \in C$: $f(x^*,y) + h(y) - h(x^*) - \langle \nabla h(x^*), y - x^* \rangle \geq 0, \quad \forall y \in C.$ (AEP)

The function

$$d(x,y) := h(y) - h(x) - \langle \nabla h(x), y - x \rangle$$

is called Bregman function. Such a function was used to define a generalized projection, called d-projection, which was used to develop algorithms for particular problems (see e.g. Censor & Lent, 1981). *An important case is* $h(x) := \frac{1}{2}\|x\|^2$. *In this case, d-projection becomes the Euclidean one.*

LEMMA 2.4 (Mastroeni, 2003) *Under assumptions (A1) and (A2), a point $x^* \in C$ is a solution of problem (AEP) if and only if*

$$x^* = \operatorname{argmin}\{f(x^*,y) + h(y) - h(x^*) - \langle \nabla h(x^*), y - x^* \rangle : y \in C\}. \quad \text{(CP)}$$

Note that since $f(x,.)$ is convex and h is strongly convex, Problem (CP) is a strongly convex program.

For each $z \in C$, by $\partial_2 f(z,z)$, we denote the subgradient of the convex function $f(z,.)$ at z, i.e.

$$\partial_2 f(z,z) := \{w \in \mathbb{R}^n : f(z,y) \geq f(z,z) + \langle w, y - z \rangle,\ \forall y \in C\}$$
$$= \{w \in \mathbb{R}^n : f(z,y) \geq \langle w, y - z \rangle,\ \forall y \in C\},$$

and each $w \in \partial_2 f(z,z)$ we define the halfspace H_z as

$$H_z := \{x \in \mathbb{R}^n : \langle w, x - z \rangle \leq 0\}. \quad (3)$$

Note that when $f(x,y) = \langle F(x), y - x \rangle$, where $F: C \to \mathbb{R}^n$, this halfspace becomes the one introduced in Solodov and Svaiter (1999). *The following lemma says that the hyperplane does not cut off any solution of problem EP(C, f).*

LEMMA 2.5 (Dinh & Muu, 2015) *Under assumptions (A2) and (A3), one has $S_f \subseteq H_z$ for every $z \in C$.*

LEMMA 2.6 (Dinh & Muu, 2015) *Under assumptions (A1) and (A2), if $\{z^k\} \subset C$ is a sequence, such that $\{z^k\}$ converges to $\bar{z}$ and the sequence $\{w^k\}$, with $w^k \in \partial_2 f(z^k, z^k)$ converges to $\bar{w}$, then $\bar{w} \in \partial_2 f(\bar{z}, \bar{z})$.*

LEMMA 2.7 *Suppose the bifunction f satisfies the assumptions (A1) and (A2), the function h satisfies the assumption (B1). If $\{x^k\} \subset C$ is bounded and $\{y^k\}$ is a sequence, such that*

$$y^k = \arg\min\left\{f(x^k,y) + \frac{1}{\rho}\left[h(y) - h(x^k) - \langle \nabla h(x^k), y - x^k \rangle\right] :\ y \in C\right\},$$

then $\{y^k\}$ is bounded.

Proof Firstly, we show that if $\{x^k\}$ converges to x^*, then $\{y^k\}$ is bounded. Indeed,

$$y^k = \arg\min\left\{f(x^k,y) + \frac{1}{\rho}\left[h(y) - h(x^k) - \langle \nabla h(x^k), y - x^k \rangle\right] : y \in C\right\},$$

and

$$f(x^k,x^k) + \frac{1}{\rho}\left[h(x^k) - h(x^k) - \langle \nabla h(x^k), x^k - x^k\rangle\right] = 0,$$

therefore

$$f(x^k,y^k) + \frac{1}{\rho}\left[h(y^k) - h(x^k) - \langle \nabla h(x^k), y^k - x^k\rangle\right] \leq 0, \quad \forall k.$$

In addition, $f(x^k,.) + \frac{1}{\rho}\left[h(.) - h(x^k) - \langle \nabla h(x^k), . - x^k\rangle\right]$ is strongly convex on C with modulus $\frac{\delta}{\rho}$; hence, for all $w^k \in \partial_2 f(x^k,x^k)$, we get

$$f(x^k,y^k) + \frac{1}{\rho}\left[h(y^k) - h(x^k) - \langle \nabla h(x^k), y^k - x^k\rangle\right] \geq \langle w^k, y^k - x^k\rangle + \frac{\delta}{\rho}\|y^k - x^k\|^2.$$

This implies $0 \geq -\|w^k\|\|y^k - x^k\| + \frac{\delta}{\rho}\|y^k - x^k\|^2$, so that

$$\|y^k - x^k\| \leq \frac{\rho}{\delta}\|w^k\|.$$

Because $\{x^k\}$ converges to x^* and $w^k \in \partial_2 f(x^k,x^k)$, by Theorem 24.5 in Rockafellar (1970), the sequence $\{w^k\}$ is bounded; combining with the boundedness of $\{x^k\}$, we have $\{y^k\}$ also bounded.

Now, we prove the Lemma 2.7. Suppose contradict that $\{y^k\}$ is unbounded, i.e. there exists an subsequence $\{y^{k_i}\} \subseteq \{y^k\}$, such that $\lim_{i\to\infty} \|y^{k_i}\| = +\infty$. By the boundedness of $\{x^k\}$, it implies $\{x^{k_i}\}$ is also bounded; without loss of gerarality, we may assume that $\lim_{i\to\infty} x^{k_i} = x^*$. By the same argument as above, we have $\{y^{k_i}\}$ is bounded, which we contradict. Therefore, $\{y^k\}$ is bounded. □

The following lemma is in Solodov and Svaiter (1999) (see also Dinh & Muu, 2015).

LEMMA 2.8 (Dinh & Muu, 2015; Solodov & Svaiter, 1999) *Suppose that* $x \in C$ *and* $u = P_{C\cap H_z}(x)$. *Then,*

$$u = P_{C\cap H_z}(\bar{x}), \quad \text{where} \quad \bar{x} = P_{H_z}(x).$$

LEMMA 2.9 (Maingé, 2008b) *Let* $T{:}C \to C$ *be an* α *demicontractive mapping, such that Fix(T) is nonempty. Then,* $T_\mu = (1-\mu)I + \mu T$ *is a quasi-nonexpansive mapping over C for every* $\mu \in [0;1-\alpha]$. *Furthermore,*

$$\|T_\mu(x) - x^*\|^2 \leq \|x - x^*\|^2 - \mu(1-\alpha-\mu)\|T(x) - x\|^2 \text{ for all} x \in C \text{ and} x^* \in Fix(T).$$

LEMMA 2.10 *(Lemma 3.1 Maingé, 2008a) Let* $\{a_k\}$ *be a sequence of real numbers that does not decrease at infinity, in the sense that there exists a subsequence* $\{a_{k_j}\}$ *of* $\{a_k\}$, *such that*

$$a_{k_j} < a_{k_j+1} \text{ for all} j \geq 0$$

Also, consider the sequence of integers $\{\sigma(k)\}_{k\geq k_0}$ *defined by*

$$\sigma(k) = \max\left\{j \leq k \mid a_j < a_{j+1}\right\}.$$

Then, $\{\sigma(k)\}_{k\geq k_0}$ *is a nondecreasing sequence verifying*

$$\lim_{k\to\infty} \sigma(k) = \infty$$

and, for all $k \geq k_0$, *the following two inequalities hold:*

$$a_{\sigma(k)} \le a_{\sigma(k)+1} \tag{4}$$

$$a_k \le a_{\sigma(k)+1} \tag{5}$$

3. A hybrid extragradient-viscosity algorithm for VIEFP (C, f, T, G)

Algorithm 1.

Initialization. Pick $x^0 \in C$, choose parameters $\eta \in (0,1)$, $\rho > 0$ and $\{\lambda_k\} \subset [0;1)$; $\{\mu_k\} \subset [0;1)$.

Iteration k**.** (k = 0, 1, 2, ...) Having x^k do the following steps:

Step 1. Solve the strongly convex program

$$\min\left\{f(x^k,y)+\frac{1}{\rho}\Big[h(y)-h(x^k)-\langle\nabla h(x^k),y-x^k\rangle\Big]\,:\,y\in C\right\} \qquad CP(x^k)$$

to obtain its unique solution y^k.

If $y^k = x^k$, take $u^k = x^k$ and go to *Step 4*. Otherwise, do Step 2.

Step 2. (Armijo linesearch rule) Find m_k as the smallest positive integer number m satisfying

$$\begin{cases} z^{k,m} = (1-\eta^m)x^k + \eta^m y^k, \\ w^{k,m} \in \partial_2 f(z^{k,m}, z^{k,m}), \\ \langle w^{k,m}, x^k - y^k\rangle \ge \frac{1}{\rho}\Big[h(y^k) - h(x^k) - \langle \nabla h(x^k), y^k - x^k\rangle\Big]. \end{cases} \tag{6}$$

Step 3. Set $\eta_k := \eta^{m_k}$, $z^k := z^{k,m_k}$, $w^k := w^{k,m_k}$. Take

$$C_k := \{x \in C : \langle w^k, x - z^k\rangle \le 0\},\ u^k := P_{C_k}(x^k). \tag{7}$$

Step 4. Compute $v^k = P_C(u^k - \lambda_k G(u^k))$.

Step 5. Set $x^{k+1} = (1-\mu_k)v^k + \mu_k T(v^k)$, and go to Step 1 with k is replaced by $k+1$.

Remark 3.1

(1) If $y^k = x^k$, then x^k is a solution to EP(C, f).

(2) $w^k \ne 0\ \forall k$, indeed, at the beginning of Step 2, $x^k \ne y^k$. By the Armijo linesearch rule and δ-strong convexity of h, we have

$$\langle w^k, x^k - y^k\rangle \ge \frac{1}{\rho}\Big[h(y^k) - h(x^k) - \langle \nabla h(x^k), y^k - x^k\rangle\Big] \ge$$
$$\ge \frac{\delta}{\rho}\|x^k - y^k\|^2 > 0.$$

Now, we are going to analyze the validity and convergence of the algorithm. Some parts in our proofs are based on the proof scheme in Maingé (2008b).

LEMMA 3.1 *Under Assumptions (A1) and (A2), the linesearch rule (6) is well defined in the sense that at each iteration k, there exists an integer number m > 0, satisfying the inequality in (6) for every* $w^{k,m} \in \partial_2 f(z^{k,m}, z^{k,m})$*, and if, in addition assumptions (A3) and (A4) are satisfied, then for every* $x^* \in S$*, one has*

$$\|x^{k+1} - x^*\|^2 \le \|x^k - x^*\|^2 - \|u^k - \bar{x}^k\|^2 - \left(\frac{\eta_k\delta}{\rho\|w^k\|}\right)^2\|x^k - y^k\|^4$$
$$- \|x^{k+1} - v^k\|^2 - 2\lambda_k\langle u^k - x^*, G(u^k)\rangle + \lambda_k^2\|G(u^k)\|^2\ \forall k, \tag{8}$$

where $\bar{x}^k = P_{H_{z^k}}(x^k)$.

Proof First, we prove that there exists a positive integer m_0 such that

$$\langle w^{k,m_0}, x^k - y^k\rangle \geq \frac{1}{\rho}\left[h(y^k) - h(x^k) - \langle \nabla h(x^k), y^k - x^k\rangle\right]$$
$$\forall w^{k,m_0} \in \partial_2 f(z^{k,m_0}, z^{k,m_0}).$$

Indeed, suppose by contradiction that for every positive integer m and $z^{k,m} = (1-\eta^m)x^k + \eta^m y^k$, there exists $w^{k,m} \in \partial_2 f(z^{k,m}, z^{k,m})$, such that

$$\langle w^{k,m}, x^k - y^k\rangle < \frac{1}{\rho}\left[h(y^k) - h(x^k) - \langle \nabla h(x^k), y^k - x^k\rangle\right].$$

Since $z^{k,m} \to x^k$ as $m \to \infty$, by Theorem 24.5 in Rockafellar (1970), the sequence $\{w^{k,m}\}_{m=1}^{\infty}$ is bounded. Thus, we may assume that $w^{k,m} \to \bar{w}$ for some $\bar{w}$. Taking the limit as $m \to \infty$, from $z^{k,m} \to x^k$ and $w^{k,m} \to \bar{w}$, by Lemma 2.6, it follows that $\bar{w} \in \partial_2 f(x^k, x^k)$ and

$$\langle \bar{w}, x^k - y^k\rangle \leq \frac{1}{\rho}\left[h(y^k) - h(x^k) - \langle \nabla h(x^k), y^k - x^k\rangle\right]. \tag{9}$$

Since $\bar{w} \in \partial_2 f(x^k, x^k)$, we have

$$f(x^k, y^k) \geq f(x^k, x^k) + \langle \bar{w}, y^k - x^k\rangle = \langle \bar{w}, y^k - x^k\rangle.$$

Combining with (9) yields

$$f(x^k, y^k) + \frac{1}{\rho}\left[h(y^k) - h(x^k) - \langle \nabla h(x^k), y^k - x^k\rangle\right] \geq 0,$$

which contradicts to the fact that

$$f(x^k, y^k) + \frac{1}{\rho}\left[h(y^k) - h(x^k) - \langle \nabla h(x^k), y^k - x^k\rangle\right] < 0.$$

Thus, the linesearch is well defined.

Now, we prove (8). For simplicity of notation, let $d^k := x^k - y^k$, $H_k := H_{z^k}$.

Since $u^k = P_{C\cap H_k}(\bar{x}^k)$ and $x^* \in S$, by Lemma 2.5, $x^* \in C \cap H_k$, we have

$$\|u^k - \bar{x}^k\|^2 \leq \langle x^* - \bar{x}^k, u^k - \bar{x}^k\rangle$$

which together with

$$\|u^k - x^*\|^2 = \|\bar{x}^k - x^*\|^2 + \|u^k - \bar{x}^k\|^2 + 2\langle u^k - \bar{x}^k, \bar{x}^k - x^*\rangle$$

implies

$$\|u^k - x^*\|^2 \leq \|\bar{x}^k - x^*\|^2 - \|u^k - \bar{x}^k\|^2. \tag{10}$$

Replacing

$$\bar{x}^k = P_{H_k}(x^k) = x^k - \frac{\langle w^k, x^k - z^k\rangle}{\|w^k\|^2} w^k$$

into (Equation 10), we obtain

$$\|u^k - x^*\|^2 \leq \|x^k - x^*\|^2 - \|u^k - \bar{x}^k\|^2 - 2\langle w^k, x^k - x^*\rangle \frac{\langle w^k, x^k - z^k\rangle}{\|w^k\|^2} + \frac{\langle w^k, x^k - z^k\rangle^2}{\|w\|^2}.$$

Substituting $x^k = z^k + \eta_k d^k$ into the last inequality, we get

$$\begin{aligned}\|u^k - x^*\|^2 &\le \|x^k - x^*\|^2 - \|u^k - \bar{x}^k\|^2 + \left(\frac{\eta_k\langle w^k, d^k\rangle}{\|w^k\|}\right)^2 - \frac{2\eta_k\langle w^k, d^k\rangle}{\|w^k\|^2}\langle w^k, x^k - x^*\rangle \\ &= \|x^k - x^*\|^2 - \|u^k - \bar{x}^k\|^2 - \left(\frac{\eta_k\langle w^k, d^k\rangle}{\|w^k\|}\right)^2 - \frac{2\eta_k\langle w^k, d^k\rangle}{\|w^k\|^2}\langle w^k, z^k - x^*\rangle.\end{aligned}$$

In addition, by the Armijo linesearch rule, using the δ-strong convexity of h, we have

$$\langle w^k, x^k - y^k\rangle \ge \frac{1}{\rho}\left[h(y^k) - h(x^k) - \langle \nabla h(x^k), y^k - x^k\rangle\right] \ge \frac{\delta}{\rho}\|x^k - y^k\|^2.$$

Note that $x^* \in H_k$ can be written as:

$$\|u^k - x^*\|^2 \le \|x^k - x^*\|^2 - \|u^k - \bar{x}^k\|^2 - \left(\frac{\eta_k\delta}{\rho\|w^k\|}\right)^2\|x^k - y^k\|^4. \tag{11}$$

From Lemma 2.9, we have

$$\|x^{k+1} - x^*\|^2 \le \|v^k - x^*\|^2 - \mu_k(1 - \mu - \mu_k)|T(v^k - v^k\|^2 \tag{12}$$

Replacing $T(v^k) - v^k = \frac{1}{\mu_k}(x^{k+1} - v^k)$ into (Equation 12), we get

$$\begin{aligned}\|x^{k+1} - x^*\|^2 &\le \|v^k - x^*\|^2 - \frac{1 - \mu - \mu_k}{\mu_k}\|x^{k+1} - v^k\|^2 \\ &\le \|v^k - x^*\|^2 - \|x^{k+1} - v^k\|^2,\end{aligned} \tag{13}$$

where the last inequality follows from $0 < \mu \le \mu_k \le \frac{1-\mu}{2}$. We have

$$\begin{aligned}\|x^{k+1} - x^*\|^2 &\le \|v^k - x^*\|^2 - \|x^{k+1} - v^k\|^2 \\ &= \|P_C(u^k - \mu_k G(u^k)) - P_C(x^*)\| - \|x^{k+1} - v^k\|^2 \\ &\le \|u^k - x^* - \mu_k G(u^k))\|^2 - \|x^{k+1} - v^k\|^2 \\ &= \|u^k - x^*\|^2 - 2\mu_k\langle u^k - x^*, G(u^k)\rangle + \mu_k^2\|G(u^k)\|^2 - \|x^{k+1} - v^k\|^2,\end{aligned}$$

which together with (11) implies

$$\begin{aligned}\|x^{k+1} - x^*\|^2 \le\ & \|x^k - x^*\|^2 - \|u^k - \bar{x}^k\|^2 - \left(\frac{\eta_k\delta}{\rho\|w^k\|}\right)^2\|x^k - y^k\|^4 \\ & - \|x^{k+1} - v^k\|^2 - 2\mu_k\langle u^k - x^*, G(u^k)\rangle + \mu_k^2\|G(u^k)\|^2\ \forall k\end{aligned} \tag{14}$$

as desired. □

LEMMA 3.2 *The sequences $\{x^k\}, \{u^k\}$, and $\{v^k\}$ generated by the Algorithm 1 are bounded under assumptions (A1), (A2), (A3), (A4), and (A5).*

Proof From (13), we get

$$\|x^{k+1} - x^*\| \le \|v^k - x^*\| \tag{15}$$

In addition,

$$\begin{aligned}
\|v^k - x^*\| &= \|P_C(u^k - \lambda_k G(u^k)) - P_C(x^*)\| \le \|u^k - \lambda_k G(u^k) - x^*\| \\
&\le \|(u^k - \lambda_k G(u^k)) - (x^* - \lambda_k G(x^*))\| + \lambda_k\|G(x^*)\| \\
&= \left\|\left(1 - L^2\frac{\lambda_k}{\beta}\right)(u^k - x^*) - L^2\frac{\lambda_k}{\beta}\left[\left(\frac{\beta}{L^2}G - I\right)u^k - \left(\frac{\beta}{L^2}G - I\right)x^*\right]\right\| \\
&\quad + \lambda_k\|G(x^*)\| \\
&\le \left(1 - L^2\frac{\lambda_k}{\beta}\right)\|u^k - x^*\| + L^2\frac{\lambda_k}{\beta}M_k + \lambda_k\|G(x^*)\|,
\end{aligned} \tag{16}$$

where $M_k = \|\left(\frac{\beta}{L^2}G - I\right)u^k - \left(\frac{\beta}{L^2}G - I\right)x^*\|$.

Since G is L-Lipschitz and β-strongly monotone, we have

$$\begin{aligned}
M_k^2 &= \left\|\frac{\beta}{L^2}(G(u^k) - G(x^*)) - (u^k - x^*)\right\|^2 \\
&= \frac{\beta^2}{L^4}\left\|G(u^k) - G(x^*)\right\|^2 - 2\frac{\beta}{L^2}\langle G(u^k) - G(x^*), u^k - x^*\rangle + \|u^k - x^*\|^2 \\
&\le \frac{\beta^2}{L^2}\|u^k - x^*\|^2 - 2\frac{\beta^2}{L^2}\|u^k - x^*\|^2 + \|u^k - x^*\|^2 \\
&= (1 - \frac{\beta^2}{L^2})\|u^k - x^*\|^2.
\end{aligned}$$

Hence, $M_k \le \sqrt{1 - \frac{\beta^2}{L^2}}\|u^k - x^*\|$. Then, combining with (16), we get

$$\begin{aligned}
\|v^k - x^*\| &\le \left(1 - \lambda_k\frac{L^2}{\beta}\left(1 - \sqrt{1 - \frac{\beta^2}{L^2}}\right)\right)\|u^k - x^*\| + \lambda_k\|G(x^*)\| \\
&= \left(1 - \lambda_k\frac{L^2}{\beta}\gamma\right)\|u^k - x^*\| + \lambda_k\|G(x^*)\| \\
&= (1 - \gamma_k)\|u^k - x^*\| + \gamma_k\left(\frac{\beta}{L^2\gamma}\|G(x^*)\|\right) \\
&\le (1 - \gamma_k)\|x^k - x^*\| + \gamma_k\left(\frac{\beta}{L^2\gamma}\|G(x^*)\|\right)
\end{aligned}$$

where, $\gamma = 1 - \sqrt{1 - \frac{\beta^2}{L^2}}$, $\gamma_k = \lambda_k\frac{L^2}{\beta}\gamma \in (0;1)$, and the last inequality deduced from (11). Combining with (15), we obtain

$$\|x^{k+1} - x^*\| \le (1 - \gamma_k)\|x^k - x^*\| + \gamma_k\left(\frac{\beta}{L^2\gamma}\|G(x^*)\|\right). \tag{17}$$

By induction, it implies $\|x^{k+1} - x^*\| \le \max\{\|x^k - x^*\|, \frac{\beta}{L^2\gamma}\|G(x^*)\|\} \le \ldots \le \max\{\|x^0 - x^*\|, \frac{\beta}{L^2\gamma}\|G(x^*)\|\}$. Hence, $\{x^k\}$ is bounded, which, from (11), implies that $\{u^k\}$ is also bounded. Since $\|v^k - u^k\| = \|P_C(u^k - \lambda_k G(u^k)) - P_C(u^k)\| \le \lambda_k\|G(u^k)\|$, and by the boundedness of $\{u^k\}$, we can conclude that $\{v^k\}$ is bounded. □

LEMMA 3.3 *If the subsequence $\{v^{k_i}\} \subset \{v^k\}$ converges to some $\bar{v}$ and*

$$\left(\frac{\eta_{k_i}}{\|w^{k_i}\|}\right)^2\|y^{k_i} - x^{k_i}\|^4 \to 0 \quad \textit{and} \quad \|x^{k_i+1} - v^{k_i}\| \to 0 \text{ as } i \to \infty, \tag{18}$$

then $\bar{x} \in S$.

Proof By definition of $\{x^k\}$ in Algorithm 1, we have $x^{k_i+1} = (1 - \mu_{k_i})v^{k_i} + \mu_{k_i}T(v^{k_i})$. Therefore, $\|T(v^{k_i}) - v^{k_i}\| = \frac{1}{\mu_{k_i}}\|x^{k_i+1} - v^{k_i}\|$. Taking the limit as $i \to \infty$ and by the closedness of mapping T, we get

$T(\bar{v}) = \bar{v}$, i.e. $\bar{v} \in Fix(T)$.

Now, we prove $\bar{v} \in S_f$. Indeed, from Lemma 3.2, $\{x^{k_i}\}$ is bounded. Without loss of generality, we may assume that $\lim_{i\to\infty} x^{k_i} = \bar{x}$. We will consider two distinct cases:

Case 1. $\text{Inf}\frac{\eta_{k_i}}{\|w^{k_i}\|} > 0$. Then, from (18), one has $\lim_{i\to\infty} \|y^{k_i} - x^{k_i}\| = 0$, thus $y^{k_i} \to \bar{x}$ and $z^{k_i} \to \bar{x}$.

According to the definition of y^{k_i}, we have

$$f(x^{k_i}, y) + \frac{1}{\rho}[h(y) - h(x^{k_i}) - \langle \nabla h(x^{k_i}), y - x^{k_i} \rangle] \geq f(x^{k_i}, y^{k_i}) + \frac{1}{\rho}[h(y^{k_i}) - h(x^{k_i}) - \langle \nabla h(x^{k_i}), y^{k_i} - x^{k_i} \rangle],\ \forall y \in C$$

by the continuity of $h, \nabla h$, we get in the limit as $i \to \infty$ that

$$f(\bar{x}, y) + \frac{1}{\rho}[h(y) - h(\bar{x}) - \langle \nabla h(\bar{x}), y - \bar{x} \rangle] \geq 0, \quad \forall y \in C$$

this fact shows that $\bar{x} \in S_f$.

Case 2. $\lim_{i\to\infty} \frac{\eta_{k_i}}{\|w^{k_i}\|} = 0$. By the linesearch rule and τ-strong convexity of h, we have

$$\begin{aligned}\langle w^{k_i}, x^{k_i} - y^{k_i} \rangle &\geq \frac{1}{\rho}\left[h(y^{k_i}) - h(x^{k_i}) - \langle \nabla h(x^{k_i}), y^{k_i} - x^{k_i} \rangle\right] \\ &\geq \frac{\tau}{\rho}\|x^{k_i} - y^{k_i}\|^2.\end{aligned}$$

Thus, $\|y^{k_i} - x^{k_i}\| \leq \sqrt{\frac{\rho}{\tau}}\|w^{k_i}\|$.

From the boundedness of $\{w^{k_i}\}$ and (18), it follows $\eta_{k_i} \to 0$, so that $z^{k_i} = (1 - \eta_{k_i})x^{k_i} + \eta_{k_i} y^{k_i} \to \bar{x}$ as $i \to \infty$. Without loss of generality, we suppose that $w^{k_i} \to \bar{w} \in \partial_2 f(\bar{x}, \bar{x})$ and $y^{k_i} \to \bar{y}$ as $i \to \infty$.

We have

$$\begin{aligned}&f(x^{k_i}, y) + \frac{1}{\rho}[h(y) - h(x^{k_i}) - \langle \nabla h(x^{k_i}), y - x^{k_i} \rangle] \\ &\quad\geq f(x^{k_i}, y^{k_i}) + \frac{1}{\rho}[h(y^{k_i}) - h(x^{k_i}) - \langle \nabla h(x^{k_i}), y^{k_i} - x^{k_i} \rangle],\ \forall y \in C\end{aligned}$$

letting $i \to \infty$, we obtain in the limit that

$$\begin{aligned}&f(\bar{x}, y) + \frac{1}{\rho}[h(y) - h(\bar{x}) - \langle \nabla h(\bar{x}), y - \bar{x} \rangle] \\ &\quad\geq f(\bar{x}, \bar{y}) + \frac{1}{\rho}[h(\bar{y}) - h(\bar{x}) - \langle \nabla h(\bar{x}), \bar{y} - \bar{x} \rangle]\ \forall y \in C.\end{aligned}$$

On the other hand, by the linesearch rule (6), for $m_{k_i} - 1$, there exists $w^{k_i, m_{k_i}-1} \in \partial_2 f(z^{k_i, m_{k_i}-1}, z^{k_i, m_{k_i}-1})$, such that

$$\langle w^{m_{k_i}-1}, x^{k_i} - y^{k_i} \rangle < \frac{1}{\rho}\left[h(y^{k_i}) - h(x^{k_i}) - \langle \nabla h(x^{k_i}), y^{k_i} - x^{k_i} \rangle\right]. \tag{19}$$

Letting $i \to \infty$ and combining with $z^{k_i, m_{k_i}-1} \to \bar{x}$, $w^{k_i, m_{k_i}-1} \to \bar{w} \in \partial_2 f(\bar{x}, \bar{x})$ we obtain in the limit from (19) that

$$\langle \bar{w}, \bar{x} - \bar{y} \rangle \leq \frac{1}{\rho}\Big[h(\bar{y}) - h(\bar{x}) - \langle \nabla h(\bar{x}), \bar{y} - \bar{x} \rangle\Big].$$

Note that $\bar{w} \in \partial f(\bar{x}, \bar{x})$; it follows from the last inequality that

$$f(\bar{x}, \bar{y}) + \frac{1}{\rho}\Big[h(\bar{y}) - h(\bar{x}) - \langle \nabla h(\bar{x}), \bar{y} - \bar{x} \rangle\Big] \geq 0.$$

Hence,

$$f(\bar{x}, y) + \frac{1}{\rho}\Big[h(y) - h(\bar{x}) - \langle \nabla h(\bar{x}), y - \bar{x} \rangle\Big] \geq 0, \ \forall y \in C,$$

which shows that $\bar{x} \in S_f$.

In addition, $\|v^{k_i} - \bar{x}\| \leq \|u^{k_i} - \bar{x}\| + \lambda_{k_i}\|G(u^{k_i})\|$ combining with (11), it implies

$$\|v^{k_i} - \bar{x}\| \leq \|x^{k_i} - \bar{x}\| + \lambda_{k_i}\|G(u^{k_i})\| \tag{20}$$

taking the limit both sides of (20), we get $\lim_{i \to \infty} v^{k_i} = \bar{x}$. Hence, $\bar{v} = \bar{x}$. Therefore, $\bar{v} \in S$. □

Now, we are in a position to prove the convergence of the proposed algorithm.

THEOREM 3.4 *Suppose that the set $S = S_f \cap Fix(T)$ is nonempty and that the function $h(.)$, the sequence $\{\lambda_k\}$, $\{\mu_k\}$ satisfies the conditions (B1), (B2), and (B3), respectively. Then, under Assumptions (A1), (A2), (A3), (A4), and (A5), the sequence $\{x^k\}$ generated by Algorithm 1 converges to the unique solution x^* of VIEFP(C, f, T, G).*

Proof By Lemma 3.1, we have

$$\begin{aligned} &\|x^{k+1} - x^*\|^2 - \|x^k - x^*\|^2 + \|x^{k+1} - v^k\|^2 + \left(\frac{\eta_k \delta}{\rho\|w^k\|}\right)^2 \|x^k - y^k\|^4 \\ &\quad \leq -2\lambda_k \langle u^k - x^*, G(u^k) \rangle + \lambda_k^2 \|G(u^k)\|^2 \ \forall k. \end{aligned} \tag{21}$$

From the boundedness of $\{u^k\}$ and $\{G(u^k)\}$, it implies that there exist positive numbers A, B, such that

$$|\langle u^k - x^*, G(u^k) \rangle| \leq A, \ \ \|G(u^k)\|^2 \leq B \ \forall k.$$

By setting $a_k = \|x^k - x^*\|^2$, and combining with the last inequalities, (21) becomes

$$a_{k+1} - a_k + \|x^{k+1} - v^k\|^2 + \left(\frac{\eta_k \delta}{\rho\|w^k\|}\right)^2 \|x^k - y^k\|^4 \leq 2\lambda_k A + \lambda_k^2 B. \tag{22}$$

We will consider two distinct cases:

Case 1. There exists k_0, such that $\{a_k\}$ is decreasing when $k \geq k_0$.

Then, there exists $\lim_{k \to \infty} a_k = a$, taking the limit on both sides of (22), we get

$$\lim_{k \to \infty} \|x^{k+1} - v^k\|^2 = 0, \text{ and } \lim_{k \to \infty} \left(\frac{\eta_k \delta}{\rho\|w^k\|}\right)^2 \|x^k - y^k\|^4 = 0. \tag{23}$$

This implies $\lim_{k \to \infty} \|v^k - x^*\| = a$. In addition,

$$\begin{aligned}\|v^k - u^k\| &= \|P_C(u^k - \lambda_k G(u^k)) - P_C(u^k)\| \\ &\le \|u^k - \lambda_k G(u^k) - u^k\| \\ &= \lambda_k \|G(u^k)\| \to 0 \text{ as } k \to \infty.\end{aligned} \tag{24}$$

Thus, $\lim_{k\to\infty} \|u^k - x^*\| = a$. From the boundedness of $\{u^k\}$, it implies that there exists $\{u^{k_i}\} \subset \{u^k\}$ and $u^{k_i} \to \bar{u} \in C$, such that $\liminf \langle u^k - x^*, G(x^*)\rangle = \lim_{i\to\infty} \langle u^{k_i} - x^*, G(x^*)\rangle$.

Combining this fact with (23) and (24), we obtain

$$v^{k_i} \to \bar{u};\ x^{k_i+1} \to \bar{u}, \text{ and } \left(\frac{\eta_{k_i+1}\delta}{\rho\|w^{k_i+1}\|}\right)^2 \|x^{k_i+1} - y^{k_i+1}\|^4 \to 0 \text{ as } i \to \infty.$$

By Lemma 3.3, we get $\bar{u} \in S$. Thus,

$$\liminf_{k\to\infty} \langle u^k - x^*, G(x^*)\rangle = \lim_{i\to\infty} \langle u^{k_i} - x^*, G(x^*)\rangle = \langle \bar{u} - x^*, G(x^*)\rangle \ge 0.$$

Since G is β-strongly monotone, one has

$$\begin{aligned}\langle u^k - x^*, G(u^k)\rangle &= \langle u^k - x^*, G(u^k) - G(x^*)\rangle + \langle u^k - x^*, G(u^*)\rangle \\ &\ge \beta\|u^k - x^*\|^2 + \langle u^k - x^*, G(u^*)\rangle.\end{aligned}$$

Taking the limit as $k \to \infty$ and remembering that $a = \lim \|u^k - x^*\|^2$, we get

$$\liminf_{k\to\infty} \langle u^k - x^*, G(u^k)\rangle \ge \beta a. \tag{25}$$

If $a > 0$, then by choosing $\epsilon = \frac{1}{2}\beta a$, from (25), it implies that there exists $k_1 > 0$, such that

$$\langle u^k - x^*, G(u^k)\rangle \ge \frac{1}{2}\beta a, \quad \forall k \ge k_1.$$

From (21), we get

$$a_{k+1} - a_k \le -\lambda_k \beta a + \lambda_k^2 B, \quad \forall k \ge k_1$$

and thus summing up from k_1 to k, we have

$$a_{k+1} - a_{k_1} \le -\sum_{j=k_1}^{K} \lambda_j \beta a + B\sum_{j=k_1}^{K} \lambda_j^2$$

combining this fact with $\sum_{k=1}^{\infty} \lambda_k = \infty$ and $\sum_{k=1}^{\infty} \lambda_k^2 < \infty$, we obtain $\liminf a_k = -\infty$, which is a contradiction.

Thus, we must have $a = 0$, i.e. $\lim_{k\to\infty} \|x^k - x^*\| = 0$.

Case 2. There exists a subsequence $\{a_{k_i}\}_{i\ge 0} \subset \{a_k\}_{k\ge 0}$, such that $a_{k_i} < a_{k_i+1}$ for all $i \ge 0$. In this situation, we consider the sequence of indices $\{\sigma(k)\}$ defined as in Lemma 2.10. It follows that $a_{\sigma(k)+1} - a_{\sigma(k)} \ge 0$, which by (22) amounts to

$$\|x^{\sigma(k)+1} - v^{\sigma(k)}\|^2 + \left(\frac{\eta_{\sigma(k)}\delta}{\rho\|w^{\sigma(k)}\|}\right)^2 \|x^{\sigma(k)} - y^{\sigma(k)}\|^4 \le 2\lambda_{\sigma(k)} A + \lambda_{\sigma(k)}^2 B.$$

Therefore,

$$\lim_{k\to\infty} \|x^{\sigma(k)+1} - v^{\sigma(k)}\|^2 = 0;\ \lim_{k\to\infty} \left(\frac{\eta_{\sigma(k)}\delta}{\rho\|w^{\sigma(k)}\|}\right)^2 \|x^{\sigma(k)} - y^{\sigma(k)}\|^4 = 0.$$

From the boundedness of $\{v^{\sigma(k)}\}$, without loss of generality, we may assume that $v^{\sigma(k)} \to \bar{v}$. By Lemma 3.3, we get $\bar{v} \in S$.

In addition,

$$\begin{aligned}\|v^{\sigma(k)} - u^{\sigma(k)}\| &= \|P_C(u^{\sigma(k)} - \lambda_{\sigma(k)} G(u^{\sigma(k)})) - P_C(u^{\sigma(k)})\| \\ &\le \lambda_{\sigma(k)} (G(u^{\sigma(k)}) \to 0 \text{ as} k \to \infty.\end{aligned}$$

Therefore, $\lim_{k\to\infty} u^{\sigma(k)} = \bar{v}$.

By (21), we get

$$\begin{aligned}2\lambda_{\sigma(k)} \langle u^{\sigma(k)} - x^*, G(u^{\sigma(k)})\rangle &\le a_{\sigma(k)} - a_{\sigma(k)+1} - \left(\frac{\eta_{\sigma(k)}\delta}{\rho\|w^{\sigma(k)}\|}\right)^2 \|x^{\sigma(k)} - y^{\sigma(k)}\|^4 \\ &\quad + \lambda^2_{\sigma(k)} \|G(u^{\sigma(k)})\|^2 \le \lambda^2_{\sigma(k)} B\end{aligned}$$

which implies

$$\langle u^{\sigma(k)} - x^*, G(u^{\sigma(k)})\rangle \le \frac{\lambda_{\sigma(k)}}{2} B. \tag{26}$$

Since G is β-strongly monotone, we have

$$\begin{aligned}\beta\|u^{\sigma(k)} - x^*\|^2 &\le \langle u^{\sigma(k)} - x^*, G(u^{\sigma(k)}) - G(x^*)\rangle \\ &= \langle u^{\sigma(k)} - x^*, G(u^{\sigma(k)})\rangle - \langle u^{\sigma(k)} - x^*, G(x^*)\rangle\end{aligned}$$

which combining with (26), we get

$$\|u^{\sigma(k)} - x^*\|^2 \le \frac{1}{\beta}\left[\frac{\lambda_{\sigma(k)}}{2} B - \langle u^{\sigma(k)} - x^*, G(x^*)\rangle\right],$$

so that

$$\lim_{k\to\infty} \|u^{\sigma(k)} - x^*\|^2 \le -\langle u^{\sigma(k)} - x^*, G(x^*)\rangle \le 0$$

which amounts to

$$\lim_{k\to\infty} \|u^{\sigma(k)} - x^*\| = 0. \tag{27}$$

In addition,

$$\begin{aligned}\|x^{\sigma(k)+1} - u^{\sigma(k)}\| &= \|P_C(u^{\sigma(k)} - \lambda_{\sigma(k)} G(u^{\sigma(k)})) - P_(u^{\sigma(k)})\| \\ &\le \lambda_{\sigma(k)} \|G(u^{\sigma(k)})\| \to 0 \text{ as} k \to \infty\end{aligned}$$

which together with (27), one has $\lim_{k\to\infty} x^{\sigma(k)+1} = x^*$, which means that $\lim_{k\to\infty} a_{\sigma(k)+1} = 0$.

By (5) in Lemma 2.10, we have

$0 \le a_k \le a_{\sigma(k)+1} \to 0$ as$k \to \infty$.

Thus, $\{x^k\}$ converges to x^* □

4. Application to variational inequalities with variational inequality and fixed point constraints

In this section, we consider the following variational inequality problem over the set that is the common elements of the solution set of a pseudomonotone variational inequality problem and the set of fixed points of a demicontractive mapping (shortly VIFP(C, F, T, G):

$$\text{Find } x^* \in S \text{ such that } \langle G(x^*), y - x^* \rangle \geq 0, \quad \forall y \in S, \tag{28}$$

where $S = S_F \cap Fix(T)$, $S_F = \{u \in C : \langle F(u), y - u \rangle \geq 0, \ \forall y \in C\}$,

i.e. S_F is the solution set of the following variational inequality problems VIP(C, F) for short)

$$\text{Find } u \in C \text{ such that } \langle F(u), y - u \rangle \geq 0, \ \forall y \in C, \tag{29}$$

and as before, $Fix(T)$ is the fixed point of the mapping T. This problem was considered by Maingé (2008b).

In the sequel, we always suppose that Assumptions (A1), (A2), (A3), (A4), and (A5) are satisfied. The algorithm for this case takes the form:

Algorithm 2.

Initialization. Pick $x^0 \in C$, choose parameters $\eta \in (0,1)$, $\rho > 0$ and $\{\lambda_k\} \subset [0;1)$; $\{\mu_k\} \subset [0;1)$.

Iteration k. (k = 0, 1, 2, ...) Having x^k do the following steps:

Step 1. $y^k = P_C(x^k - \frac{\rho}{2}F(x^k))$

If $y^k = x^k$, take $u^k = x^k$ and go to *Step 4*. Otherwise, do Step 2.

Step 2. (Armijo linesearch rule) Find m_k as the smallest positive integer number m satisfying

$$\begin{cases} z^{k,m} = (1 - \eta^m)x^k + \eta^m y^k, \\ \langle F(z^{k,m}), x^k - y^k \rangle \geq \frac{1}{\rho}\|y^k - x^k\|^2. \end{cases} \tag{30}$$

Step 3. Set $\eta_k := \eta^{m_k}$, $z^k := z^{k,m_k}$. Take

$$C_k := \{x \in C : \langle F(z^k), x - z^k \rangle \leq 0\}, \ u^k := P_{C_k}(x^k). \tag{31}$$

Step 4. Compute $v^k = P_C(u^k - \lambda_k G(u^k))$.

Step 5. Set $x^{k+1} = (1 - \mu_k)v^k + \mu_k T(v^k)$, and go to Step 1 with k is replaced by $k + 1$.

Similar to Theorem 3.1, we have the following theorem

THEOREM 4.1 *Under assumptions (A1), (A2), (A3), (A4), and (A5) and (B1), (B2), and (B3), the sequence $\{x^k\}$ generated by Algorithm 2 converges to the unique solution x^* of VIFP(C, F, T, G).*

5. Conclusion

We have proposed a hybrid extragradient-viscosity algorithm for solving strongly monotone variational inequality problems over the set that is common points of the set of solutions of a pseudomonotone equilibrium problem and the set of fixed points of a demicontractive mapping. The convergence of the proposed algorithm is obtained, and a special case of this problem is also considered.

Funding
This work is supported in part by RFIT@LQDTU.

Author details
Bui Van Dinh[1]
E-mail: vandinhb@gmail.com

[1] Faculty of Information Technology, Department of Mathematics, Le Quy Don Technical University, No. 236 Hoang Quoc Viet Road, Hanoi, Vietnam.

References

Anh, P. N., Kim, J. K., & Muu, L. D. (2012). An extragradient algorithm for solving bilevel variational inequalities. *Journal of Global Optimization, 52*, 627–39.

Blum, E., & Oettli, W. (1994). From optimization and variational inequalities to equilibrium problems. *The Mathematics Student, 63*, 127–149.

Censor, Y., & Lent, A. (1981). An iterative row-action method for interval convex programming. *Journal of Optimization Theory and Applications, 34*, 321–353.

Ding, X. P. (2010). Auxiliary principle and algorithm for mixed equilibrium problems and bilevel equilibrium problems in Banach spaces. *Journal of Optimization Theory and Applications, 146*, 347–357.

Dinh, B. V., & Muu, L. D. (2015). Projection algorithm for solving pseudomonotone equilibrium problems and it's application to a class of bilevel equilibria. *Optimization, 64*, 559–575.

Facchinei, F., & Pang, J. S. (2003). *Finite-dimensional variational inequalities and complementarity problems*. New York, NY: Springer.

Konnov, I. V. (2001). *Combined relaxation methods for variational inequalities* (Lecture notes in economics and mathematical systems). Berlin: Springer.

Luo, J. Q., Pang, J. S., & Ralph, D. (1996). *Mathematical programs with equilibrium constraints*. Cambridge: Cambridge University Press.

Maingé, P. E. (2008a). Strong convergence of projected subgradient methods for nonsmooth and nonstrictly convex minimization. *Set-Valued Analysis, 16*, 899–912.

Maingé, P. E. (2008b). A hybrid extragradient viscosity methods for monotone operators and fixed point problems. *SIAM Journal on Control and Optimization, 47*, 1499–1515.

Mastroeni, G. (2003). On auxiliary principle for equilibrium problems. *Journal of Global Optimization, 27*, 411–426.

Migdalas, M. A., Pardalos, P., & Varbrand, P. (Eds.). (1988). *Multilevel optimization: Algorithms and applications*. Dordrecht: Kluwer.

Moudafi, A. (2010). Proximal methods for a class of bilevel monotone equilibrium problems. *Journal of Global Optimization, 47*, 287–292.

Muu, L. D., & Oettli, W. (1992). Convergence of an adaptive penalty scheme for finding constrained equilibria. *Nonlinear Analysis, Theory, Methods & Applications, 18*, 1159–1166.

Rockafellar, R. T. (1970). *Convex analysis*. Princeton, NJ: Princeton University Press.

Solodov, M. V., & Svaiter, B. F. (1999). A new projection method for variational inequality problems. *SIAM Journal on Control and Optimization, 37*, 765–776.

Yao, Y., Liou, Y. C., & Kang, S. M. (2010). Minimization of equilibrium problems, variational inequality problems and fixed point problems. *Journal of Global Optimization, 48*, 643–656.

7

On the eigenvalue asymptotics of Zonal Schrödinger operators in even metric and non-even metric

Lung-Hui Chen[1*]

*Correponding author: Lung-Hui Chen, Department of Mathematics, National Chung Cheng University, 168 University Rd. Min-Hsiung, Chia-Yi County 621, Taiwan

E-mails: mr.lunghuichen@gmail.com, lhchen@math.ccu.edu.tw.

Reviewing editor: Carlo Cattani, University of Tuscia, Italy

Abstract: We discuss a spectral asymptotics theory of an even zonal metric and a Schrödinger operator with zonal potentials on a sphere. We decompose the eigenvalue problem into a series of one-dimensional problems. We consider the individual behavior of this series of one-dimensional problems. We find certain Weyl's type of asymptotics on the eigenvalues.

Subjects: Advanced Mathematics; Analysis - Mathematics; Differential Equations; Mathematics & Statistics; Science

Keywords: zonal potential; zonal metric; eigenvalue asymptotics; Weyl asymptotics; Sturm–Liouville theory; Cartwright–Levinson theory

AMS subject classifications: 35P25; 35R30; 34B24; 34K23; 35J10

1. Introduction

In this paper, we compute the eigenvalue asymptotics of the operator Δ_g with a metric g that satisfies the following even zonal metric assumption: Let $(x_1, \dots, x_{N+1}, z)$ be the standard coordinate on $\mathbb{R}^{N+2}$. We consider the hypersurface defined by the equation

$$\sum_{n=1}^{N+1} x_n^2 = r^2(z), \quad -1 \leq z \leq 1, \tag{1}$$

in which we assume r is an even function of z, $r(-1) = r(1) = 0$ and $0 < r(z) < \infty$ for $-1 \leq z \leq 1$. Following the framework of Carlson (1997), we let $s(z)$ denote the arc length

$$s(z) = \int_{-1}^{z} \sqrt{1 + \{\frac{dr(t)}{dt}\}^2} dt, \tag{2}$$

ABOUT THE AUTHOR

The author grew up in Taiwan, and received his PhD in mathematics at Purdue University, 2007. He held a postdoctoral position at National Taiwan University in 2008. Currently, he is teaching at National Chung-Cheng University in Chia-Yi. The author's research interests are scattering theory and spectral analysis of differential operators. Recently, he has been involved with the inverse problems concerning wave propagation. It sounds a bit cliched, but when he is not with math, he's with his family and wine tasting. He is an audiophile, and has collected a few thousands of music CDs across all genres.

PUBLIC INTEREST STATEMENT

In this paper, we describe a connection between the eigenvalue distribution theorems and the geometric characteristics for a class of manifolds: eigenvalues are frequencies. It is believed that one can figure out some of the physical characteristics of a wave-reflecting object by hitting it with a band of frequencies of testing waves. It is a research interest among many disciplines, e.g. acoustics, optics, remote sensing, medical imaging, national defense, astrophysics, and quantum mechanics. Wherever there is a wave propagating through the media, we ask if one can analyze the perturbed wave and figure out the perturbation. As asked by Mark Kac, "Can you hear the shape of a drum?".

in which we set $L:=s(1)$. Hence, a metric is induced on the hypersurface. Now, we calculate Δ_g using the method provided in Shubin (1987, p. 157), and then we deduce that

$$\Delta_g = r^{-N}(s)\partial_s r^N(s)\partial_s + r^{-2}(s)\Delta_S, \tag{3}$$

in which S is the N sphere. We put the operator in the Liouville form with respect to the variable s:

$$r^{N/2}(s)\Delta_g r^{-N/2} = \partial_s^2 - \frac{N^2}{4}(\frac{r'}{r})^2 - \frac{N}{2}(\frac{r'}{r})' + r^{-2}(s)\Delta_S, \tag{4}$$

where we observe that the function $-\frac{N^2}{4}\left(\frac{r'}{r}\right)^2 - \frac{N}{2}\left(\frac{r'}{r}\right)'$ is again an even function about the midpoint of [0, L]. For the standard $N+1$ sphere, we note that $r(s)=\sin(s)$. Accordingly, we are required to assume that $r(s)\in C^2[0,L]$ and for some $p_0\in C^2[0,L/2]$, we have the following properties:

$$r(s) = s[1+p_0(s)],\ \lim_{s\to 0} p_0(s) = 0,\ \lim_{s\to 0} p_0'(s)/s \text{ exists.} \tag{5}$$

Here,(1.5) is to assure the hypersurface behaves like $N+1$ sphere near $z=-1$. Then, for $0<s<L/2$,$r^{-1}(s) = s^{-1}\left[\frac{1}{1+p_0(s)}\right]^{-1}$, and from which we deduce that

$$r^{-2}(s) = s^{-2} + p_1(s),\ p_1(s)\in C[0,L/2]. \tag{6}$$

Accordingly,

$$-\frac{N^2}{4}\left(\frac{r'}{r}\right)^2 - \frac{N}{2}\left(\frac{r'}{r}\right)' = \frac{-N(N-2)}{4s^2} + p_2(s),\ p_2(s)\in C[0,L/2]. \tag{7}$$

The function p_1, p_2 depends on p_0 and its derivatives.

Let $\{\beta_k\}$ be the eigenvalues of Δ_S. Then, the eigenvalue problem (4) is reduced to be

$$\partial_s^2 y + \left[-\frac{N^2}{4}\left(\frac{r'}{r}\right)^2 - \frac{N}{2}\left(\frac{r'}{r}\right)' + r^{-2}(s)\beta_k\right] y = \lambda y; \tag{8}$$

For a Schrödinger operator with a zonal potential, we are dealing with an equation in the following form:

$$\partial_s^2 y + \left[-\frac{N^2}{4}\left(\frac{r'}{r}\right)^2 - \frac{N}{2}\left(\frac{r'}{r}\right)' + r^{-2}(s)\beta_k\right] y + p(s)y = \lambda y, \tag{9}$$

where $p\in\mathcal{L}^2[0,L]$ is assumed to be an even function with midpoint $L/2$, and $r(s)=\sin(s)$. We say $p(s)$ is zonal because p is a function of z as in the hypersurfaces(1.1) and (1.2) (Gurarie, 1988,1990). A potential of this type has various applications in mathematical physics. We refer more introduction on potentials of this class to (Gurarie, 1990, p. 567). The differential equations (1.8) and (1.9) require the following regularization conditions:

$$\lim_{s\to 0}\frac{y(s)}{s^{m+1}} < \infty; \tag{10}$$

$$\lim_{s\to L} L - \frac{y(s)}{s^{m+1}} < \infty. \tag{11}$$

Moreover, (1.6) and (1.7) imply that

$$-\frac{N^2}{4}\left(\frac{r'}{r}\right)^2 - \frac{N}{2}\left(\frac{r'}{r}\right)' + r^{-2}(s)\beta_k = \frac{4\beta_k - N(N-2)}{4s^2} + p(s),\ 0<s<L/2, \tag{12}$$

in which r(s) is even, $p(s) = p_2(s) + \beta_k p_1(s) \in C[0, L]$ by its construction and its solution satisfies the initial condition $r'(L/2) = 0$. Theorem 1.4 in Carlson (1993) says that p(s) is uniquely determined by the spectral data of Δ_g. According to (1.8) and (1.9), we set that

$$m(m+1) = \frac{N^2 - 2N - 4\beta_k}{4}. \tag{13}$$

Now, we are studying the eigenvalue asymptotics of the equation of the following form:

$$\partial_s^2 y + [\frac{-m(m+1)}{s^2} + p(s)]y = z^2 y,\ z \in \mathbb{C}, \tag{14}$$

in which the asymptotic expansion of y(s; z) is analyzed in Carlson (1993,1997,1994). Setting the solution $y = y(s;z)$, y(s; z) is an entire function of exponential type (Carlson, 1993; Pöschel & Trubowitz, 1987). The union of all eigenvalues of (1.14) over $\{\beta_k\}$ gives the collection of the total eigenvalues of Δ_g in (1.4) and vice versa. The cluster structure of the eigenvalues for each β_k is known among the work in Gurarie (1988,1990) and many others. Most important of all (Carlson, 1993, p. 23), due to the evenness assumption on Δ_g and $\Delta_g + p$, the eigenvalues of (1.14) is split into two kinds for each β_k: the zeros of y(L / 2; z) and $y'(L/2;z)$. The zeros of y(L / 2; z) and of $y'(L/2;z)$ correspond to the Dirichlet and Neumann spectral data of (1.14) at $r = L/2$, respectively. Hence, in the first part of this paper, we collect all zeros of y(L / 2; z) and $y'(L/2;z)$ for each β_k.

In this paper, we consider the Weyl's type of eigenvalue asymptotics of (1.8) and (1.9) on surfaces of type (1.4).

THEOREM 1.1 *Let N(v) be the eigenvalue counting function in an interval of length v starting at the origin. Then, the following asymptotics holds:*

$$N(v) \sim \frac{2Lv^{\frac{N+1}{2}}}{\pi N!}. \tag{15}$$

A Weyl's theorem of this kind is classic in many perturbations (Chen, 2015a; Gurarie, 1988,1990; Shubin, 1987). We provide an extra information on the arc length L. The new ingredient in this paper is an analysis in entire function theory and its extension to non-even metrics.

2. Zeros of y(L / 2; z) and $y'(L/2;z)$

We apply the entire function in complex analysis (Koosis, 1997; Levin, 1972,1996) to study the distribution of its zeros.

Definition 2.1 Let f(z) be an entire function. Let $M_f(r) := \max_{|z|=r} |f(z)|$. An entire function of f(z) is said to be a function of finite order if there exists a positive constant k, such that the inequality

$$M_f(r) < e^{r^k} \tag{16}$$

is valid for all sufficiently large values of r. The greatest lower bound of such numbers k is called the order of the entire function f(z). By the type σ of an entire function f(z) of order ρ, we mean the greatest lower bound of positive number A for which asymptotically we have

$$M_f(r) < e^{Ar^\rho}. \tag{17}$$

That is,

$$\sigma := \limsup_{r\to\infty} \frac{\ln M_f(r)}{r^\rho}. \tag{18}$$

If $0 < \sigma < \infty$, then we say f(z) is of normal type or mean type.

We note that

$$e^{(\sigma-\epsilon)r^\rho} \underset{n}{<} M_f(r) \underset{as}{<} e^{(\sigma+\epsilon)r^\rho}, \tag{19}$$

where we mean the first inequality holds for some sequence going to infinity and the second one holds asymptotically.

Definition 2.2 If an entire function *f(z)* is of order one and of normal type, then we say it is an entire function of exponential type σ.

LEMMA 2.3 *Let f and g be two entire functions. Then, the following two inequalities hold.*

$$\begin{aligned} &h_{fg}(\theta) \leq h_f(\theta) + h_g(\theta), \text{ if one limit exists;} \\ &h_{f+g}(\theta) \leq \max_\theta\{h_f(\theta), h_g(\theta)\}, \end{aligned} \tag{20}$$

where if the indicator of the two summands is not equal at some θ_0, then the equality holds in (2.6).

The equality in (2.5) holds if one function is of completely regular growth. This is classic and we refer these to Levin (1972, p. 51, 159). □

Definition 2.4 Let *f(z)* be an integral function of finite order ρ in the angle $[\theta_1, \theta_2]$. We call the following quantity the indicator of the function *f(z)*.

$$h_f(\theta) := \limsup_{r\to\infty} \frac{\ln|f(re^{i\theta})|}{r^\rho}, \ \theta_1 \leq \theta \leq \theta_2. \tag{21}$$

Definition 2.5 The following quantity is called the width of the indicator diagram of the entire function *f*:

$$d := h_f\left(\frac{\pi}{2}\right) + h_f\left(-\frac{\pi}{2}\right). \tag{22}$$

The order and the type of an integral function in an angle can be defined similarly. The connection between the indicator $h_f(\theta)$ and its type σ is specified by the following theorem.

Definition 2.6 Let *f(z)* be an entire function of order ρ. We use $N(f, [\alpha, \beta], r)$ to denote the number of the zeros of *f(z)* inside $[\alpha, \beta]$ and $|z| \leq r$; we define the density function

$$\Delta_f(\alpha, \beta) := \limsup_{r\to\infty} \frac{N(f, [\alpha, \beta], r)}{r^\rho} \tag{23}$$

and

$$\Delta(\beta) := \Delta(\alpha_0, \beta), \tag{24}$$

with fixed $\alpha_0 \notin E$, with E as an at most countable set.

The two definitions above are necessary vocabularies to apply Cartwright–Levinson theory (Levin, 1972, p. 251) in complex analysis.

LEMMA 2.7 *(Levin, 1972, p. 72) The maximal value of the indicator $h_f(\theta)$ of the function f(z) on the interval $\alpha \leq \theta \leq \beta$ is equal to the type σ of this function inside the angle $\alpha \leq \arg z \leq \beta$.*

LEMMA 2.8 *(Levin, 1972) Let α and β be real constants.*

$$h_{\sin\{\alpha z+\beta\}}(\theta) = |\alpha \sin\theta|. \tag{25}$$

Let $y(s;z)$ be the solution of the following problem in Carlson (1997):

$$\begin{cases} -y''(s;z) + \frac{m(m+1)y(s;z)}{s^2} + p(s)y(s;z) = z^2 y(s;z); \\ \lim_{s\downarrow 0}\{\frac{y(s;z)}{s} - j_m(sz)\} = 0. \end{cases} \tag{26}$$

For the initial conditions, we actually have

$$\hat{y}_0(0) = 0,\ \hat{y}_0'(0) = 1; \hat{y}(0) = 0,\ \hat{y}'(0) = 0,\ m > 0. \tag{27}$$

The following result is well known for the classic case (Pöschel & Trubowitz, 1987, p. 14) and for singular potentials without a lower bound (Faddeev, 1960, (14.14),(14.15)). However, we give the more precise first-order asymptotics from a point of view from Carlson (1993,1997).

Proposition 2.9 $y(L/2;z)$ and $y'(L/2;z)$ are entire functions of exponential type $L/2$.

Firstly, we apply the estimates in Carlson (1993,1997,1994).

$$|y(s;z) - \frac{\sin sz}{z}| \le \frac{K\log(1+|z|)\exp\{|\Im z|s\}}{|z|^2}, \tag{28}$$

so we have

$$y(s;z) = \frac{\sin sz}{z} + O(\frac{\log(1+|z|)\exp\{|\Im z|s\}}{|z|^2}). \tag{29}$$

The classic result (Pöschel & Trubowitz, 1987, p. 27) shows that there exists a constant C_δ depending on the distance to the zeros of $\sin sz$, such that the following inequality holds away from the zeros of $\sin sz$:

$$\exp\{|s\Im z|\} < C_\delta |\sin sz|. \tag{30}$$

Hence,

$$\begin{aligned} y(s;z) &= \frac{\sin sz}{z}\left[1 + \frac{z}{\sin sz}O\left(\frac{\log(1+|z|)\exp\{s|\Im z|\}}{|z|^2}\right)\right] \\ &= \frac{\sin sz}{z}\left[1 + O\left(\frac{1}{|z|^{1-\epsilon}}\right)\right],\ z \notin \frac{\pi\mathbb{Z}}{s}. \end{aligned} \tag{31}$$

Thus, the indicator function of $y(s;z)$ is

$$h_{y(s;z)}(\theta) = s\sin\theta,\ \theta \neq 0, \pi. \tag{32}$$

Because $y(s;z)$ is entire in z for a fixed s, $h_{y(s;z)}(\theta)$ is a continuous function of θ (Levin, 1972). Thus,

$$h_{y(s;z)}(\theta) = s\sin\theta. \tag{33}$$

Lemma 2.7 implies that $y(s;z)$ is an entire function of exponential type s. A similar argument holds for $y'(s;z)$. That is,

$$y'(s;z) = \cos sz[1 + O(\frac{1}{|z|^{1-\epsilon}})],\ z \notin \frac{\pi\mathbb{Z} + \frac{\pi}{2}}{z}; \tag{34}$$

$$h_{y'(s;z)}(\theta) = s\sin\theta. \tag{35}$$

This proves the lemma setting $s = L/2$. □

LEMMA 2.10 *Let z_j be the zeros of $y(s; z)$ and z_j' be the zeros of $y'(s;z)$. Then,*

$$z_j = \frac{j\pi}{s} + O(\frac{1}{j}), j \in \mathbb{Z};$$
$$z_j' = \frac{(j-\frac{1}{2})\pi}{s} + O(\frac{1}{j}), j \in \mathbb{Z}. \tag{36}$$

This is a direct consequence of Rouché's theorem on the boundary of a suitable sequence of neighborhoods containing the zeros of $\frac{\sin sz}{z}$ or $\cos sz$ under the estimates (2.9) and (2.12), respectively, by considering the following inequality:

$$|y(s;z) - \frac{\sin\{sz\}}{z}| = O(\frac{1}{|z|^{1-\epsilon}})|\frac{\sin\{sz\}}{z}|. \tag{37}$$

We refer the detailed proof to Carlson (1993,1997), Chen (2015a,2015b), Pöschel and Trubowitz (1987). □

Therefore, there is an asymptotically uniform structure of eigenvalues of (1.3) for each β_k-eigenvalue. The first term in the asymptotics is independent of p(s). They overlay asymptotically periodically from one β_k to another to give each cluster of eigenvalues of (1.3). Let $N_k(v)$ be denoted as the counting function for eigenvalues in interval [0, v] for each β_k-eigenvalues. We refer the structure of the eigenvalues of a zonal eigenvalues to (Gurarie, 1990, p. 576). We collect two kinds of spectra for each β_k by applying Lemmas 2.10 and (2.9):

$$N_k(v) \sim \frac{Lv}{\pi}, \text{ as} v \to \infty. \tag{38}$$

The locations of $\{\beta_k\}$ of N-sphere are well known in Gurarie (1988) and Shubin (1987):

$$\beta_k = k(k+N-1), k = 0, 1, \dots, \tag{39}$$

with increasing multiplicity $d_k = \begin{pmatrix} N+k \\ N \end{pmatrix} = O(k^{N-1})$. Given an interval of length v starting at the origin, we have the quantity of

$$\sum_{k=0}^{1-N+\sqrt{(N-1)^2+4v}/2} d_k \sim \frac{2v^{N/2}}{N!} \tag{40}$$

of eigenvalues of Δ_S (Shubin, 1987, p. 165). Hence,

$$N(v) = \sum_{\beta_k} N_k(v) \sim \frac{2Lv^{\frac{N+1}{2}}}{\pi N!}, \text{ as} v \to \infty. \tag{41}$$

This proves Theorem 1.1.

3. Non-even zonal potentials

Now, we drop the assumption that r and p are even functions in s in Theorem 1.1. We note that (1.7), (1.8), and (1.9) hold in [0, L]. Accordingly, we are considering a Schrödinger operator with a zonal potential $p(s)$; we are dealing with the equation

$$\partial_s^2 y + [-\frac{N^2}{4}(\frac{r'}{r})^2 - \frac{N}{2}(\frac{r'}{r})' + r^{-2}(s)\beta_k]y + p(s)y = z^2 y, \tag{42}$$

where $p \in \mathcal{L}_c^2[0, L]$, $r(-1) = r(1) = 0, 0 < r(z) < \infty$ for $-1 \leq z \leq 1$ and $r(s) = \sin(s\pi/L)$. Without the symmetry at $s = L/2$, we do not consider the zeros of $y(L/2; z)$ and of $y'(L/2;z)$ any more. After the linearization near $s = 0$ in (1.5) in Section 1, we consider a differential equation of the following form similar to (1.13) and (1.14).

$$\partial_s^2 y^{(l)} + [\frac{-l(l+1)}{s^2} + p(s)]y^{(l)} = z^2 y^{(l)}, \ l > 0, \tag{43}$$

whose solutions are spanned by the Jost solutions $\{f^{(l)}(s;z), f^{(l)}(s; -z)\}$ Faddeev (1960, (14.11)) and Reed and Simon (1979, p. 140), which satisfy the following integral equation

$$f^{(l)}(s;z) = f_0^{(l)}(s;z) - \int_s^{\infty} J_l(s, t;z)p(t)f(t;z)dt, \tag{44}$$

in which $J_l(s, t;z)$ is defined as in (Faddeev, 1960, (14.12)). Thus,

$$f^{(l)}(s;z) = f_0^{(l)}(s;z) := i^{l+1}h_l(sz), \tag{45}$$

whenever s is beyond the support of the perturbation. In our case, $[0, L]$; $h_l(sz)$ is the spherical Bessel function of second kind. Therefore, we write $y^{(l)}(s;z)$ as

$$y^{(l)}(s;z) = \alpha_l(z)f^{(l)}(s;z) + \beta_l(z)f^{(l)}(s; -z). \tag{46}$$

In general, we recall the scattering formula in Faddeev (1960, (14.17)) for $z \in 0i + \mathbb{R}$:

$$y^{(l)}(s;z) = \frac{1}{2iz}(\frac{1}{iz})^l[f^{(l)}(s;z)M^{(l)}(-z) - (-1)^l f^{(l)}(s; -z)M^{(l)}(z)], \tag{47}$$

where $S^{(l)}(z) := \frac{M^{(l)}(-z)}{M^{(l)}(z)}$ gives the scattering matrix to equation (3.2). For any z that solves (3.6) is an eigenvalue of (3.2). Most important of all, the M-function $M^{(l)}(z)$ and $S^{(l)}(z)$ are independent of l (Faddeev, 1960, Theorem 14.1). The proof is given in Faddeev (1960, p. 90) and carries to continuous l. Moreover, $S^{(l)}(z)$ can be meromorphically extended from the upper half plane to the complex z-plane or $\log z$-plane without poles on the real axis except for the origin (Melrose, 1995, p. 16) depending on dimension parity. For our case, the dimension is one. Therefore, $M^{(l)}(z)$ and $S^{(l)}(z)$ can be defined in

$$\Lambda := \{z \in \mathbb{C} | -\epsilon \leq \arg z \leq \pi + \epsilon, \ \text{for some} \epsilon > 0\}. \tag{48}$$

For some extension and uniqueness theory of $M(z)$, we refer to Faddeev (1960, p. 42). Furthermore, the constants $\alpha_l(z)$ and $\beta_l(z)$ are independent of the space variable s and can be solved by the scattering theory in half line Aktosun, Gintides, and Papanicolaou (2013, p. 13) and Freiling and Yurko (2001) as follows:

$$\begin{bmatrix} y^{(l)}(s;z) \\ y^{(l)\prime}(s;z) \end{bmatrix} = \begin{bmatrix} f^{(l)}(s;z) & f^{(l)}(s; -z) \\ f^{(l)\prime}(s;z) & f^{(l)\prime}(s; -z) \end{bmatrix} \begin{bmatrix} \alpha_l(z) \\ \beta_l(z) \end{bmatrix}. \tag{49}$$

Evaluating (3.7) at $s = 0$, (2.13) implies that

$$\begin{aligned} \begin{bmatrix} 0 \\ 0 \end{bmatrix} &= \begin{bmatrix} f^{(l)}(0;z) & f^{(l)}(0; -z) \\ f^{(l)\prime}(0;z) & f^{(l)\prime}(0; -z) \end{bmatrix} \begin{bmatrix} \alpha_l(z) \\ \beta_l(z) \end{bmatrix}, \ l > 0; \\ \begin{bmatrix} 0 \\ 1 \end{bmatrix} &= \begin{bmatrix} f^{(l)}(0;z) & f^{(l)}(0; -z) \\ f^{(l)\prime}(0;z) & f^{(l)\prime}(0; -z) \end{bmatrix} \begin{bmatrix} \alpha_l(z) \\ \beta_l(z) \end{bmatrix}, \ l = 0. \end{aligned} \tag{50}$$

Evaluating (3.7) at $s = L$, we have

$$\begin{bmatrix} y^{(l)}(L;z) \\ y^{(l)\prime}(L;z) \end{bmatrix} = \begin{bmatrix} f_0^{(l)}(L;z) & f_0^{(l)}(L;-z) \\ f_0^{(l)\prime}(L;z) & f^{(l)\prime}_0(L;-z) \end{bmatrix} \begin{bmatrix} \alpha_l(z) \\ \beta_l(z) \end{bmatrix}. \tag{51}$$

Only the case $l = 0$ is solvable which is found in Aktosun et al. (2013, p. 13). Equations (3.9) and (3.10) imply that

$$y_0^{(0)}(L;z) = \frac{1}{2iz} f^{(0)}(0;-z)h_0(Lz) - \frac{1}{2iz} f^{(0)}(0;z)h_0(-Lz), \tag{52}$$

in which $h_0(sz) = \exp\{isz\}$. *Therefore, the eigenvalues of (14) are z-solutions to the equation* (3.12). This is a compatibility condition. From complex analysis, we have a zero distribution theory for the exponential polynomials in the form of (3.12) (Koosis, 1997; Levin, 1972,1996).

In general, by referring to Faddeev (1960, Lemma 1.5),

$$f^{(0)}(s;z) = e^{isz} + o(e^{-\Im zs}),\ \Im z \geq 0, \tag{53}$$

uniformly for any $s \geq 0$. However, we need the behavior of $M^{(0)}(z)$ slightly below the real axis (see Figure 1).

We recall the well-known integral equation (Faddeev, 1960, p. 38):

$$M^{(0)}(k) = 1 + \int_0^\infty A(0,t)e^{ikt}dt,\ k \in 0i + \mathbb{R}, \tag{54}$$

where $A(0, t)$ is compactly supported for our potential $p(s)$. We refer the construction of $A(0, t)$ to Faddeev (1960, p. 30). Thus, we observe that

$$M^{(0)}(k - \delta i) = 1 + \int_0^\infty [A(0,t)e^{\delta t}]e^{ikt}dt,\ \delta > 0, \tag{55}$$

to which we apply the Riemann–Lebesgue Lemma. Hence, the integral vanishes for large $|k|$.

$$M^{(0)}(k - \delta i) = 1 + o(1),\ \delta > 0,\ \text{as}|k| \to \infty. \tag{56}$$

Moreover, $e^{\pm izL}$ are the exponential functions of type L by applying (2.4). We also infer from (2.14) and the complex analysis in section 2 that the right-hand side of (3.12) is an analytic function of order one and at most of type L.

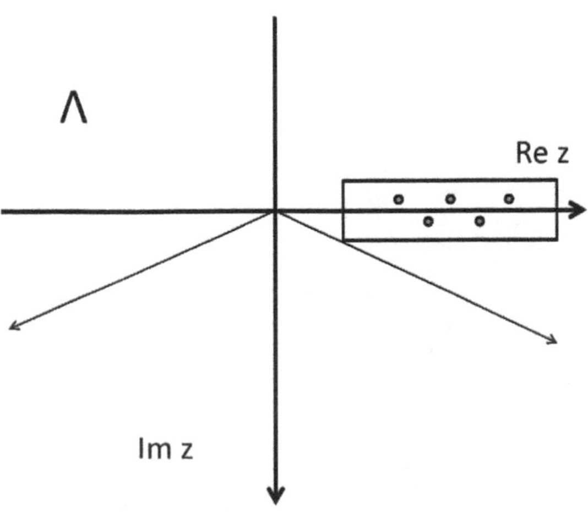

Figure 1. Rouchés Theorem.

Because $M^{(0)} = M^{(l)}$ for any l, we take $f^{(0)}(0;z) = M^{(l)}(z)$ for any l (Faddeev, 1960, Theorem 14.1). By (3.6) and (3.12), it suffices to study the zeros of the following analytic function

$$F(z) := y^{(l)}(L;z) - \frac{1}{2iz}(\frac{1}{iz})^l f^{(0)}(0;-z)e^{izL} + \frac{1}{2iz}(\frac{1}{iz})^l f^{(0)}(0;z)e^{-izL}, \tag{57}$$

wherein (2.6) suggests that (3.17) is an analytic function of type at most L in $\{z \in \mathbb{C} | \Im z \geq -\delta\}$. We need a lower bound. We use (2.17) to describe the asymptotic behaviors of the solution $y^{(l)}(L;z)$. More specifically, we apply Prof. Carlson's result (1997, Lemma 2.4, Lemma 3.2).

$$F(z) = \frac{\sin Lz}{z^{l+1}}[1 + O(\frac{1}{|z|^{1-\epsilon}})] - \frac{1}{2iz}(\frac{1}{iz})^l f^{(0)}(0;-z)e^{izL} + \frac{1}{2iz}(\frac{1}{iz})^l f^{(0)}(0;z)e^{-izL},\ z \notin \frac{\pi\mathbb{Z}}{s}, \tag{58}$$

to which we use (2.16) again with (3.16) and deduce that

$$\begin{aligned} F(z) &= \frac{\sin Lz}{z^{l+1}}\{1 + O(\frac{1}{|z|^{1-\epsilon}}) + o(1)\} \\ &= \frac{\sin Lz}{z^{l+1}}\{1 + o(1)\},\ z \notin \frac{\pi\mathbb{Z}}{s},\ z \in \Lambda. \end{aligned} \tag{59}$$

Hence, we apply Rouche's theorem in a strip of width less than δ containing $0i + \mathbb{R} \setminus \{0\}$ in the interior in Figure 1. There are only finitely many zeros outside this strip when examined (3.19) by Rouche's theorem. Let $N_k(v)$ be the eigenvalue counting function inside the strip $\{z | 0 < \Re z \leq v,\ |\Im z| \leq 2\delta\}$ for some $\delta > 0$ for each β_k-eigenvalue. Rouche's theorem also suggests that the zeros of $F(z)$ asymptotically periodically approach to the ones of $\sin Lz$ with the eigenvalue density described by Lemma 2.10. Then,

$$N_k(v) \sim \frac{Lv}{\pi},\ \text{as} v \to \infty. \tag{60}$$

Previously, we have obtained this in (2.25) for even metrics. We repeat the same argument using (2.26), (2.27), and (2.28). Thus, (1.15) follows again for the asymptotics for non-even metrics.

Funding
The author received no direct funding for this research.

Author details
Lung-Hui Chen[1]
E-mails: mr.lunghuichen@gmail.com, lhchen@math.ccu.edu.tw.
[1] Department of Mathematics, National Chung Cheng University, 168 University Rd. Min-Hsiung, Chia-Yi County, 621, Taiwan.

References

Aktosun, T., Gintides, D., & Papanicolaou, V. G. (2013). Reconstruction of the wave speed from transmission eigenvalues for the spherically symmetric variable-speed wave equation. *Inverse Problems, 29*(6), 065007.

Carlson, R. (1993). Inverse spectral theory for some singular Sturm--Liouville problems. *Journal of Differential Equations, 106*, 121–140.

Carlson, R. (1997). A Borg--Levinson theorem for Bessel operators. *Pacific Journal of Mathematics, 177*(1), 1–26.

Carlson, R. (1994). Inverse Sturm-Liouville problems with a singularity at zero. *Inverse Problems, 10*, 851–864.

Chen, L-H. (2015a). An uniqueness result with some density theorems with interior transmission eigenvalues. *Applicable Analysis*. doi:10.1080/00036811.2014.936403

Chen, L.-H. (2015b). A uniqueness theorem on the eigenvalues of spherically symmetric interior transmission problem in absorbing medium. *Complex Variables and Elliptic Equations, 60*, 145–167.

Faddeev, L. D. (1960). *The quantum theory of scattering* (Research report, No. Em-165). New York: Institute of Mathematical Science, Division of Mathematical Research, New York University.

Freiling, G., & Yurko, V. (2001). *Inverse Sturm--Liouville problems and their applications*. New York, NY: Nova Science Publishers.

Gurarie, D. (1988). Inverse spectral problem for the 2-sphere Schroedinger operators with zonal potentials. *Letters in Mathematical Physics, 16*, 313–323.

Gurarie, D. (1990). Zonal Schroedinger operator on the n-sphere inverse spectral problem and rigidity. *Communications in Mathematical Physics, 131*, 571–603.

Koosis, P. (1997). *The logarithmic integral I*. Cambridge: Cambridge University Press.

Levin, B. J. (1972). *Distribution of zeros of entire functions* (Revised ed., Translations of Mathematical Monographs). Providence, RI: American Mathematical Society.

Levin, B. J. (1996). *Lectures on entire functions* (Translation of Mathematical Monographs, Vol. 150). Providence, RI: AMS.

Melrose, R. B. (1995). *Geometric scattering theory.* New York, NY: Cambridge University Press.
Pöschel, J., & Trubowitz, E. (1987). *Inverse spectral theory.* Orlando: Academic Press.
Reed, M., & Simon, B. (1979). *Methods of modern mathematical physics,* Vol. 3. London: Academic Press.
Shubin, M. A. (1987). *Pseudodifferential operators and spectral theory.* Berlin: Springer-Verlag.

8

Games of entangled agents

Michail Zak[1*]

*Corresponding author: Michail Zak, California Institute of Technology, 9386 Cambridge Street, Cypress, CA 90630, USA
E-mail: michail.zak@gmail.com

Reviewing editor: Igor Yakov Subbotin, National University, USA

Abstract: The paper extends the mathematical formalism of quantum physics to include games of intelligent agents that communicate due to entanglement. The novelty of the approach is based upon a human factor-based ***behavioral*** model of an intelligent agent. The model is quantum inspired: it is represented by a modified Madelung equation in which the gradient of quantum potential is replaced by a specially chosen information force. It consists of motor dynamics simulating actual behavior of the agent, and mental dynamics representing evolution of the corresponding knowledge base and incorporating this knowledge in the form of information flows into the motor dynamics. Due to feedback from mental dynamics, the motor dynamics attains quantum-like properties: its trajectory splits into a family of different trajectories, and each of those trajectories can be chosen with the probability prescribed by the mental dynamics; each agent is entangled (in a quantum way) to other agents and makes calculated predictions for future actions; human factor is associated with violation of the second law of thermodynamics: the system can move from disorder to order without external help, and that represents intrinsic intelligence. All of these departures actually extend and complement the classical methods making them especially successful in analysis of communications of agents represented by new mathematical formalism, and in particular, in agent-based economics with a human factor.

ABOUT THE AUTHOR

This work opens up the way to development of a fundamentally new approach to mathematical formalism of ***economics*** that is inspired by the formalism of quantum mechanics. Mathematical treatment of economics has a relatively short history. Recently developed Newtonian models have been criticized for application to human behavior, arguing that some human choices are irreducible to Newtonian physics. Actually, the alert was expressed much earlier by Newton who stated "I can calculate the motion of heavenly bodies, but not the madness of people." Limitations of modern mathematical methods in economics are especially transparent in the area of agent-based computational economics. In this context, the presented mathematical formalism incorporates a human factor in agent-based economics while departing from physics formalism toward physics of life.
The area of recent research activity of the author is: Mathematical physics, artificial intelligence, and theory of turbulence.

PUBLIC INTEREST STATEMENT

The novelties introduced in the manuscript, and in particular, departures from Newtonian approach actually extend and complement the classical methods making them especially successful in analysis of communications of agents represented by new mathematical formalism, and in particular, in agent-based economics with a human factor.

Subjects: Bioscience; Mathematics & Statistics; Science

Keywords: randomness; stability; attractors

1. Introduction

This paper is devoted to a new approach to differential games, (Isaacs, 1965), i.e. to a group of problems related to the modeling and analysis of conflict in the context of a dynamical system. In corresponding agent-based models, the "agents" are "computational objects modeled as interacting according to rules" over space and time, not real people. The rules are formulated to model behavior and social interactions based on incentives and information. Such rules could also be the result of optimization, realized through use of AI methods.

We will concentrate on the non-Newtonian properties of dynamics describing a psychology-based behavior of intelligent agents as players. In other words, we will introduce dynamical systems with a human factor.

1.1. Justification for non-Newtonian approach

All the previous attempts to develop models for so-called active systems (i.e. systems that possess certain degree of autonomy from the environment that allows them to perform motions that are not directly controlled from outside) have been based upon the principles of Newtonian and statistical mechanics. These models appear to be so general that they predict not only physical, but also some biological and economical, as well as social patterns of behavior exploiting such fundamental properties of non-linear dynamics as attractors. Notwithstanding indisputable successes of that approach (neural networks, distributed active systems, etc.), there is still a fundamental limitation that characterizes these models on a dynamical level of description: they propose no difference between a solar system, a swarm of insects, and a stock market. Such a phenomenological reductionism is incompatible with the first principle of progressive biological evolution associated with Darwin. According to this principle, the evolution of living systems is directed toward the highest levels of complexity if the complexity is measured by an irreducible number of different parts, which interact in a well-regulated fashion (although in some particular cases deviations from this general tendency are possible). At the same time, the solutions to the models based upon dissipative Newtonian dynamics eventually approach attractors where the evolution stops while these attractors dwell on the subspaces of lower dimensionality, and therefore, of the lower complexity (until a "master" reprograms the model). Therefore, such models fail to provide an autonomous progressive evolution of living systems (i.e. evolution leading to increase in complexity). Let us now extend the dynamical picture to include thermal forces. That will correspond to the stochastic extension of Newtonian models, while the Liouville equation will extend to the Fokker–Planck equation that includes thermal force effects through the diffusion term. Actually, it is a well-established fact that evolution of life has a diffusion-based stochastic nature as a result of the multi-choice character of behavior of living systems. Such an extended thermodynamics-based approach is more relevant to model of living systems, and therefore, the simplest living species must obey the second law of thermodynamics as physical particles do. However, then the evolution of living systems (during periods of their isolation) will be regressive since their entropy will increase. Therefore, Newtonian physics is not sufficient for simulation the specific properties typical for intelligence.

There is another argument in favor of a non-Newtonian approach to modeling intelligence. As pointed out by Penrose (1955), the Gödel's famous theorem has the clear implication that mathematical understanding cannot be reduced to a set of known computational rules. That means that no knowable set of purely computational procedures could lead to a computer-control robot that possesses genuine mathematical understanding. In other words, such privileged properties of intelligent systems as common sense, intuition, or consciousness are non-computable within the framework of classical models. That is why a fundamentally new physics is needed to capture these "mysterious" aspects of intelligence, and in particular, to decision-making process.

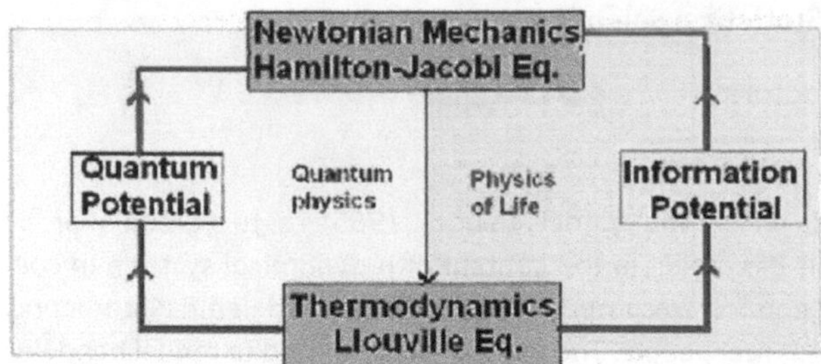

Figure 1. Classical physics, quantum physics, and physics of life.

2. Dynamical model for simulations

In this section, we review and discuss a ***behavioral*** model of intelligent agents, or players. The model is based upon departure from Newtonian dynamics to quantum inspired dynamics that was first introduced in Zak (1998, 2007, 2008, 2014a, 2014b), Figure 1.

2.1. Destabilizing effect of Liouville feedback

We will start with the derivation of an auxiliary result that illuminates departure from Newtonian dynamics. For mathematical clarity, we will consider here a one-dimensional motion of a unit mass under action of a force f depending upon the *velocity* v and time t and present it in a dimensionless form

$$\dot{v} = f(v, t) \tag{1}$$

referring all the variables to their representative values v_0, t_0, etc.

If initial conditions are not deterministic, and their probability density is given in the form

$$\rho_0 = \rho_0(V), \quad \text{where} \quad \rho \geq 0, \quad \text{and} \quad \int_{-\infty}^{\infty} \rho dV = 1 \tag{2}$$

while ρ is a *single-valued* function, then the evolution of this density is expressed by the corresponding Liouville equation

$$\frac{\partial \rho}{\partial t} + \frac{\partial}{\partial V}(\rho f) = 0 \tag{3}$$

The solution of this equation subject to initial conditions and normalization constraints (2) determines probability density as a function of V and t:

$$\rho = \rho(V, t) \tag{4}$$

Remark. Here and below we make distinction between the random variable v(t) and its values V in probability space.

In order to deal with the constraint (2), let us integrate Equation (3) over the whole space assuming that $\rho \to 0$ at $|V| \to \infty$ and $|f| < \infty$. Then

$$\frac{\partial}{\partial t}\int_{-\infty}^{\infty} \rho dV = 0, \quad \int_{-\infty}^{\infty} \rho dV = \text{const}, \tag{5}$$

Hence, the constraint (2) is satisfied for $t > 0$ if it is satisfied for $t = 0$.

Let us now specify the force f as a feedback from the Liouville equation

$$f(v, t) = \phi[\rho(v, t)] \tag{6}$$

and analyze the motion after substituting the force (6) into Equation (2)

$$\dot{v} = \phi[\rho(v,t)], \tag{7}$$

This is a fundamental step in our approach. Although the theory of ODE does not impose any restrictions upon the force as a function of space coordinates, the Newtonian physics does: equations of motion are never coupled with the corresponding Liouville equation. Moreover, it can be shown that such a coupling leads to non-Newtonian properties of the underlying model. Indeed, substituting the force f from Equation (6) into Equation (3), one arrives at the *non-linear* equation of evolution of the probability density

$$\frac{\partial \rho}{\partial t} + \frac{\partial}{\partial V}\{\rho\phi[\rho(V,t)]\} = 0 \tag{8}$$

Let us now demonstrate the destabilizing effect of the feedback (6). For that purpose, it should be noticed that the derivative $\partial\rho/\partial v$ must change its sign at least once, within the interval $-\infty < v < \infty$, in order to satisfy the normalization constraint (2).

But since

$$Sign\frac{\partial \dot{v}}{\partial v} = Sign\frac{d\phi}{d\rho}Sign\frac{\partial \rho}{\partial v} \tag{9}$$

there will be regions of v where the motion is unstable, and this instability generates randomness with the probability distribution guided by the Liouville equation (8). It should be noticed that the condition (9) may lead to exponential or polynomial growth of v (in the last case the motion is called neutrally stable, however, as will be shown below, it causes the emergence of randomness as well if prior to the polynomial growth, the Lipcshitz condition is violated).

2.2. Emergence of self-generated stochasticity

In order to illustrate mathematical aspects of the concepts of Liouville feedback in systems under consideration as well as associated with it instability and randomness, let us take the feedback (6) in the form

$$f = -\sigma^2\frac{\partial}{\partial v}\ln\rho, \tag{10}$$

to obtain the following equation of motion

$$\dot{v} = -\sigma^2\frac{\partial}{\partial v}\ln\rho, \tag{11}$$

This equation should be complemented by the corresponding Liouville equation (in this particular case, the Liouville equation takes the form of the Fokker–Planck equation)

$$\frac{\partial \rho}{\partial t} = \sigma^2\frac{\partial^2 \rho}{\partial V^2} \tag{12}$$

Here v stands for a particle velocity, and σ^2 is the constant diffusion coefficient.

The solution of Equation (12) subjects to the sharp initial condition

$$\rho = \frac{1}{2\sigma\sqrt{\pi t}}\exp\left(-\frac{V^2}{4\sigma^2 t}\right) \tag{13}$$

describes diffusion of the probability density, and that is why the feedback (10) will be called a diffusion feedback.

Substituting this solution with Equation (11) at **V = v** one arrives at the differential equation with respect to *v (t)*

$$\dot{v} = \frac{v}{2t} \tag{14}$$

and therefore,

$$v = C\sqrt{t} \tag{15}$$

where *C* is an arbitrary constant. Since *v = 0* at *t = 0* for any value of *C*, the solution (15) is consistent with the sharp initial condition for the solution (13) of the corresponding Liouvile equation (12). The solution (15) describes the simplest irreversible motion: it is characterized by the "beginning of time" where all the trajectories intersect (that results from the violation of Lipcsitz condition at *t = 0*, Figure 2), while the backward motion obtained by replacement of *t* with (*−t*) leads to imaginary values of velocities. One can notice that the probability density (13) possesses the same properties.

It is easily verifiable that the solution (15) has the same structure as the solution of the Madelung equation (Zak, 2014b), although the dynamical system (11), (12) is not quantum! The explanation of such a "coincidence" is very simple: the system (11), (12) has the same dynamical topology as that of the Madelung equation where the equation of conservation of the probability is coupled with the equation of conservation of the momentum. As will be shown below, the systems (11), (12) neither quantum nor Newtonian, and we will call such systems quantum-inspired, or self-supervised.

Further analysis of the solution (15) demonstrates that the solution (15) is *unstable* since

$$\frac{d\dot{v}}{dv} = \frac{1}{2t} > 0 \tag{16}$$

and therefore, an initial error always grows generating *randomness*. Initially, at *t = 0*, this growth is of infinite rate since the Lipchitz condition at this point is violated

$$\frac{d\dot{v}}{dv} \to \infty \quad \text{at} \quad t \to 0 \tag{17}$$

This type of instability has been introduced and analyzed in Zak (1992). The unstable equilibrium point (*v* = 0) has been called a terminal attractor, and the instability triggered by the violation of the Lipchitz condition—a non-Lipchitz instability. The basic property of the non- Lipchitz instability is the following: if the initial condition is infinitely close to the repeller, the transient solution will escape the repeller during a *bounded* time while for a regular repeller the time would be *unbounded*. Indeed, an escape from the simplest regular repeller can be described by the exponent $v = v_0 e^t$. Obviously, $v \to 0$ if $v_0 \to 0$, unless the time period is unbounded. On the contrary, the period of escape from the terminal attractor (15) is bounded (and even infinitesimal) if the initial condition is infinitely small, (see Equation (17)).

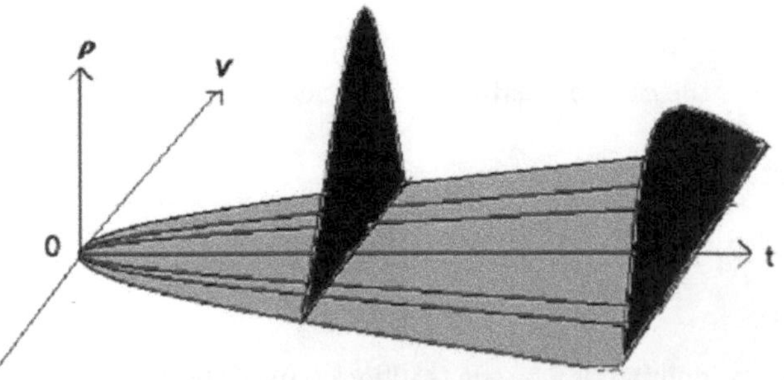

Figure 2. Stochastic process and probability density.

Considering first Equation (15) at fixed *C* as a sample of the underlying stochastic process (13), and then varying *C*, one arrives at the whole ensemble characterizing that process (see Figure 2). One can verify that, as follows from Equation (13), (Risken, 1989), the expectation and the variance of this process are, respectively,

$$\bar{v} = 0, \quad \tilde{v} = 2\sigma^2 t \tag{18}$$

The same results follow from the ensemble (15) at $-\infty \leq C \leq \infty$. Indeed, the first equality in (18) results from symmetry of the ensemble with respect to $v = 0$; the second one follows from the fact that

$$\tilde{v} \propto v^2 \propto t \tag{19}$$

It is interesting to notice that the stochastic process (15) is an alternative to the following Langevin equation, (Risken, 1989)

$$\dot{v} = \Gamma(t), \quad \bar{\Gamma} = 0, \quad \tilde{\Gamma} = \sigma \tag{20}$$

that corresponds to the *same* Fokker–Planck Equation (12). Here, $\Gamma(t)$ is the Langevin (random) force with zero mean and constant variance σ.

Thus, the emergence of self-generated stochasticity is the first basic non-Newtonian property of the dynamics with the Liouville feedback.

2.3. Second law of thermodynamics

In order to demonstrate another non-Newtonian property of the systems considered above, let us start with the dimensionless form of the Langevin equation for a one-dimensional Brownian motion of a particle subjected to a random force, (Risken, 1989)

$$\dot{v} = -kv + \Gamma(t), \quad < \Gamma(t) >= 0, \quad < \Gamma(t)\Gamma(t') >= 2\sigma\delta(t - t'), [\Gamma] = 1/s \tag{21}$$

Here, v is the dimensionless velocity of the particle (referred to a representative velocity v_0), k is the coefficient of a linear damping force, $\Gamma(t)$ is the Langevin (random) force per unit mass, $\sigma > 0$ is the noise strength. The representative velocity v_0 can be chosen, for instance, as the initial velocity of the motion under consideration.

The corresponding continuity equation for the probability density ρ is the following Fokker–Planck equation

$$\frac{\partial \rho}{\partial t} = k\frac{\partial (V\rho)}{\partial V} + \sigma\frac{\partial^2 \rho}{\partial V^2}, \int_{-\infty}^{\infty} \rho dV = 1 \tag{22}$$

Obviously without external control, the particle cannot escape from the Brownian motion.

Let us now introduce a new force (referred to unit mass and divided by v_0) as a Liouville feedback

$$f = \sigma \exp \sqrt{D}\frac{\partial}{\partial v} \ln \rho, \quad [f] = 1/s \tag{23}$$

Here D is the dimensionless variance of the stochastic process $D(t) = \int_{-\infty}^{\infty} \rho V^2 dV$,

Then the new equation of motion takes the form

$$\dot{v} = -kv + \Gamma(t) + \sigma \exp \sqrt{D}\frac{\partial}{\partial v} \ln \rho, \tag{24}$$

and the corresponding Fokker–Planck equation becomes non-linear

$$\frac{\partial \rho}{\partial t} = k\frac{\partial (V\rho)}{\partial V} + \sigma(1 - \exp\sqrt{D})\frac{\partial^2 \rho}{\partial V^2}, \quad \int_{-\infty}^{\infty} \rho dV = 1 \tag{25}$$

Obviously, the diffusion coefficient in Equation (25) is *negative.* Multiplying Equation (25) by V^2, then integrating it with respect to V over the whole space, one arrives at ODE for the variance $\tilde{V}(t)$

$$\dot{D} = 2[\sigma(1 - \exp\sqrt{D}) - kD] \tag{26}$$

Thus, as a result of *negative* diffusion, the variance D monotonously vanishes regardless of the initial value $D(0)$. It is interesting to note that the time T of approaching $D = 0$ is finite

$$T = \frac{1}{2}\int_{D(0)}^{0} \frac{dD}{\sigma(1 - \exp\sqrt{D}) - kD} \leq \frac{1}{2\sigma}\int_{0}^{\infty} \frac{dD}{\exp\sqrt{D} - 1} = \frac{\pi}{6\sigma} \tag{27}$$

This terminal effect is due to violation of the Lipchitz condition, at $D = 0$, (Zak, 2014a).

Let us review the structure of the force (23): it is composed only of the probability density and its variance, i.e. out of the components of the conservation Equation (25); at the same time, Equation (25) itself is generated by the equation of motion (24). Consequently, the force (23) is *not* an external force. Nevertheless, it allows the particle escape from the Brownian motion using its own "internal effort." It would be reasonable to call the force (23) an *information force* since it links to information rather than to energy.

Thus, we came across the phenomenon that violates the second law of thermodynamics when the dynamical system moves from disorder to the order without external interactions due to a feedback from the equation of conservation of the probability to the equation of conservation of the momentum. One may ask why the negative diffusion was chosen to be non-linear. Let us turn to a linear version of Equation (26)

$$\frac{\partial \rho}{\partial t} = -\sigma^2\frac{\partial^2 \rho}{\partial V^2}, \quad \int_{-\infty}^{\infty} \rho dV = 1 \tag{28}$$

and discuss a negative diffusion in more detail. As follows from the linear equivalent of Equation (26)

$$\dot{D} = -2\sigma, \text{ i.e. } \quad D = D_0 - 2\sigma t < 0 \quad \text{at} \quad t < D_0/(2\sigma) \tag{29}$$

Thus, eventually the variance becomes negative, and that disqualifies Equation (29) from being meaningful. As shown in Zak (2014a), the initial value problem for this equation is ill-posed: its solution is not differentiable at any point. Therefore, a *negative diffusion must be non-linear* in order to protect the variance from becoming negative, Figure 3.

It should be emphasized that negative diffusion represents a major departure from both Newtonian mechanics and classical thermodynamics by providing a progressive evolution of complexity against the second law of thermodynamics.

In the next subsection, we will demonstrate again that formally the dynamics introduced above does not belong to the Newtonian world; nevertheless, its self-supervising capability may associate such a dynamics with a potential model for intelligent behavior. For that purpose, we will turn to even simpler version of this dynamics by removing the external Langevin force and simplifying the information force.

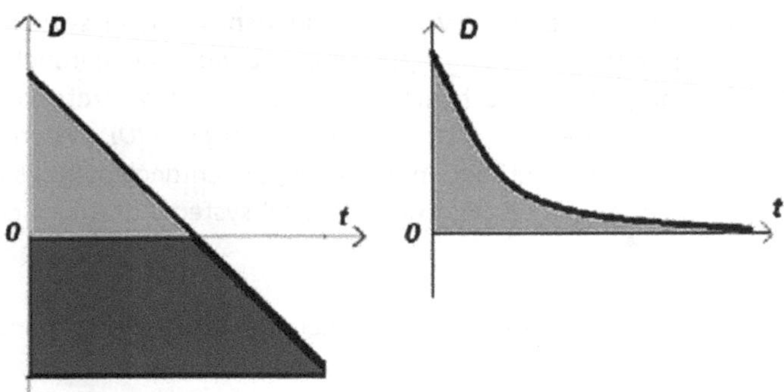

Linear negative diffusion. *Nonlinear negative diffusion.*

Figure 3. Negative diffusion.

In 1945, Schrödinger wrote in his book "What is life": *"Life is to create order in the disordered environment against the second law of thermodynamics."* The self-supervised dynamical system introduced above is fully consistent with this statement. Indeed, consider a simplified version of Equations (21) and (22)

$$\dot{v} = \sigma\sqrt{D}\frac{\partial}{\partial v}\ln\rho, \tag{30}$$

$$\frac{\partial\rho}{\partial t} = -\sigma\sqrt{D}\frac{\partial^2\rho}{\partial V^2}, \quad \int_{-\infty}^{\infty}\rho dV = 1 \tag{31}$$

Removal of the Langevin forces makes the particle *isolated*. Nevertheless, the particle has a capability of moving from disorder to order. For demonstration of this property, we will assume that the Langevin force was suddenly removed at $t = 0$ so that the initial variance $D_0 > 0$. Then,

$$\dot{D} = -2\sigma\sqrt{D} \tag{32}$$

$$\text{whence} \quad D = \left(\sqrt{D_0} - \sigma t\right)^2 \tag{33}$$

As follows from Equation (33), as a result of *internal, self-generated* force

$$F = \sigma\sqrt{D}\frac{\partial}{\partial v}\ln\rho, \tag{34}$$

the Brownian motion gradually disappears and then vanishes abruptly:

$$D \to 0, \quad \dot{D} \to 0, \quad \frac{d\dot{D}}{dD} \to \infty \quad at \quad t \to \frac{\sqrt{D_0}}{\sigma} \tag{35}$$

Thus, the probability density shrinks to a delta-function at $t = \frac{\sqrt{D_0}}{\sigma}$. Consequently, the entropy $H(t) = -\int_V \rho \ln \rho dV$ decreases down to zero, and that violates the second law of thermodynamics.

Another non-Newtonian property is entanglement.

2.4. Entanglement

In this subsection, we will introduce a fundamental and still mysterious property that was predicted theoretically and corroborated experimentally in quantum systems: entanglement. Quantum entanglement is a phenomenon in which the quantum states of two or more objects have to be described with reference to each other, even though the individual objects may be spatially separated. This leads to correlations between observable physical properties of the systems. As a result, measurements performed on one system seem to be instantaneously influencing other systems entangled with it. Different views of what is actually occurring in the process of quantum entanglement give rise to different interpretations of quantum mechanics. Here, we will demonstrate that entanglement is not a prerogative of quantum systems: it occurs in *quantum-inspired (QI)* systems that are under consideration in this paper. That will shed light on the concept of entanglement as a special type of global constraint imposed upon a broad class of dynamical systems that include quantum as well as *QI* systems.

In order to introduce entanglement in *QI* system, we will start with Equations (11) and (12) and generalize them to the two-dimensional case

$$\dot{v}_1 = -a_{11}\frac{\partial}{\partial v_1}\ln\rho - a_{12}\frac{\partial}{\partial v_2}\ln\rho, \tag{36}$$

$$\dot{v}_2 = -a_{21}\frac{\partial}{\partial v_1}\ln\rho - a_{22}\frac{\partial}{\partial v_2}\ln\rho, \tag{37}$$

$$\frac{\partial\rho}{\partial t} = a_{11}\frac{\partial^2\rho}{\partial V^2} + (a_{12}+a_{21})\frac{\partial^2\rho}{\partial V_1\partial V_2} + a_{22}\frac{\partial^2\rho}{\partial V_2}, \tag{38}$$

As in the one- dimensional case, this system describes diffusion without a drift

The solution to Equation (38) has a closed form

$$\rho = \frac{1}{\sqrt{2\pi\det[\hat{a}_{ij}]t}}\exp(-\frac{1}{4t}b'_{ij}V_iV_j), \quad i = 1,2. \tag{39}$$

Here,

$$[b'_{ij}] = [\hat{a}_{ij}]^{-1}, \hat{a}_{11} = a_{11}, \hat{a}_{22} = a_{22}, \hat{a}_{12} = \hat{a}_{21} = a_{12} + a_{21}, \hat{a}_{ij} = \hat{a}_{ji}, b'_{ij} = b'_{ji}, \tag{40}$$

Substituting the solution (39) with Equations (36) and (37), one obtains

$$\dot{v}_1 = \frac{b_{11}v_1 + b_{12}v_2}{2t} \tag{41}$$

$$\dot{v}_2 = \frac{b_{21}v_1 + b_{22}v_2}{2t}, \qquad b_{ij} = b'_{ij}\hat{a}_{ij} \tag{42}$$

Eliminating *t* from these equations, one arrives at an ODE in the configuration space

$$\frac{dv_2}{dv_1} = \frac{b_{21}v_1 + b_{22}v_2}{b_{11}v_1 + b_{12}v_2}, \quad v_2 \to 0 \quad at \quad v_1 \to 0, \tag{43}$$

This is a classical singular point treated in text books on ODE.

Its solution depends upon the roots of the characteristic equation

$$\lambda^2 - 2b_{12}\lambda + b_{12}^2 - b_{11}b_{22} = 0 \tag{44}$$

Since both the roots are real in our case, let us assume for concreteness that they are of the same sign, for instance, $\lambda_1 = 1,\ \lambda_2 = 1$. Then the solution to Equation (43) is represented by the family of straight lines

$$v_2 = \tilde{C}v_1, \quad \tilde{C} = const. \tag{45}$$

Substituting this solution into Equation (41) yields (45)

$$v_1 = Ct^{\frac{1}{2}(b_{11}+\tilde{C}b_{12})} \quad v_2 = \tilde{C}Ct^{\frac{1}{2}(b_{11}+\tilde{C}b_{12})} \tag{46}$$

Thus, the solutions to Equations (36) and (37) are represented by two-parametrical families of random samples, as expected, while the randomness enters through the time-independent parameters C and $\tilde{C}$that can take any real numbers. Let us now find such a combination of the variables that is deterministic. Obviously, such a combination should not include the random parameters C or $\tilde{C}$. It is easily verifiable that

$$\frac{d}{dt}(\ln v_1) = \frac{d}{dt}(\ln v_2) = \frac{b_{11} + \tilde{C}b_{12}}{2t} \tag{47}$$

and therefore,

$$(\frac{d}{dt}\ln v_1)/(\frac{d}{dt}\ln v_2) \equiv 1 \tag{48}$$

Thus, the ratio (48) is deterministic although both the numerator and denominator are random, (see Equation (47)). This is a fundamental non-classical effect representing a global constraint. Indeed, in theory of stochastic processes, two random functions are considered statistically equal if they have the same statistical invariants, but their point-to-point equalities are not required (although it can happen with a vanishingly small probability). As demonstrated above, the *diversion of determinism into randomness* via *instability (due to a Liouville feedback), and then conversion of randomness to partial determinism (or coordinated randomness)* via *entanglement* is the fundamental non-classical paradigm that may lead to instantaneous transmission of *conditional* information on remote distance that to be discussed below.

2.5. Relevance to model of intelligent agents

The model under discussion was inspired by E. Schrödinger, the creator of quantum mechanics who wrote in his book "What is Life": "Life is to create order in the disordered environment against the second law of thermodynamics." The proposed model illuminates the "border line" between living and non-living systems. The model introduces a biological particle that, in addition to Newtonian properties, possesses the ability to process information. The probability density can be associated with the *self-image* of the biological particle as a member of the class to which this particle belongs, while its ability to convert the density into the information force—with the *self-awareness* (both these concepts are adopted from psychology). Continuing this line of associations, the equation of motion (such as Equations (1) or (7)) can be identified with a motor dynamics, while the evolution of density (see Equations (3) or (8)—with a mental dynamics. Actually, the mental dynamics plays the role of the Maxwell sorting demon: it rearranges the probability distribution by creating the information potential and converting it into a force that is applied to the particle. One should notice that mental dynamics describes evolution of the whole class of state variables (differed from each other only by initial conditions), and that can be associated with the ability to generalize that is a privilege of living systems. Continuing our biologically inspired interpretation, it should be recalled that the second law of thermodynamics states that the entropy of an isolated system can only increase. This law has a clear probabilistic interpretation: increase of entropy corresponds to the passage of the system from less probable to more probable states, while the highest probability of the most disordered state (that is the state with the highest entropy) follows from a simple combinatorial analysis. However, this statement is correct only if there is no Maxwell' sorting demon, i.e. nobody inside the system is rearranging the probability distributions. But this is precisely what the Liouville feedback is doing: it takes the probability density ρ from Equation (3), creates functionals and functions of this density, converts them into a force and applies this force to the equation of motion (1). As already mentioned above, because of that property of the model, the evolution of the probability density becomes non-linear, and the entropy may decrease "against the second law of thermodynamics," Figure 4.

Obviously the last statement should not be taken literary; indeed, the proposed model captures only those aspects of the living systems that are associated with their ***behavior***, and in particular, with their motor-mental dynamics, since other properties are beyond the dynamical formalism. Therefore, such physiological processes that are needed for the metabolism, reproduction, est., are not included into the model. That is why this model is in a formal disagreement with the first and second laws of thermodynamics while the living systems are not. Indeed, applying the first law of thermodynamics we imply violation of conservation of ***mechanical*** energy since other types of energies (chemical, electro-magnetic, etc.) are beyond our mathematical formalism. Applying the second law of thermodynamics, we consider our system as isolated one while the underlying real system is open due to other activities of livings that were not included in our model. Nevertheless, despite these limitations, the proposed model captures the "magic" of life: the ability to create self-image and self-awareness, and that fits perfectly to the concept of ***intelligent agent.*** Actually, the proposed model represents governing equations for interactions of intelligent agents. In order to emphasize the autonomy of the agents' decision-making process, we will associate the proposed models with ***self-supervised (SS) active systems.***

By an active system, we will understand here a set of interacting intelligent agents capable of processing information, while an intelligent agent is an autonomous entity, which observes and acts upon an environment and directs its activity toward achieving goals. The active system is not derivable from the Lagrange or Hamilton principles, but it is rather created for information processing. One of the specific differences between active and physical systems is that the former are supposed to act in uncertainties originated from incompleteness of information. Indeed, an intelligent agent almost never has access to the whole truth of its environment. Uncertainty can also arise because of incompleteness and incorrectness in the agent's understanding of the properties of the environment. That is why *QI SS* systems are well suited for representation of active systems.

2.6. Self-supervised active systems with integral feedback

In this subsection, we will introduce a feedback from the mental to motor dynamics that is different from the feedback (6) discussed above. This feedback will make easier to formulate new principles of the competitive mode of agents associated with game theory.

Let us introduce the following feedback, (Zak, 2008)

$$f = \frac{1}{\rho(V,t)} \int_{-\infty}^{V} [\rho(\zeta,t) - \rho^{*}(\zeta)]dv \tag{49}$$

With the feedback (49), Equations (7) and (8) take the form, respectively,

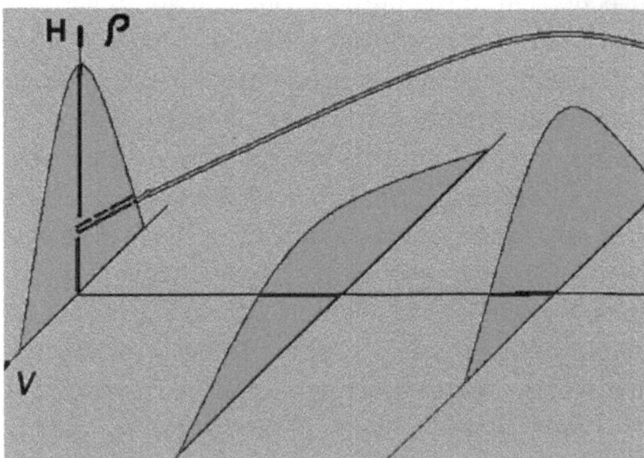

Figure 4. Living system: Deviation from classical thermodynamics.

$$f = \frac{1}{\rho(v,t)} \int_{-\infty}^{v} [\rho(\zeta,t) - \rho^*(\zeta)]dv \tag{50}$$

$$\frac{\partial \rho}{\partial t} + \rho(t) - \rho^* = 0 \tag{51}$$

The last equation has the analytical solution

$$\rho = (\rho_0 - \rho^*)e^{-t} + \rho^* \tag{52}$$

Subject to the initial condition

$$\rho(t = 0) = \rho_0 \tag{53}$$

This solution converges to a prescribed, or target, stationary distribution $\rho^*(V)$. Obviously the normalization condition for ρ is satisfied if it is satisfied for ρ_0 and ρ^*. This means that Equation (51) has an attractor in the probability space, and this attractor is stochastic. Substituting the solution (52) with Equation (50), one arrives at the ODE that simulates the stochastic process with the probability distribution (52)

$$\dot{v} = \frac{e^{-t}}{[\rho_0(v) - \rho^*(v)]e^{-t} + \rho^*(v)} \int_{-\infty}^{v} [\rho_0(\zeta) - \rho^*(\zeta)]d\zeta \tag{54}$$

As notices above, the randomness of the solution to Equation (54) is caused by instability that is controlled by the corresponding Liouville equation. It should be emphasized that in order to run the stochastic process started with the initial distribution ρ_0 and approaching a stationary process with the distribution ρ^*, one should substitute into Equation (54) the *analytical expressions* for these functions.

It is reasonable to assume that the solution (4) starts with sharp initial condition

$$\rho_0(V) = \delta(V) \tag{55}$$

As a result of that assumption, all the randomness is supposed to be generated *only* by the controlled instability of Equation (54). Substitution of Equation (55) into Equation (54) leads to two different domains of *v*: $v \neq 0$ and $v = 0$ where the solution has two different forms, respectively,

$$\int_{-\infty}^{v} \rho^*(\zeta)d\zeta = \frac{C_1}{e^{-t} - 1}, \quad v \neq 0 \tag{56}$$

$$v \equiv 0 \tag{57}$$

Equation (57) represents a singular solution, while Equation (56) is a regular solution that include arbitrary constant *C*. The regular solutions is unstable at $t = 0$, $|v| \to 0$ where the Lipschitz condition is violated

$$\frac{d\dot{v}}{dv} \to \infty \quad at \quad t \to 0 \quad , |v| \to 0 \tag{58}$$

and therefore, an initial error always grows generating *randomness*.

Let us analyze the behavior of the solution (56) in more detail. As follows from this solution, all the particular solutions intersect at the same point $v = 0$ at $t = 0$, and that leads to non-uniqueness of the solution due to violation of the Lipcshitz condition. Therefore, the same initial condition $v = 0$ at $t = 0$

yields infinite number of different solutions forming a family (56); each solution of this family appears with a certain probability guided by the corresponding Liouville equation (51). For instance, in cases plotted in Figures 5 and 6, the "winner" solution is, respectively,

$$v_1 = \varepsilon \to 0, \quad \rho(v_1) = \rho_{max}, \quad \text{and} \quad v = v_2, \quad \rho(v_2) = \sup\{\rho\} \tag{59}$$

since it passes through the maximum of the probability density (51). However, with lower probabilities, other solutions of the family (53) can appear as well. Obviously, this is a non-classical effect. Qualitatively, this property is similar to those of quantum mechanics: the system keeps all the solutions simultaneously and displays each of them "by a chance," while that chance is controlled by the evolution of probability density (51).

The approach is generalized to n-dimensional case simply by replacing v with a vector $v = v_1, v_2, \ldots v_n$ since Equation (51) does not include space derivatives.

Examples. Let us start with the following normal distribution

$$\rho^*(V) = \frac{1}{\sqrt{2\pi}} e^{-\frac{V^2}{2}} \tag{60}$$

Substituting the expression (60) and (55) with Equation (56) at $V = v$, one obtains

$$v = erf^{-1}\left(\frac{C_1}{e^{-t} - 1}\right), \quad v \neq 0 \tag{61}$$

As another example, let us choose the target density ρ^* as the student's distribution, or so called power law distribution

$$\rho^*(V) = \frac{\Gamma(\frac{\nu+1}{2})}{\sqrt{\nu\pi}\Gamma(\frac{\nu}{2})}\left(1 + \frac{V^2}{\nu}\right)^{-(\nu+1)/2} \tag{62}$$

Substituting the expression (62) with Equation (56) at $V = v$, and $\nu = 1$, one obtains

$$v = \tan\left(\frac{C}{e^{-t} - 1}\right) \quad for \quad v \neq 0 \tag{63}$$

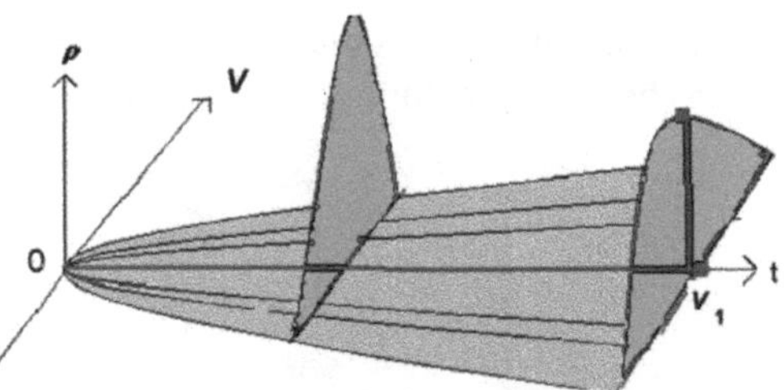

Figure 5. Stochastic process and probability density.

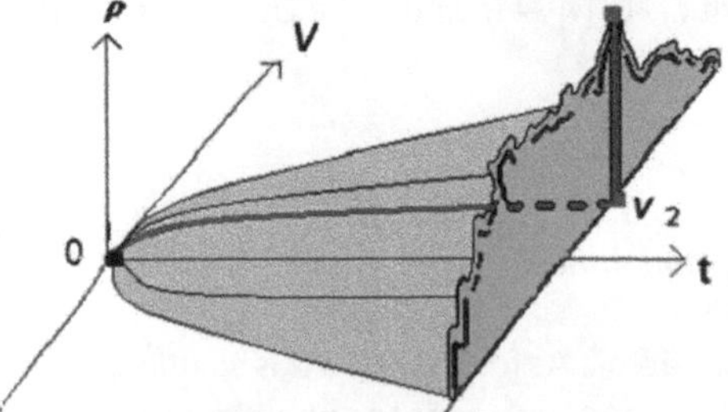

Figure 6. Global maximum.

The 3D plot of the solutions of Equations (61) and (63), are presented in Figures 7, and 8 respectively.

2.7. Finding global maximum

Based upon the proposed model with integral feedback, a simple algorithm for finding a global maximum of an n-dimensional function can be formulated. The idea of the proposed algorithm is very simple: based upon the model with integral feedback (50), and (51), introduce a *positive* function $\psi(v_1, v_2, ...v_n)$, $|v_i| < \infty$ to be maximized as the probability density $\rho^*(v_1, v_2, ...v_n)$ to which the solution of Equation (50) is attracted. Then the larger value of this function will have the higher probability to appear. The following steps are needed to implement this algorithm.

(1) Build and implement the n-dimensional version of the model Equations (50), and (51), as an analog devise

$$\dot{v}_i = \frac{e^{-t}}{n\{[\rho_0(v) - \rho^*(v)]e^{-t} + \rho^*(v)\}} \int_{-\infty}^{v_i} [\rho_0(\zeta) - \rho^*(\zeta)]d\zeta \quad , \quad i = 1, 2, \dots n. \tag{64}$$

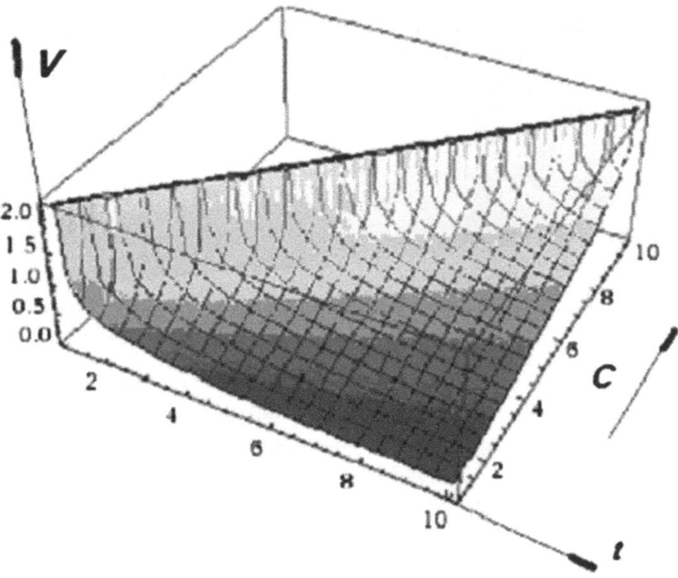

Figure 7. Dynamics driving random events to normal distribution.

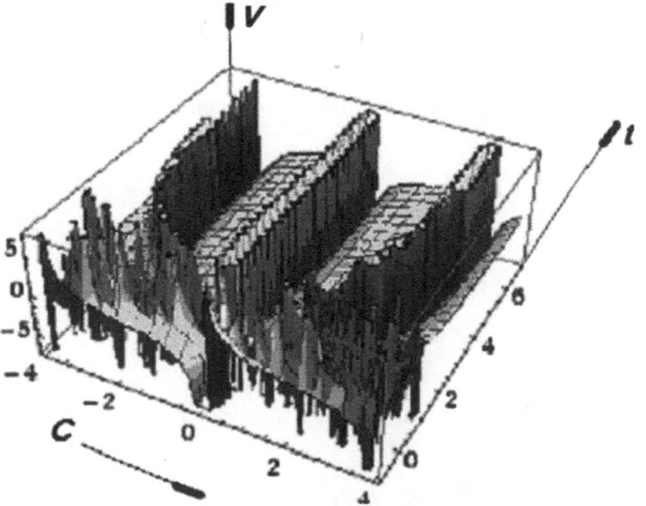

Figure 8. Dynamics driving random events to power law.

(2) Normalize the function to be maximized

$$\bar{\psi}(\{v\}) = \frac{\psi(\{v\})}{\int_{-\infty}^{\infty} \psi(\{v\}) d\{v\}} \tag{65}$$

(3) Using Equation (51), evaluate time τ of approaching the stationary process to accuracy ε

$$\tau \approx \ln \frac{1-\bar{\psi}}{\varepsilon \bar{\psi}} \tag{66}$$

(4) Substitute $\bar{\psi}$ instead of ρ^* with Equations (64) and run the system during the time interval τ.

(5) The solution will "collapse" into one of possible solutions with the probability $\bar{\psi}$.Observing (measuring) the corresponding values of {v'}, find the first approximation to the optimal solution.

(6) Switching the device to the initial state and then starting again, arrive at the next approximations.

(7) The sequence of the approximations represents Bernoulli trials that exponentially improve the chances of the optimal solution to become a winner. Indeed, the probability of success ρ_s and failure ρ_f after the first trial are, respectively,

$$\rho_s = \bar{\psi}_1, \quad \rho_f = 1 - \bar{\psi}_1 \tag{67}$$

Then the probability of success after *M* trials is

$$\rho_{sM} = 1 - (1 - \bar{\psi})^M \to 1 \quad at \quad M \to \infty \tag{68}$$

Therefore, after *polynomial* number of trials, one arrived at the solution to the problem (unless the function ψ is flat).

The main advantage of the proposed methodology is in a weak restriction imposed upon the *space structure* of the function $\bar{\psi}(\{v\})$: it should be only integrable since there is no space derivatives included in the model (64). This means that $\bar{\psi}(\{v\})$ is not necessarily to be differentiable. For instance, it can be represented by a Weierstrass-like function $f(v) = \sum_0^\infty a^n \cos(b^n \pi v)$, where $0 < a < 1$, *b* is a positive odd integer, and $ab > 1 + 1.5\pi$.

In a particular case when $\bar{\psi}(\{x\})$ is twice differentiable, the algorithm is insensitive to local maxima because it is driven *not by gradients*, but by the *values* of this function.

2.8. Entanglement in QI active systems with integral feedback

We will continue the analysis of the QI system with integral feedback introduced above proceeding with the two-dimensional case

$$\dot{v}_1 = \frac{1}{2\rho(v_1, v_2, t)} \int_{-\infty}^{v_1} [\rho(\eta, v_2, t) - \rho^*(\eta, v_2)] d\eta \tag{69}$$

$$\dot{v}_2 = \frac{1}{2\rho(v_1, v_2, t)} \int_{-\infty}^{v_2} [\rho(v_1, \eta, t) - \rho^*(v_1, \eta)] d\eta \tag{70}$$

$$\frac{\partial \rho(V,t)}{\partial t} + \rho(V,t) - \rho^*(V) = 0 \tag{71}$$

The solution of Equation (71) has the same form as for one-dimensional case, (see Equation (52),

$$\rho = \rho_0 + \rho^*(1 - e^{-t}) \tag{72}$$

Substitution this solution into Equations (69) and (70), yields, respectively,

$$\dot{v}_1 = \frac{e^{-t}}{2[\rho_0(v_1, v_2) - \rho^*(v_1, v_2)]e^{-t} + \rho^*(v_1, v_2)} \int_{-\infty}^{v_1} [\rho_0(\eta, v_2) - \rho^*(\eta, v_2)]d\eta \tag{73}$$

$$\dot{v}_2 = \frac{e^{-t}}{2[\rho_0(v_1, v_2) - \rho^*(v_1, v_2)]e^{-t} + \rho^*(v_1, v_2)} \int_{-\infty}^{v_2} [\rho_0(v_1, \eta) - \rho^*(v_1, \eta)]d\eta \tag{74}$$

that are similar to Equation (54). Following the same steps as in one-dimensional case, one arrives at the following solutions of Equations (73) and (74), respectively,

$$\int_{-\infty}^{v_1} \rho^*(\eta, v_2)d\eta = \frac{C_1}{e^{-t} - 1}, \quad v_1 \neq 0 \tag{75}$$

$$\int_{-\infty}^{v_2} \rho^*(\eta, v_1)d\eta = \frac{C_2}{e^{-t} - 1}, \quad v_2 \neq 0 \tag{76}$$

that are similar to the solution (56). Since $\rho(v_1, v_2)$ is the known (preset) function, Equations (75) and (76) implicitly define v_1 and v_2 as functions of time. Eliminating time t and orbitary constants C_1, C_2, one obtains

$$\frac{d}{dt}[ln \int_{-\infty}^{v_1} \rho^*(\eta, v_2)d\eta]/\frac{d}{dt}[ln \int_{-\infty}^{v_2} \rho^*(v_1, \eta)d\eta] \equiv 1 \tag{77}$$

Thus, the ratio (77) is deterministic although both the numerator and denominator are random, (see Equations (75) and (66)).

3. Application to games of entangled agents

In this section, we will address a situation when agents are competing. That means that they have *different* objectives. Turning to Equation (75), (76), one can rewrite them for the case of competing agents

$$\dot{v}_i = \frac{1}{n\rho(v_{1,} \dots v_n, t)} \int_{-\infty}^{v_i} [\rho(\zeta_i, v_{j\neq i}, t) - \rho_i^*(\zeta_i, v_{j\neq i})]d\zeta_i \tag{78}$$

$$\frac{d\rho}{dt} + \rho(t) - \frac{1}{n}\sum_{1}^{n} a_k\rho_k^* = 0, \sum_{k=1}^{n} a_k = n\rho_i^* \neq \rho_j^* if i \neq j. \tag{79}$$

where ρ_k^* is the preset density of the kth agent that can be considered as his objective, a_k is a constant weight of the kth agent's effort to approach his objective.

Thus, each kth agent is trying to establish his own static attractor ρ_k^*, but due to entanglement, the whole system will approach the weighted average

$$\rho = [\rho^0 - \frac{1}{n}\sum_{i=1}^{n}(a_i\rho_i^*)]e^{-t} + \frac{1}{n}\sum_{i=1}^{n} a_i\rho_i^* \tag{80}$$

$$\rho(t) = \frac{1}{n}\sum_{1}^{n} a_k \rho_k^* \text{ at } t \to \infty$$

Substituting the solution (80) with Equation (78), one arrives at a coupled system of n ODE with respect to n state variables v_i. Although a closed form analytical solution of the system (78) and (79) is not available, its property of the Lipcshitz instability at $t = 0$ could be verified. This means that the solution to the system (78) and (79) is random, and if the system is run many times, the statistical properties of the whole ensemble will be described by Equation (80). Obviously, those agents who have chosen density with a sharp maximum are playing more risky game. Here, we have assumed that competing agents are still entangled, and therefore, their information about each other is complete. More complex situation when the agents are not entangled, and exchanged information is incomplete is address in the next section. The simplest way to formalize the incompleteness of information possessed by competing agent is to include the "vortex" terms into Equation (77): these terms could change each particular trajectory of the agent motion, but they would not change the statistical invariants that remain available to the competing agents

$$\dot{v}_i = \frac{1}{n\rho(v_{1,} \dots v_n, t)}[\int_{-\infty}^{v_i} [\rho(\zeta_i, v_{j\neq i}, t) - \rho^*(\zeta_i, v_{j\neq i})]d\zeta_i + \sum_{j\neq i}^{n} T_{ij} \tanh v_j] \tag{81}$$

It is easily verifiable that the augmented neural net-like terms do not effect the corresponding Liouville equation, and therefore, they do not change the static attractor in the probability space described by Equation (76). However, they may significantly change the configuration of the random trajectories in physical space making their entanglement more sophisticated. Another way to formalize uncertainty is to introduce a complex joint probability density where its imaginary part represents a measure of uncertainty in density distribution. This case will be considered below in more details.

3.1. Problem formulation

In this section, we will present a draft of application of self-supervised *active* dynamical systems to differential games. Following von-Neuman, and Isaaks, (Isaacs, 1965), we will introduce a two-player zero-sum (antagonistic) differential game that is described by dynamical Equation (64) rewritten for $i = 1,2$

$$\dot{v}_i = \frac{1}{2\rho(v_{1,} \dots v_n, t)} \int_{-\infty}^{v_i} [\rho(\zeta_i, v_{j\neq i}, t) - a_i C_i f_i(\zeta_i, v_{j\neq i})]d\zeta_i \quad f_1 = f, \quad f_2 = f^{-1}, a_1 + a_2 = 2. \tag{82}$$

where v is the state vector, f_1 is the control vector of the maximizing player E (evader), f_2 is the control vector of the minimizing player P (pursuer), and C_1, C_2 are the normalizing factors. Obviously, f is a known function of both state variables.

However, the rules of the game we propose is slightly different from those introduced by Isaaks, namely: the player P tries to minimize the function f, (i.e. maximize the function f^{-1}) while the player E tries to maximize f in the same manner as it was described in the previous subsection i.e. via entanglement. The Liouville equation for the system (69) follows from Equation (79)

$$\frac{d\rho}{dt} + \rho(t) - \frac{1}{2}(a_1 C_1 f + a_2 C_2 f^{-1}) = 0 \tag{83}$$

whence

$$\rho = [\rho^0 - \frac{1}{2}(a_1 C_1 f + a_2 C_2 f^{-1})]e^{-t} + \frac{1}{2}(a_1 C_1 f + a_2 C_2 f^{-1}), \tag{84}$$

We will now give a description of the game.

The game starts with zero initial conditions:

$$v_1 = 0, \quad v_2 = 0, \quad \rho^0 = \delta(v_1, v_2) \quad at \quad t = 0 \tag{85}$$

It is assumed that each player has access to the systems (82), (84), and therefore, he has complete information about its state. The substitution of Equation (84) with Equation (82) closes the system (82). However, because of a failure of the Lipcshitz condition at $t = 0$ (see Equation (58)), the solution of Equation (82) is random, and each player can predict it only in terms of probability. As follows from Equation (84), the highest probability to appear has the solution that delivers the global maximum to the payoff function

$$F = a_1 C_1 f + a_2 C_2 f^{-1} \tag{86}$$

Obviously, the player that has higher weight a_i would have better chances to win since the global maximum of Equation (86) is closer to the global maximum of his goal function. With reference to Equation (86), a player can evaluate time τ of approaching the stationary process to accuracy ε as

$$\tau \approx \ln \frac{1-F}{\varepsilon F} \tag{87}$$

and introduce

$$v_1^1 = v_1(\tau), \quad v_2^1(\tau), \quad f_1^1 = f_1(v_1^1, v_2^2), \quad f_2^1 = f_2(v_1^1, v_2^1) \tag{88}$$

This is the end of the first move. After that, each player updates his weight as following

$$a_1^1 = a_1 + \frac{f_1^1 - f_1^0}{f_1^0}, a_2^1 = a_2 - \frac{f_1^1 - f_1^0}{f_1^0}, f_1^0 \neq 0, \tag{89}$$

and starts the next move with the same initial conditions (85). But the system (82), (84) is different now: the control functions f_1, f_2 are to be replaced by their updated values f_1^1, f_2^1, respectively. Thus, during the first move, the potential winner is selected by a chance, and during the next move, his chances are increased due to favorable update of the weights. However, the role of the chance is still significant even during the subsequent moves; indeed, if the global maximum of the control function F is sharp, the initially selected potential winner still can lose.

The game ends when one of the players achieves his goal my maximizing his control function to a preset level, for instance, if

$$f_1 - f_2 > A^2 \tag{90}$$

3.2. Games with incomplete information

The theory presented above includes applications to such problems as battle games, games with moving craft, pursuit games, etc. However, the main limitation of this theory, as well as the most of the game theories, is that it requires complete information about the state variables available to both players. This limitation significantly diminishes the applicability of the theory to real-life games where the complete information is not available. That is why the extension of this theory to cases of incomplete information is of vital importance. In our application, we will assume that each player knows only his own state variables, while he has to guess about the state variables of his adversary. For that case, the mathematical formalism of *QI* systems can offer a convenient tool to replace unknown value of a state variable by its expected value. Such a possibility is available due to players' dependence (but not necessarily entanglement) via the joint probability density: since each player possesses the joint density, he can, at any moment, compute the expected value of the state variable of the other player.

We assume that the players follow the strategy: "what do you think I think you think ...?" and we will start with the assumption that each player takes a conservative view by thinking that although he does not know the values of the state variable of his adversary, the adversary *does know* the values of his state variable. Then the governing equation for the Evader will be

$$\dot{v}_1 = \frac{1}{2\rho(v_1,\bar{V}_2,t)}\int_{-\infty}^{v_1} \rho(\zeta_1,\bar{V}_2,t) - a_1C_1f_1(\zeta_1,\bar{V}_2)d\zeta_1 \tag{91}$$

Here $\bar{V}_2$ is the expected value of V_2

$$\bar{V}_2 = \int_{-\infty}^{\infty} V_2\rho(V_1,V_2)dV_1dV_2 \tag{92}$$

Now the Evader has to create the *image* of the Pursuer using the expected value of his state variable

$$\dot{v}_{2|1} = \frac{1}{2\rho(v_1,\bar{V}_2,t)}\int_{-\infty}^{v_2} [\rho(v_1,\zeta_2,t) - a_2C_2f_2(v_1,\zeta_2)]d\zeta_2 \tag{93}$$

where $v_{2|1}$ is the state variable of the Pursuer in view of the Evader.

The corresponding Liouville equation that governs the joint probability equation is not changed: it is still given by Equation (79). Its solution (84) should be substituted with Equations (91) and (93) along with the Equation (92). Obviously, the expected value (92) is found from the solution (84)

$$\bar{V}_i = [\bar{V}_i^0 - \int_{-\infty}^{\infty} d\zeta_1 \int_{-\infty}^{\infty} d\zeta_2\{[\frac{1}{2}(a_1C_1f + a_2C_2f^{-1})]e^{-t} + \frac{1}{2}(a_1C_1f + a_2C_2f^{-1})]\bar{V}_i\}, \quad i = 1,2. \tag{94}$$

The system of Equations (91), (93), and (79) with reference to Equations (92), (94) is closed.

Similar system can be obtained for governing equation of the Pursuer coupled with the governing equation of his *image* of the Evador:

$$\dot{v}_2 = \frac{1}{2\rho(\bar{V}_1,v_2,t)}\int_{-\infty}^{v_1} \rho(\bar{V}_1,\zeta_2,t) - a_2C_2f_2(\bar{V}_1,\zeta_2)d\zeta_2 \tag{95}$$

$$\dot{v}_{1|2} = \frac{1}{2\rho(\bar{V}_1,v_2,t)}\int_{-\infty}^{v_1} \rho(\zeta_1,v_2,t) - a_2C_2f_2(\zeta_1,v_2)d\zeta_1 \tag{96}$$

$$\bar{V}_1 = \int_{-\infty}^{\infty} V_1\rho(V_1,V_2)dV_1dV_2 \tag{97}$$

After substitution of Equation (79) with Equations (95) and (96), with reference to Equations (97) and (94), one arrives at the closed system.

Thus, we obtain two independent systems of ODE describing *entanglement* of the player with the image of his adversary. Each system has random solutions that appear with the probability described by Equation (79). After time interval τ (see Equation (87)), each player gets access to the real values of the functions f_i to be maximized, and based upon that, he can update the state variables and weights for the next move, (see Equations (88) and (89)).

Let us consider now the case when the players do not know the values of state variables of their adversary.

Then instead of the systems (91), (93), and (95), (96) we have, respectively

$$\dot{v}_1 = \frac{1}{2\rho(v_1,\bar{V}_2,t)} \int_{-\infty}^{v_1} \rho(\zeta_1,\bar{V}_2,t) - a_1C_1f_1(\zeta_1,\bar{V}_2)d\zeta_1 \quad (98)$$

$$\dot{v}_{2|1|1} = \frac{1}{2\rho(\bar{V}_1,\bar{V}_2,t)} \int_{-\infty}^{v_2} [\rho(\bar{V}_1,\zeta_2,t) - a_2C_2f_2(\bar{V}_1,\zeta_2)]d\zeta_2 \quad (99)$$

$$\dot{v}_2 = \frac{1}{2\rho(\bar{V}_1,v_2,t)} \int_{-\infty}^{v_1} \rho(\bar{V}_1,\zeta_2,t) - a_2C_2f_2(\bar{V}_1,\zeta_2)d\zeta_2 \quad (100)$$

$$\dot{v}_{1||2|2} = \frac{1}{2\rho(\bar{V}_1,\bar{V}_2,t)} \int_{-\infty}^{v_1} \rho(\zeta_1,\bar{V}_2,t) - a_2C_2f_2(\zeta_1,\bar{V}_2)d\zeta_1 \quad (101)$$

Here, $v_{2|1|1}$ is the state variable of the Pursuer's view on the Evader in view of the Pursuer, and $v_{1|2|2}$ is the state variable of the Evader's view on the Pursuer in view of the Evader.

It is easy to conclude that the image Equations (99) and (101) can be solved independently

$$v_{2|1|1} = \int_0^t dt \frac{1}{2\rho(\bar{V}_1,\bar{V}_2,t)} \int_{-\infty}^{v_2} [\rho(\bar{V}_1,\zeta_2,t) - a_2C_2f_2(\bar{V}_1,\zeta_2)]d\zeta_2 \quad (102)$$

$$v_{1|2|2} = \int_0^t dt \frac{1}{2\rho(\bar{V}_1,\bar{V}_2,t)} \int_{-\infty}^{v_1} \rho(\zeta_1,\bar{V}_2,t) - a_2C_2f_2(\zeta_1,\bar{V}_2)d\zeta_1 \quad (103)$$

Now replacing $\bar{V}_2, \bar{V}_1$ in Equations (98) and (100) by the solutions for $v_{2|1|1}$ and $v_{1|2|2}$, respectively, one arrives at two independent ODE describing behaviors of the players. Therefore, at this level of incompleteness of information, the entanglement disappears.

The games with incomplete information give a reason to distinguish two type of dependence between the agents described by the variables v_i in the *iQ* systems. The first type of dependence is entanglement that has been introduced and discussed above. One should recall that in order to be entangled, the agents are supposed to run the system jointly during some initial period of time. But what happens if the agents had never been in contact? Obviously they are not entangled, i.e. they cannot predict each other's motions. However, they are not completely independent: they can make random decisions, but the probability of these decisions will be correlated via the joint probability. As a result, the agents will be able to predict expected decisions of each other. We will call such correlation a ***weak*** entanglement. As follows from the games with incomplete information considered above, weak entanglement was presented as entanglement of an agent with the probabilistic image of another agent.

4. Games of partially entangled agents

In this section we introduce a new, more sophisticated entanglement that does not exist in quantum mechanics, but can be found in QI models. This finding is based upon existence of incompatible stochastic processes that are considered below.

4.1. Incompatible stochastic processes

Classical probability theory defines conditional probability densities based upon the existence of a joint probability density. However, one can construct correlated stochastic processes that are represented only by conditional densities since a joint probability density does not exist. For that purpose, consider two coupled Langevin equations (Risken, 1989)

$$\dot{x}_1 = g_{11}(x_2)L_1(t) \tag{104}$$

$$\dot{x}_2 = g_{22}(x_1)L_2(t) \tag{105}$$

where the Langevin forces $L_1(t)$ and $L_2(t)$ satisfy the conditions

$$\langle L_i(t)\rangle = 0, \quad \langle L_i(t)L_j(t')\rangle = 2g_{ii}\delta(t-t') \tag{106}$$

Then the joint probability density $\rho(X_1, X_2)$ describing uncertainties in values of the random variables $x_1$and x_2 evolves according to the following Fokker–Planck equation

$$\frac{\partial \rho}{\partial t} = g_{11}^2(X_2)\frac{\partial^2 \rho}{\partial X_1^2} + g_{22}^2(X_1)\frac{\partial^2 \rho}{\partial X_2^2} \tag{107}$$

Let us now modify Equations (104) and (105) as following

$$\dot{x}_1 = g_{11}^2(x_2^*)L_1(t) \tag{108}$$

$$\dot{x}_2 = g_{22}^2(x_1^*)L_2(t) \tag{109}$$

where x_1^* and x_1^* are fixed values of x_1 and x_2 that play role of parameters in Equations (108) and (109), respectively. Now the uncertainties of $x_1$and x_2 are characterized by conditional probability densities $\rho_1(X_1|X_2)$ and $\rho_2(X_2|X_1)$ while each of these densities is governed by its own Fokker–Planck equation

$$\frac{\partial \rho_1}{\partial t} = g_{11}^2(X_2)\frac{\partial^2 \rho_1}{\partial X_1^2} \tag{110}$$

$$\frac{\partial \rho_2}{\partial t} = g_{22}^2(X_1)\frac{\partial^2 \rho_2}{\partial X_2} \tag{111}$$

The solutions of these equations subject to sharp initial conditions

$$\rho_i(X_i, t|X_i', t') = \delta(X_i - X_i'), \qquad i = 1, 2. \tag{112}$$

for $t > t'$ read

$$\rho_1(X_1|X_2) = \frac{1}{\sqrt{4\pi g_{11}^2(X_2)(t-t')}}\exp(-\frac{(X_1 - X_1')^2}{4g_{11}^2(X_2)(t-t')} \tag{113}$$

$$\rho_2(X_2|X_1) = \frac{1}{\sqrt{4\pi g_{22}^2(X_1)(t-t')}}\exp(-\frac{(X_2 - X_2')^2}{4g_{22}^2(X_1)(t-t')} \tag{114}$$

As shown in Zak (1998), a joint density for the conditional densities (113) and (114) exists only in special cases of the diffusion coefficients g_{11} and g_{22} when the conditional probabilities are compatible. These conditions are

$$ink(\rho_1, \rho_2) = \frac{\partial}{\partial X_1 \partial X_2}\ln\frac{\rho_1(X_1|X_2)}{\rho_2(X_2|X_1)} \equiv 0 \tag{115}$$

Indeed

$$\rho(X_1,X_2) = \rho_1(X_1|X_2)\int_{-\infty}^{\infty}\rho(\zeta,X_2)d\zeta = \rho_2(X_2|X_1)\int_{-\infty}^{\infty}\rho(X_1,\zeta)d\zeta, \tag{116}$$

whence

$$\ln\frac{\rho_1(X_1|X_2)}{\rho_2(X_2|X_1)} = \ln\int_{-\infty}^{\infty}\rho(X_1,\zeta)d\zeta - \ln\int_{-\infty}^{\infty}\rho(\zeta,X_2)d\zeta \tag{117}$$

and that leads to Equation (115).

Thus, the existence of the join density $\rho(X_1, X_2)$ for the conditional densities $\rho_1(X_1|X_2)$, and $\rho_2(X_2|X_1)$ requires that

$$\frac{\partial^2}{\partial X_1 \partial X_2}[\frac{(X_1 - X_1')^2}{4g_{11}^2(X_2)} - \frac{(X_2 - X_2')}{4g_{22}^2(X_1)}] \equiv 0 \tag{118}$$

Obviously the identity (118) holds only for specially selected functions $g_{11}(X_2)$ and $g_{22}(X_1)$, and therefore, existence of the joint density is an exception rather than a rule.

4.2. Partial entanglement

In order to prove existence of a new form of entanglement, let us modify the system Equations (36), (37), and (38) as following

$$\dot{V}_1 = -a_{11}(V_2)\frac{\partial}{\partial V_1}\ln\rho_1(V_1|V_2) \tag{119}$$

$$\frac{\partial\rho_1(V_1|V_2)}{\partial t} = a_{11}(V_2)\frac{\partial^2\rho_1(V_1|V_2)}{\partial V_1^2} \tag{120}$$

$$\dot{V}_2 = -a_{22}(V_1)\frac{\partial}{\partial V_1}\ln\rho_2(V_2|V_1) \tag{121}$$

$$\frac{\partial\rho_2(V_2|V_1)}{\partial t} = a_{22}(V_1)\frac{\partial^2\rho_2(V_2|V_1)}{\partial V_2^2} \tag{122}$$

Since here we do not postulate existence of a joint density, the system is written in terms of conditional densities, while Equations (120) and (121) are similar to Equations (110) and (111). The solutions of these PDE can be written in the form similar to the solutions (113) and (114)

$$\rho_1(V_1|V_2) = \frac{1}{\sqrt{4\pi a_{11}(V_2)(t-t')}}\exp(-\frac{(V_1 - V_1')^2}{4a_{11}(V_2)(t-t')} \tag{123}$$

$$\rho_2(V_2|V_1) = \frac{1}{\sqrt{4\pi a_{22}(V_1)(t-t')}}\exp(-\frac{(V_2 - V_2')^2}{4a_{11}(V_1)(t-t')} \tag{124}$$

As noticed in the previous subsection, the existence of the joint density $\rho(V_1, V_2)$ for the conditional densities $\rho_1(V_1|V_2)$ and $\rho_2(V_2|V_1)$ require that

$$\frac{\partial^2}{\partial V_1 \partial V_2}[\frac{(V_1 - V_1')^2}{4a_{11}(V_2)} - \frac{(V_2 - V_2')}{4a_{22}(V_1)}] \equiv 0 \tag{125}$$

In this case, the joint density exists (although its finding is not trivial, Zak, 1998), and the system (119)–(122) can be reduced to a system similar to (36)–(38). But here we will be interested in case when the joint density does ***not*** exist. It is much easier to find such functions $a_{11}(V_2)$, $a_{22}(V_1)$ for which the identity (125) does ***not*** hold, and we assume that

$$\frac{\partial^2}{\partial V_1 \partial V_2}[\frac{(V_1 - V_1')^2}{4a_{11}(V_2)} - \frac{(V_2 - V_2')}{4a_{22}(V_1)}] \neq 0 \tag{126}$$

In this case, the system (119)–(122) cannot be simplified. In order to analyze this system in detail, lets substitute the solutions (123) and (124) with Equations (119) and (121), respectively. Then with reference to Equation (14), one obtains

$$\dot{v}_1 = \frac{v_1}{2t} \tag{127}$$

$$\dot{v}_2 = \frac{v_2}{2t} \tag{128}$$

and therefore

$$v_1 = C_1\sqrt{t} \tag{129}$$

$$v_2 = C_2\sqrt{t} \tag{130}$$

It should be recalled that according to the terminology introduced in Section 3, the systems (119)–(120) and the systems (121)–(122) can be considered as dynamical models for interaction of two *communicating* agents where Equations (119) and (121) describe their motor dynamics, and Equations (120) and (122)—mental dynamics, respectively. Also, it should be reminded that the solutions (129) and (130) are represented by one parametrical family of random samples, as in Equation (15), while the randomness enters through the time-independent parameters C_1 and C_2 that can take any real numbers. As follows from Figure 2, all the particular solutions (129) and (130) intersect at the same point $v_{1,2} = 0$ at $t = 0$, and that leads to non-uniqueness of the solution due to violation of the Lipcshitz condition. Therefore, the same initial condition $v_{1,2} = 0$ at $t = 0$ yields infinite number of different solutions forming a family; each solution of this family appears with a certain probability guided by the corresponding Fokker-Planck Equations (120) and (122), respectively. Similar scenario was described in the Section 2 of this paper. But what is unusual in the system (119)–(121) is correlations: although Equations (120) and (122) are correlated, and therefore, mental dynamics are entangled, Equations (119) and (121) are *not* correlated (since they can be presented in the form of independent Equations (127) and (128), respectively), and therefore, the motor dynamics are *not* entangled. This means that in the course of communications, each agent "selects" a certain pattern of behavior from the family of solutions (129) and (130), respectively, and these patterns are independent; but the *probabilities* of these "selections" are entangled via Equations (120) and (122). Such sophisticated correlations *cannot* be found in physical world, and they obviously represent a "human touch." Unlike the entanglement in system with joint density (such as that in Equations (36)–(38)) here the agents do not share any deterministic invariants (compare to Equation 48). Instead, the agents can communicate via "best guesses" based upon known conditional probability densities distributions.

In order to quantify the amount of uncertainty due to incompatibility of the conditional probability densities (123) and (124), let us introduce a concept of complex probability, (Zak, 1998),

$$f(V_1, V_2) = a(V_1, V_2) + ib(V_1, V_2) \tag{131}$$

Then the marginal densities are

$$f_1(V_1) = \int_{-\infty}^{\infty} a(V_1, V_2)dV_2 + i\int_{-\infty}^{\infty} b(V_1, V_2)dV_2 = a_1(V_1) + ib_1(V_1) \tag{132}$$

$$f_2(V_2) = \int_{-\infty}^{\infty} a(V_1, V_2)dV_1 + i\int_{-\infty}^{\infty} b(V_1, V_2)dV_1 = a_2(V_2) + ib_2(V_2) \tag{133}$$

Following the formalism of conditional probabilities, the conditional densities will be defined as

$$f_{1|2} = \frac{f(V_1, V_2)}{f_2(V_2)} = \frac{a(V_1, V_2) + ib(V_1, V_2)}{a_2(V_2) + ib_2(V_2)} = \frac{aa_2 + bb_2}{a_2^2 + b_2^2} + i\frac{a_2 b - ab_2}{a_2^2 + b_2^2} \tag{134}$$

$$f_{2|1} = \frac{f(V_1, V_2)}{f_1(V_1)} = \frac{a(V_1, V_2) + ib(V_1, V_2)}{a_1(V_1) + ib_1(V_1)} = \frac{aa_1 + bb_1}{a_1^2 + b_1^2} + i\frac{a_1 b - ab_1}{a_1^2 + b_1^2} \tag{135}$$

with the normalization constraint

$$\int_{-\infty}^{\infty}\int_{-\infty}^{\infty} (a^2 + b^2)^{1/2} dV_1 dV_2 = 1 \tag{136}$$

This constraint can be enforced by introducing a normalizing multiplier in Equation (131) which will not affect the conditional densities (134) and (135).

Clearly,

$$A \leq (a^2 + b^2)^{1/2}, \text{and} \int_{-\infty}^{\infty}\int_{-\infty}^{\infty} a dV_1 dV_2 \leq 1 \tag{137}$$

Now our problem can be reformulated in the following manner: given two conditional probability densities (123) and (124), and considering them as real parts of (unknown) complex densities (134) and (135), find the corresponding complex joint density (131), and therefore, all the marginal (132) and (133), as well as the imaginary parts of the conditional densities. In this case, one arrives at two coupled integral equations with respect to two unknowns $a(V_1, V_2)$ and $b(V_1, V_2)$ (while the formulations of $a_1(V_1, V_2)$, $a_2(V_1, V_2)$, $b_1(V_1, V_2)$ and $b_2(V_1, V_2)$ follow from Equations (132) and (133)). These equations are

$$\rho_1(V_1, V_2) = \frac{aa_2 + bb_2}{a_2^2 + b_2^2}, \rho_2(V_1, V_2) = \frac{aa_1 + bb_1}{a_2^2 + b_2^2}, \tag{138}$$

The system (138) is non-linear, and very little can be said about general property of its solution without detailed analysis. Omitting such an analysis, let us start with a trivial case when

$$b = 0 \tag{139}$$

In this case, the system (138) reduces to the following two integral equations with respect to *one* unknown that is $a(V_1, V_2)$:

$$\rho_1(V_1, V_2) = \frac{a(V_1, V_2)}{\int_{-\infty}^{\infty} a(V_1, V_2)dV_2}, \rho_2(V_1, V_2) = \frac{a(V_1, V_2)}{\int_{-\infty}^{\infty} a(V_1, V_2)dV_1}, \tag{140}$$

This system is overdetermined unless the compatibility conditions (115) are satisfied.

As known from classical mechanics, the incompatibility conditions are usually associated with a fundamentally new concept or a physical phenomenon. For instance, incompatibility of velocities in fluid (caused by non-existence of velocity potential) introduces vorticity in rotational flows, and incompatibility in strains describes continua with dislocations. In order to interpret the incompatibility (115), let us return to the system (138). Discretizing the functions in Equation (138) and replacing the integrals by the corresponding sums, one reduces Equation (138) to a system of n algebraic equations with respect to n unknowns. This means that the system is closed, and cases when a solution does not exist are exceptions rather than a rule. Therefore, in most cases, for any arbitrarily chosen conditional densities, for instant, for those given by Equations (123) and (124), the system (138) defines the complex joint density in the form (131).

Now we are ready to discuss a physical meaning of the imaginary component of the complex probability density. Firstly, as follows from comparison of Equations (138) and (140), the imaginary part of the probability density appears as a response to incompatibility of the conditional probabilities, and therefore, it can be considered as a "compensation" for the incompatibility. Secondly, as follows from the inequalities (137), the imaginary part consumes a portion of the "probability mass" increasing thereby the degree of uncertainty in the real part of the complex probability density. Hence, the imaginary part of the probability density can be defined as a measure of the uncertainty "inflicted" by the incompatibility into the real part of this density.

In order to avoid solving the system of integral equations (138), we can reformulate the problem in an inverse fashion by assuming that the *complex* joint density is given. Then the real parts of the conditional probabilities that drive Equations (119) and (120) can be found from simple formulas (134) and (135).

Let us illustrate this new paradigm, and consider two players assuming that each player knows his own state but does not know the state of his adversary. In order to formalize the degree of initial incompleteness of information, introduce the complex joint probability density,

$$\rho_0(V_1, V_2) = a_0\delta(V_1, V_2) + ib_0\delta(V_1, V_2)$$

that shows how much the players know and how much they do not know about each other when the game starts. With reference to the normalization constraint (136),

$$(a_0^2 + b_0^2)^{1/2} = 1 \tag{142}$$

The structure of the real part of the joint probability density can be chosen the same as in Equation (84)

$$\text{Re } \rho = a_0\{[\delta - \frac{1}{2}(\alpha_1 C_1 f + \alpha_2 C_2 f^{-1})]e^{-t} + \frac{1}{2}(\alpha_1 C_1 f + \alpha_2 C_2 f^{-1})\}, \tag{143}$$

However, since here $a_0 < 1$, the real part of the joint probability density is reduced due to a "leak" of the probability "mass" from the real to the imaginary part, and this makes predictions less certain for the both players. Otherwise, the formal structure of the motor dynamics is similar to that described by Equations (83) and (84).

The imaginary part can be preset as

$$\text{Im}\rho = b_0[(\delta - C_3\rho^*)e^{-t} + C_3\rho^*] \tag{144}$$

where ρ^* is the probability density characterizing the degree of uncertainty of information that the players have about each other, while the larger ρ^* the more the probability leak from the real to imaginary part of the complex probability density. The arbitrary constants C_1, C_2, and C_3 couples the real and the imaginary parts via the normalization constraint (136)

$$\int_{-\infty}^{\infty}\int_{-\infty}^{\infty} [(\mathrm{Re}\rho)^2 + (\mathrm{Im}\rho^2)]^{1/2} dV_1 dV_2 = 1 \tag{145}$$

The motor dynamics has a slight change compare to Equation (82)

$$\dot{v}_1 = \mathrm{Re}\{\frac{e^{-t}}{2[\delta(v_1,v_2) - \rho^*(v_1,v_2)]e^{-t} + \rho^*(v_1,v_2)}\int_{-\infty}^{v_1}[\delta(\zeta,v_2) - \rho^*(\zeta,v_2)]d\zeta\} \tag{146}$$

$$\dot{v}_2 = \mathrm{Re}\{\frac{e^{-t}}{2[\delta(v_1,v_2) - \rho^*(v_1,v_2)]e^{-t} + \rho^*(v_1,v_2)}\int_{-\infty}^{v_2}[\delta(v_1,\zeta) - \rho^*(v_1,\zeta)]d\zeta\} \tag{147}$$

Thus, both players rely only upon the real part of the *complex* joint density instead of a *real* joint density (that may not exist in this case). But as follows from the inequalities (137), the values of density of the *real part* are lowered due to loss of the probability mass, and this increases the amount of uncertainty in player's predictions. In order to minimize that limitation, the players can invoke the imaginary part of the joint density that gives them *qualitative* information about the amount of uncertainty at the selected maxima.

It should be noticed that the game starts with a significant amount of uncertainties that will grow with next moves. Such subtle and sophisticated relationship is typical for communications between humans, and the proposed model captures it via partial entanglement introduced above.

Remark. So far we considered the imaginary part of a joint probability density as a result of incompatibility of conditional densities of the players. However, this part can have a different origin: it can also represent a degree of deception that the players apply in real-life games. As in the previous example, in games with deception the imaginary part of the joint probability density increases uncertainty of the players' prediction capabilities. The mathematical formalism of the game with deception is similar to that discussed above.

5. Passive period of players' performance

In this section, we will discuss the capacity of ***mathematical formalism*** that provides an extension of the proposed model to a new space with imaginary time where players exhibit virtual motions such as dreams and memories. In order to demonstrate that, let us replace Equation (54) by the following

$$f = \frac{\zeta}{\rho(v,t)}\int_{-\infty}^{v}[\rho(\zeta,t) - \rho^*(\zeta)]dv \tag{148}$$

where

$$\zeta = \frac{\sqrt{T-t}}{\sqrt{|T-t|}} \tag{149}$$

and T is the period of active performance of the player.

Then at $0 < t < T$

$$\zeta = 1 \tag{150}$$

the player is active, and its activity is described by the governing Equations (50) and (51).

For $t = T$

$\zeta = 0$

the player is at rest, and its state is described by a simple Newtonian state.

But for $t > T$ the feedback (148)

$$f = \frac{i}{\rho(v,t)} \int_{-\infty}^{v} [\rho(\eta,t) - \rho^*(\eta)]d\eta \tag{151}$$

as well as Equations (50), (61) and (52)

$$\dot{v} = i\frac{\zeta_0}{\rho(v,t)} \int_{-\infty}^{v} [\rho(\eta,t) - \rho^*(\eta)]d\eta \tag{152}$$

$$\frac{\partial \rho}{\partial t} + i\zeta_0[\rho(t) - \rho^*] = 0 \tag{153}$$

become complex. For better interpretation, it will be more convenient to introduce an imaginary time

$$\tilde{t} = it \tag{154}$$

Then, the formal solutions of these equations are

$$\rho = [(\rho_0 - \rho^*)e^{-\tilde{t}} + \rho^*] \tag{155}$$

$$\int_{-\infty}^{v} \rho^*(\eta)d\eta = (\frac{C}{e^{-\tilde{t}} - 1}), \quad v \neq 0 \tag{166}$$

Thus, the velocity v and the probability density ρ become real functions of imaginary time. It is reasonable to assume that the family of trajectories in the solution (63) describes virtual motions evolving in imaginary time with the probability (62), while the time scale of these motions could be different from the real one. Such a surrealistic activity can be associated with memories and dreams, i.e. with the period of passive performance during which a player has an opportunity to enrich his information with help of memories, and to plan and test his future performance.

6. Discussion and conclusion

We start the discussion with outlining the mathematical novelties of this work. Actually a new class of ODE that are coupled with their Liouville equation is introduced. The leading ideas came from the Madelung equation that is a hydrodynamics version of the Schrödinger equation

$$\frac{\partial \rho}{\partial t} + \nabla \bullet (\frac{\rho}{m}\nabla S) = 0 \tag{167}$$

$$\frac{\partial S}{\partial t} + (\nabla S)^2 + F - \frac{\hbar^2 \nabla^2 \sqrt{\rho}}{2m\sqrt{\rho}} = 0 \tag{168}$$

Here ρ and S are the components of the wave function $\psi = \sqrt{\rho}e^{iS/\hbar}$, and $\hbar$ is the Planck constant divided by 2π. The last term in Equation (2) is known as quantum potential. From the viewpoint of Newtonian mechanics, Equation (167) expresses continuity of the flow of probability density, and

Equation (168) is the Hamilton–Jacobi equation for the action S of the particle. Actually the quantum potential in Equation (168), as a feedback from Equation (167) to Equation (168), represents the difference between the Newtonian and quantum mechanics, and therefore, it is solely responsible for fundamental quantum properties.

Our approach is based upon replacing the quantum potential with a different Liouville feedback, Figure 1.

In Newtonian physics, the concept of probability ρ is introduced via the Liouville equation

$$\frac{\partial \rho}{\partial t} + \nabla \bullet (\rho \mathbf{F}) = 0 \tag{169}$$

generated by the system of ODE

$$\frac{d\mathbf{v}}{dt} = \mathbf{F}[\mathbf{v}_1(t), \dots \mathbf{v}_n(t), t] \tag{170}$$

where $\mathbf{v}$ is velocity vector.

It describes the continuity of the probability density flow originated by the error distribution

$$\rho_0 = \rho(t = 0) \tag{171}$$

in the initial condition of ODE (171).

Let us rewrite Equation (168) in the following form

$$\frac{d\mathbf{v}}{dt} = \mathbf{F}[\rho(\mathbf{v})] \tag{172}$$

where $\mathbf{v}$ is a velocity of a hypothetical particle.

This is a fundamental step in our approach: in Newtonian dynamics, the probability never explicitly enters the equation of motion. In addition to that, the Liouville equation generated by Equation (172) is non-linear with respect to the probability density ρ

$$\frac{\partial \rho}{\partial t} + \nabla \bullet \{\rho \mathbf{F}[\rho(\mathbf{V})]\} = 0 \tag{173}$$

and therefore, the system (172), (173) departs from Newtonian dynamics. However, although it has the same topology as quantum mechanics (since now the equation of motion is coupled with the equation of continuity of probability density), it does not belong to it either. Indeed Equation (172) is more general than the Hamilton–Jacoby Equation (2): it is not necessarily conservative, and **F** is not necessarily the quantum potential although further we will impose some restriction upon it that links **F** to the concept of information. The relation of the system (172), (123) to Newtonian and quantum physics is illustrated in Figure 1. Two different types of feedbacks replacing the quantum potential (see Equation (23) and Equation (49)) are introduced and analyzed. Both of these feedbacks lead to different approaches to psychological games. As demonstrated, this new class of ODE includes mathematical formalization of human factor. Therefore, the paper extends the mathematical formalism of quantum physics to include games and economics with a component of psychology. The novelty of the approach is based upon a human factor-based ***behavioral*** model of an intelligent agent. The model is quantum inspired: it is represented by a modified Madelung equation in which the gradient of quantum potential is replaced by a specially chosen information force. It consists of motor dynamics simulating actual behavior of the agent, and mental dynamics representing evolution of the corresponding knowledge base and incorporating this knowledge in the form of information flows into the motor dynamics. Due to feedback from mental dynamics, the motor dynamics attains quantum-like properties: its trajectory splits into a family of different trajectories, and each of those trajectories can be chosen with the probability prescribed by the mental dynamics; each agent entangled (in a quantum way) to other agents and makes calculated predictions for future actions. In case of a complex density, its imaginary part represents a measure of uncertainty of the density distribution. Human factor is associated with violation of the second law of thermodynamics: the system can move from disorder to order without external help, and that represent intrinsic intelligence. All of these departures actually extend and complement the classical methods making them especially successful in analysis of communications of agents represented by new mathematical formalism. Special attention is concentrated on new approach to theory of differential games, and in particular, on behavioral properties of players as intelligent subjects possessing self-image and self-awareness. Due to quantum-like entanglement they are capable to predict and influence actions of their adversaries. The model addresses a new type of entanglement that correlates the probabilities of actions of livings rather than the actions themselves.

There are several differences between the proposed and conventional game theories. Firstly, in the proposed game, the players are entangled: they cannot make independent deterministic decisions; instead, they make coordinated random decisions such that, at least, the probabilities of these decisions are dependent. Therefore, the proposed game represents a special case of non-determine symmetric simultaneous zero-sum game. Secondly, the maximization of the pay-off function here does not require any special methods (like gradient ascend) since it is "built-in" into the dynamical model. Indeed, the pay-off function (86) is represented by the probability density of the stochastic attractor, and therefore, its maximum value has the highest probability to appear as a random solution of the underlying dynamical model (82). Moreover, the payoff function (86) is not required to be differentiable at all (although it must be integrable).

This work opens up the way to development of a fundamentally new approach to mathematical formalism of ***economics*** that is inspired by the formalism of quantum mechanics. Mathematical treatment of economics has a relatively short history. Formal economic modeling began in the nineteenth century with the use of differential calculus to represent and explain economic behavior, such as utility maximization, an early economic application of mathematical optimization. Economics became more mathematical as a discipline throughout the first half of the twentieth century, but introduction of new and generalized techniques in the period around the World War II, as in game theory, (Isaacs, 1965), would greatly broaden the use of mathematical formulations in economics. However, not withstanding undisputable success of mathematical methods in economics, there were alarmed critics of the discipline, as well as some noted economists. John Maynard Keynes, Robert Heilbroner, Friedrich Hayek, and others have criticized the broad use of mathematical models for human behavior, arguing that some human choices are irreducible to mathematics. Actually the alert was expressed much earlier by Newton who stated "I can calculate the motion of heavenly bodies, but not the madness of people."

Limitations of modern mathematical methods in economics are especially transparent in the area of agent-based computational economics. It is a relatively recent field, dating from about the 1990s as to published work. It studies economic processes, including whole economies, as dynamic systems of interacting agents over time. As such, it falls in the paradigm of complex adaptive systems. In corresponding agent-based models, agents are not real people but "computational objects modeled as interacting according to rules" ... "whose micro-level interactions create emergent patterns" in space and time. The rules are formulated to predict behavior and social interactions based on incentives and information. The theoretical assumption of mathematical optimization by agents markets is replaced by the less restrictive postulate of agents with bounded rationality adapting to market forces.

In this context, the presented mathematical formalism incorporates a human factor in agent-based economics while departing from physics formalism toward physics of life.

Funding
The author received no direct funding for this research.

Author details
Michail Zak[1]
E-mail: michail.zak@gmail.com
[1] California Institute of Technology, 9386 Cambridge street, Cypress, CA 90630, USA.

References

Isaacs, R. (1965). *Differential games*. New York, NY: Wiley.

Penrose, R. (1955). *Why new physics is needed to understand the mind*. Cambridge: University Press.

Risken, H. (1989). *The Fokker-Planck equation*. New York, NY: Springer. http://dx.doi.org/10.1007/978-3-642-61544-3

Zak, M. (1992). Terminal model of Newtonian dynamics. *International Journal of Theoretical Physics, 32*, 159–190.

Zak, M. (1998). Incompatible stochastic processes and complex probabilities. *Physics Letters A, 238*, 1–7. http://dx.doi.org/10.1016/S0375-9601(97)00763-9

Zak, M. (2007). Physics of life from first principles. *EJTP 4, 16*, 11–96.

Zak, M. (2008). Quantum-inspired maximizer. *Journal of Mathematical Physics, 49*, 042702. http://dx.doi.org/10.1063/1.2908281

Zak, M. (2014a). *Particle of life: Mathematical abstraction or reality?* New York, NY: Nova.

Zak, M. (2014b). Origin of randomness in quantum mechanics. *Electronic Journal of Theoretical Physics, 11*, 149–164.

9

Picard's iterative method for nonlinear multicomponent transport equations

Juergen Geiser[1]*

*Corresponding author: Juergen Geiser, Department of Electrical Engineering and Information Technology, Ruhr University of Bochum, Universitätsstrasse 150, D-44801 Bochum, Germany

E-mail: juergen.geiser@ruhr-uni-bochum.de

Reviewing editor: Jen-Chih Yao, National Sun Yat-sen University, Taiwan

Abstract: In this paper, we present a Picard's iterative method for the solution of nonlinear multicomponent transport equations. The multicomponent transport equations are important for mixture models of the ionized and neutral particles in plasma simulations. Such mixtures deal with the so-called Stefan–Maxwell approaches for the multicomponent diffusion. The underlying nonlinearities are delicate and it is not necessary to be an analytical function of the dependent variables. The proposed solver method is based on Banach's contraction fix-point principle that allows to solve such nonlinearities without making any use to Lagrange multipliers and constrained variations. Such an improvement allows to solve delicate nonlinear problems and we test the application to model with multicomponent transport equations.

Subjects: Algorithms & Complexity; Computer Science; Computer Science (General); Engineering & Technology; Mathematics & Statistics for Engineers; Technology

Keywords: Picard's iterative method; variational iterative method; multicomponent transport; multilevel iterative method; plasma models; multidiffusion model; nonlinear solver methods

AMS Subject classifications: 35K25; 35K20; 74S10; 70G65

1. Introduction

We are motivated to apply nonlinear multicomponent transport equations, which are given by delicate plasma processes.

ABOUT THE AUTHOR

Juergen Geiser, researcher and lecturer at the Ruhr-University of Bochum, Germany, has been involved in teaching and research projects and has collaborated with engineering and physicist groups on numerical modeling of technical and physical models. The research activity refers to the mathematical modeling, numerics, and analysis of transport and flow problems in engineering applications, e.g. groundwater modeling and plasma modeling. He is a specialist in multiscale solvers and iterative solvers and most of the topics of the special issue. Moreover, Juergen Geiser is the author of scientific books and editor of various scientific journals and has thus able to manage the editorial activity. He is also a visiting professor at the Centrale Supelec in the laboratory: Mathematics in Interaction with Computer Sciences, Chatenay-Malabry, Cedex, France.

PUBLIC INTEREST STATEMENT

In this paper a multi component model is presented to simulate ionized and neutral particles in plasma simulations. An improved diffusion operator is presented based on the Stefan–Maxwell approach. Such nonlinear diffusion operators are solved with novel solver methods based on the Banach's contraction fix-point principle. For higher accuracy, multilevel iterative methods are presented to solve the nonlinearities. An extension to exponential Picard's iterative methods is discussed, which allows to split into linear and nonlinear parts. The novel schemes are tested in blow-up, Bernoulli examples.

A real-life example based on a multicomponent diffusion model is solved with the new methods. The paper also discusses future work to the novel iterative solvers.

We deal with problems of normal pressure, room temperature plasma applications, which are used for medical and technical processes. Here, the increasing importance of plasma chemistry based on the multicomponent plasma is a key factor for such a trend, see for low pressure plasma (Senega & Brinkmann, 2006) and for atmospheric pressure regimes (Tanaka, 2004). Due to the fact of the influence of the mass transfer in the multicomponent mixture, also the standard conservation laws have to be improved, such improvements are well-known in fusion research, see the modeling of in high-ionized plasmas (Igitkhanov, 2011), but only few works are done for weak-ionized plasma for atmospheric pressure regimes.

We concentrate on a diffusion reaction model for a mixture of H, H_2, H_2^+ species. The model equation results in a delicate nonlinear multidiffusion equation. Such nonlinear equations are often solved with standard fix-point iteration schemes. Here, we propose a novel higher order Picard's iterative method, which allows to accelerate the nonlinear solver based on intermediate level. Such a treatment allows to save computational time and we obtained higher order accurate results.

The paper is outlined as follows.

In Section 2, we present our mathematical model. The Picard's iterative methods are discussed in Section 3. The numerical experiments are presented in Section 4. In the contents, that are given in Section 5, we summarize our results.

2. Mathematical model

In the following, a model is presented due to the motivation in Senega and Brinkmann (2006), which deals with a fluid dynamical description of a plasma model based on a small Knudsen number.

The Knudsen number is the ratio of the mean free path λ over the typical domain size L of the apparatus. For small Knudsen numbers $Kn \approx 0.01 - 1.0$, we deal with a Navier–Stokes equation, where for large Knudsen numbers $Kn \geq 1.0$, we deal with the Boltzmann equations.

We deal with the following plasma model of a mixture of H, H_2, H_2^+.

We take into account the dissociation and ionization reactions, which are given as:

$$H_2 + e \underset{\longleftrightarrow}{\lambda_1} H_2^+ + 2\,e, \tag{1}$$

$$H_2 + e \underset{\longleftrightarrow}{\lambda_2} 2H + e, \tag{2}$$

where the electron temperature is given as $T_e = 17,400$ K and the gas temperature values remain constant $T_h = 600$ K.

Further, we have $\lambda_1 = 1.58 \times 10^{-15}\, T_e^{0.5} \exp\left(\frac{-15.378}{T_e}\right) = 2.082 \times 10^{-13}$ and $\lambda_2 = 1.413 \times 10^{-15}\, T_e^{2} \exp\left(\frac{-4.48}{T_e}\right) = 4.276 \times 10^{-7}$.

The diffusion coefficients are given as in the following formula:

$$D_{ij} = \frac{3}{16} \frac{f_{ij} k_B^2 T_i T_j}{p\, m_{ij} \Omega_{ij}^{(1,1)}(T_{ij})}, \tag{3}$$

where the parameters are: f_{ij} is a correction factor of order unity, $m_{ij} = \frac{m_i m_j}{m_i + m_j}$ is the reduced mass, m_i mass of species i, m_j mass of species j, p pressure, t_i, T_j the temperature of the corresponding species, and $\Omega_{ij}^{(1,1)}$ a collision integral (Hirschfelder, Curtiss, & Bird, 1966).

We assume the following binary diffusion parameters for our experiments:

$$D_{H_2, H_2^+} = 0.34\,[\,\mathrm{cm}^2/\,\mathrm{sec}], \tag{4}$$
$$D_{H_2, H} = 0.21\,[\,\mathrm{cm}^2/\,\mathrm{sec}], \tag{5}$$
$$D_{H_2^+, H} = 0.21\,[\,\mathrm{cm}^2/\,\mathrm{sec}]. \tag{6}$$

We have the following underlying model: We deal with:

$$\partial_t \xi_i + \nabla \cdot N_i = S_i, \quad 1 \leq i \leq 3, \tag{7}$$
$$\sum_{j=1}^{3} N_j = 0, \tag{8}$$
$$\frac{\xi_2 N_1 - \xi_1 N_2}{D_{12}} + \frac{\xi_3 N_1 - \xi_1 N_3}{D_{13}} = -\nabla \xi_1, \tag{9}$$
$$\frac{\xi_1 N_2 - \xi_2 N_1}{D_{12}} + \frac{\xi_3 N_2 - \xi_2 N_3}{D_{23}} = -\nabla \xi_2, \tag{10}$$

and the kinetic term S_i is given as:

$$S_i = \sum_{j=1}^{3} \lambda_{i,j} \xi_j, \tag{11}$$

where $\lambda_{i,j}$ are the reaction rates. Further, the domain is given as $\Omega \in \mathbb{R}^d, d \in N^+$ with $\xi_i \in C^2$.

We decompose the diffusion and the reaction parts and apply the following: a splitting approach to our problem; we compute $n = 1, \ldots, N, t_0, t_1, \ldots, t_n$ time steps:

The first step is given as (Diffusion step):

$$\partial_t \tilde{\xi}_i + \nabla \cdot N_i = 0, \quad 1 \leq i \leq 3, \tag{12}$$
$$\sum_{j=1}^{3} N_j = 0, \tag{13}$$
$$\frac{\tilde{\xi}_2 N_1 - \tilde{\xi}_1 N_2}{D_{12}} + \frac{\tilde{\xi}_3 N_1 - \xi_1 N_3}{D_{13}} = -\nabla \xi_1, \tag{14}$$
$$\frac{\tilde{\xi}_1 N_2 - \tilde{\xi}_2 N_1}{D_{12}} + \frac{\tilde{\xi}_3 N_2 - \tilde{\xi}_2 N_3}{D_{23}} = -\nabla \tilde{\xi}_2, \text{ for } t \in [t^n, t^{n+1}], \tag{15}$$
$$\tilde{\xi}_i(t^n) = \xi_i(t^n),\ i = 1, 2, 3, \tag{16}$$

and the next step (Reaction step):

$$\partial_t \xi_i = S_i,\ 1 \leq i \leq 3, \text{ for } t \in [t^n, t^{n+1}], \tag{17}$$
$$\xi_i(t^n) = \tilde{\xi}_i(t^{n+1}),\ i = 1, 2, 3. \tag{18}$$

Remark 1 Based on the derived model, we discuss the application of the novel nonlinear solvers. We can also generalize the schemes for more than three species.

3. Iterative method

In the following, we discuss Picard's iterative method with different levels. Picard's iterative methods are known to solve delicate nonlinear problems (see Ramos, 2008, 2009).

In the following, we discuss the basic ideas and we develop improved Picard's iterative methods, e.g. exponential schemes, for our special treatments.

The basic Picard's iterative method is given as:

$$u_{k+1}(x,t) = (Pu_k)(x,t), \quad k = 0,1,2,\dots, \tag{19}$$

where the Picard's operator is given as:

$$(Pu)(x,t) = u(x,t_0) + \int_{t_0}^{t} F(x,s,u(x,s),\nabla u(x,s),\Delta u(x,s))\,ds. \tag{20}$$

Picard's operator is applied for the two-level method as:

$$\begin{aligned} u_{k+1}(x,t) &= u(x,t_0) \\ &+ \int_{t_0}^{t} F(x,s,u_k(x,s),\nabla u_k(x,s),\Delta u_k(x,s))\,ds, \quad k = 0,1,2,\dots. \end{aligned} \tag{21}$$

Further, the operator is applied for the three-level method as:

$$\begin{aligned} u_{k+1}(x,t) &= u_k(x,t) \\ &+ \int_{t_0}^{t} F_k(s) - F_{k-1}(s)\,ds, \quad k = 0,1,2,\dots, \end{aligned} \tag{22}$$

where $F_k(s) = F(x,s,u_k(x,s),\nabla u_k(x,s),\Delta u_k(x,s))$.

Proof We combine two two-level methods, given as:

$$\begin{aligned} u_{k+1}(x,t) &= u(x,t_0) \\ &+ \int_{t_0}^{t} F_k(s)\,ds, \quad k = 0,1,2,\dots, \end{aligned} \tag{23}$$

and

$$\begin{aligned} u_k(x,t) &= u(x,t_0) \\ &+ \int_{t_0}^{t} F_{k-1}(s)\,ds, \quad k = 0,1,2,\dots, \end{aligned} \tag{24}$$

where if we subtract Equation (23) with (24), we obtain: □

$$\begin{aligned} u_{k+1}(x,t) &= u_k(x,t) \\ &+ \int_{t_0}^{t} F_k(s) - F_{k-1}(s)\,ds, \quad k = 0,1,2,\dots, \end{aligned} \tag{25}$$

3.1. Multilevel iterative method based on Picard's iterative method

The multilevel iterative method is given as:

THEOREM 1 *We have the following construction formula for the k-level iterative method based on Picard's iterative method:*

- Two-level method is given as:

$$u_i(x,t) - u(x,t_0) = \int_{t_0}^{t} F_{i-1}(s)\,ds. \tag{26}$$

- i-level method with $i \geq 3$ is given as:

$$\sum_{j=0}^{\tilde{k}} (-1)^j \binom{\tilde{k}}{j} u_{i-j}(x,t) = \int_{t_0}^{t} \sum_{j=0}^{\tilde{k}} (-1)^j \binom{\tilde{k}}{j} F_{i-j-1}(s)\,ds. \tag{27}$$

where $\tilde{k} = i - 2$.

Proof The case $i = 2$ is the standard Picard method and clear.

For the case $i \geq 3$ we have the following complete induction given as:

We start from $i = 3$ (and $\tilde{k} = i - 2$) and have:

$$\sum_{j=0}^{1}(-1)^j \begin{pmatrix} 1 \\ j \end{pmatrix} u_{i-j}(x,t) = \int_{t_0}^{t} \sum_{j=0}^{1}(-1)^j \begin{pmatrix} 1 \\ j \end{pmatrix} F_{i-j-1}(s)\, ds, \tag{28}$$

$$u_i(x,t) - u_{i-1}(x,t) = \int_{t_0}^{t} (F_{i-1}(s) - F_{i-2}(s))\, ds, \tag{29}$$

The induction step is given as $i \to i + 1$, where we have to multiply $(u_1 - u_0)$ to the i-level method as:

$$(u_1 - u_0)\sum_{j=0}^{\tilde{k}}(-1)^j \begin{pmatrix} \tilde{k} \\ j \end{pmatrix} u_{i-j}(x,t) \tag{30}$$

$$= \sum_{j=0}^{\tilde{k}}(-1)^j \begin{pmatrix} \tilde{k} \\ j \end{pmatrix} u_{i+1-j}(x,t) + \sum_{j=0}^{\tilde{k}}(-1)^j \begin{pmatrix} \tilde{k} \\ j \end{pmatrix} u_{i-j}(x,t) \tag{31}$$

$$= \sum_{j=0}^{\tilde{k}+1}(-1)^j \begin{pmatrix} \tilde{k} \\ j-1 \end{pmatrix} u_{i-j}(x,t) + \sum_{j=0}^{\tilde{k}+1}(-1)^j \begin{pmatrix} \tilde{k} \\ j \end{pmatrix} u_{i-j}(x,t) \tag{32}$$

$$= \sum_{j=0}^{\tilde{k}+1}(-1)^j \begin{pmatrix} \tilde{k}+1 \\ j \end{pmatrix} u_{i-j}(x,t) \tag{33}$$

where $\tilde{k} = i - 2$.

The same is done for the right hand side and we have proven the formula. □

Remark 2 The next level methods are given as:

- Four-level method:

$$u_k(x,t) = 2u_{k-1}(x,t) - u_{k-2}(x,t) + \int_{t_0}^{t} \left(F_{k-1}(s) - 2F_{k-2}(s) + F_{k-3}(s)\right) ds, \; k = 0, 1, 2, \ldots, \tag{34}$$

- Five-level method:

$$u_k(x,t) = 3u_{k-1}(x,t) - 3u_{k-2}(x,t) + u_{k-3}(x,t) + \int_{t_0}^{t} \left(F_{k-1}(s) - 3F_{k-2}(s) + 3F_{k-3}(s) - 3F_{k-4}(s)\right) ds, \quad k = 0, 1, 2, \ldots, \tag{35}$$

3.2. Exponential Picard's iterative method

In the following, we discuss an extension of Picard's iterative method with a linear operator (exponential idea).

The exponential Picard's iterative method is given as:

$$\frac{\partial u_{k+1}(x,t)}{\partial t} = Au_{k+1} + F_k(t), \quad k = 0, 1, 2, \ldots, \tag{36}$$

where the variation of constant formula is applied and we obtain:

$$u_{k+1}(x,t) = \exp(At)u(x,t_0) + \int_{t_0}^{t} \exp(A(t-s))F_k(s)\,ds, \quad k = 0,1,2,\ldots \tag{37}$$

So the two-level method is given as:

$$\begin{aligned} u_{k+1}(x,t) &= \exp(At)u(x,t_0) \\ &\quad + \int_{t_0}^{t} \exp(A(t-s))F(x,s,u_k(x,s),\nabla u_k(x,s),\Delta u_k(x,s))\,ds, \quad k = 0,1,2,\ldots, \end{aligned} \tag{38}$$

Further, the three-level method is given as:

$$\begin{aligned} u_{k+1}(x,t) &= u_k(x,t) \\ &\quad + \int_{t_0}^{t} \exp(A(t-s))\big(F_k(s) - F_{k-1}(s)\big)\,ds, \quad k = 0,1,2,\ldots, \end{aligned} \tag{39}$$

where $F_k(s) = F(x,s,u_k(x,s),\nabla u_k(x,s),\Delta u_k(x,s))$.

The multilevel iterative method is given as:

THEOREM 2 *We have the following construction formula for the k-level iterative method based on Picard's iterative method:*

- Two-level method is given as:

$$u_i(x,t) - \exp(At)\,u(x,t_0) = \int_{t_0}^{t} \exp(A(t-s))\,F_{i-1}(s)\,ds. \tag{40}$$

- k-level method with $k \geq 3$ is given as:

$$\begin{aligned} &\sum_{j=0}^{\tilde{k}} (-1)^j \binom{\tilde{k}}{i} u_{i-j}(x,t) \\ &= \int_{t_0}^{t} \exp(A(t-s)) \left(\sum_{j=0}^{\tilde{k}} (-1)^j \binom{\tilde{k}}{j} F_{i-j-1}(s) \right) ds, \end{aligned} \tag{41}$$

where $\tilde{k} = k - 2$.

Proof The case $k = 2$ is the variation of constants with the standard Picard method and is clear.

For the case $k \geq 3$, we have the following complete induction given as in the proof of Theorem 1. □

Remark 3 The next level methods are given as:

- Four-level method:

$$\begin{aligned} u_k(x,t) &= 2u_{k-1}(x,t) - u_{k-2}(x,t) \\ &\quad + \int_{t_0}^{t} \exp(A(t-s))\big(F_{k-1}(s) - 2F_{k-2}(s) + F_{k-3}(s)\big)\,ds, \quad k = 0,1,2,\ldots, \end{aligned} \tag{42}$$

- Five-level method:

$$\begin{aligned} u_k(x,t) &= 3u_{k-1}(x,t) - 3u_{k-2}(x,t) + u_{k-3}(x,t) \\ &\quad + \int_{t_0}^{t} \exp(A(t-s))\big(F_{k-1}(s) - 3F_{k-2}(s) + 3F_{k-3}(s) - 3F_{k-4}(s)\big)\,ds \quad k = 0,1,2,\ldots, \end{aligned} \tag{43}$$

4. Numerical experiments

In the following experiments, we discuss the improvements of our novel Picard's iterative methods to the standard approaches. We start with test examples of blow-up and Bernoulli's equations, where we could compare to analytical solutions. A multicomponent model is discussed in the next steps and the benefits to the higher order multilevel methods are presented.

4.1. First numerical example: blow-up equation

As a nonlinear differential example, we chose the Bernoulli's equation:

$$\frac{\partial u(t)}{\partial t} = \lambda(u(t)^p, \; t \in [0, T], \text{ with } u(0) = 1, \tag{44}$$

where the analytical solution can be derived as (see also Geiser, 2008):

$$u(t) = \left(\frac{p-1}{-C-\lambda t}\right)^{\frac{1}{p-1}}.$$

When we apply $p = 2$, we have

$$u(t) = \frac{1}{1-\lambda t}$$

We choose $p = 2, \lambda = -0.2, T = 0.2$ and $u(0) = 1$.

For the numerical experiments, we have the following analytical solution:

$$u_1(t) = 1 + \lambda t, \tag{45}$$

$$u_2(t) = 1 + \lambda t + \lambda^2 t^2 + \frac{1}{3}\lambda^3 t^3, \tag{46}$$

$$u_3(t) = 1 + \lambda t + \lambda^2 t^2 + \lambda^3 t^3 + \frac{5}{12}\lambda^4 t^4 + \frac{1}{3}\lambda^5 t^5 + \frac{1}{9}\lambda^6 t^6 + \frac{1}{63}\lambda^7 t^7. \tag{47}$$

We apply the Picard's iterative method, which is given in the following as a two-level method. Here, we approximate in each sub-interval $[t^n, t^{n+1}]$, $n = 0, 1, \dots, N$, the integral:

$$u_{k+1,2}(t^{n+1}) = u_0(t^n) + \int_{t^n}^{t^{n+1}} \lambda u_{k,2}^2(s)\, ds, \quad k = 0, 1, 2, \dots. \tag{48}$$

For the numerical integration, we apply the different Simpson's rules of higher order.

- Simpson's Rule: We apply the Simpson's rule for the two-level method:

$$u_{k+1,2}(t^{n+1}) = u_0(t^n) + \lambda\frac{\Delta t}{6}\left(u_{k,2}^2(t^{n+1}) + 4u_{k,2}^2(t^{n+1/2}) + u_{k,2}^2(t^n)\right), \quad k = 0, 1, 2, \dots, \tag{49}$$

where $u_{k,2}(t^{n+1/2}) = u_{k,2}(^{n+1}) - \frac{1}{2}\Delta t \,\frac{du_{k,2}}{dt}(t^{n+1}) = u_{k,2}(^{n+1}) - \lambda\frac{1}{2}\Delta t\, u_{k,2}^2(t^{n+1})$.

- 3/8 Simpson's rule: We apply the 3/8-Simpson's rule for the two-level method:

$$u_{k+1,2}(t^{n+1}) = u_0(t^n) + \lambda\frac{\Delta t}{8}\left(u_{k,2}^2(t^{n+1}) + 3u_{k,2}^2(t^{n+1/3}) + 3u_{k,2}^2(t^{n+2/3}) + u_{k,2}^2(t^n)\right), \quad k = 0, 1, 2, \dots, \tag{50}$$

where

$$u_{k,2}(t^{n+1/3}) = u_{k,2}(^{n+1}) - \frac{2}{3}\Delta t\,\frac{du_{k,2}}{dt}(t^{n+1}) = u_{k,2}(^{n+1}) - \lambda\frac{2}{3}\Delta t\, u_{k,2}^2(t^{n+1}),$$

$$u_{k,2}(t^{n+2/3}) = u_{k,2}(^{n+1}) - \frac{1}{3}\Delta t\,\frac{du_{k,2}}{dt}(t^{n+1}) = u_{k,2}(^{n+1}) - \lambda\frac{1}{3}\Delta t\, u_{k,2}^2(t^{n+1}).$$

Further, for the next time step, we have $u(t^{n+1}) = u_{K,2}(t^{n+1})$ and K is the number of iterative steps, e.g. K= 4.

Further, the three-level method is given as a combination of two two-level methods' means, and we have:

$$\Delta u_{k+1,k,3}(t^{n+1}) = u_{k+1,3}(t^{n+1}) - u_{k,3}(t^{n+1})$$

$$= \int_{t^n}^{t^{n+1}} \lambda\left(u_{k,3}^2(s) - u_{k-1,3}^2(s)\right) ds, \tag{51}$$

$$u_{k+1,3}(t^{n+1}) = u_{k,3}(t^{n+1}) + \Delta u_{k+1,k,3}(t^{n+1}) \tag{52}$$

$$k = 0, 1, 2, \ldots,$$

where $u_0(t) = u(t^n)$ and for $u_{1,3}(t) = u_{1,2}(t)$, we have the two-level method. Further, we apply the trapezoidal or Simpson's rule for the numerical integration of the integrals. Further, for the next time step, we have $u(t^{n+1}) = u_{K,3}(t^{n+1})$ and K is the number of iterative steps, e.g. K = 4.

Remark 4 The three-level method has the benefit of applying the numerical integration for the differences between k and $k-1$ solutions of the right-hand side, such that we can accelerate the solver process, while we skip an additional numerical integration in each step.

Based on the reference solution, we deal with the following errors:

$$E_{L_1,k,\Delta t,[0,T]} = |u_{method,k,\Delta t}(t) - u_{ref,\Delta t}(t)|$$

$$= \sum_{n=0}^{N} \Delta t |u_{method,k,\Delta t}(t^n) - u_{ref,\Delta t}(t^n)|, \tag{53}$$

where *method* is the different Picard's methods and k is the number of iterative steps.

We apply the Picard's iteration with the two-level method (see Figure 1 and Table 1).

Remark 5 Here, we see the improvement till $K = 5$, while here the accuracy is reached. The reduction of the errors is done with each iterative step.

We apply the Picard's iteration with the three-level method (see Figure 2 and Table 2).

Remark 6 Here, we see the improvement till $K = 4$, while here the accuracy is reached. The reduction of the errors is done with each iterative step.

We apply the blow-up experiment with $\lambda = 0.999$ and obtain the following solutions. The Picard's iteration with the two-Level method is given in Figure 3 for $\lambda = 0.999$.

The Picard's iteration with the three-level method is given in Figure 4 for $\lambda = 0.999$.

Remark 7 We discussed the multilevel Picard's method with different iterative steps to a blow-up problem. We saw that we obtain high accurate solutions with $k \approx 3-5$ iterative steps and that a three-level method can obtain faster numerical results.

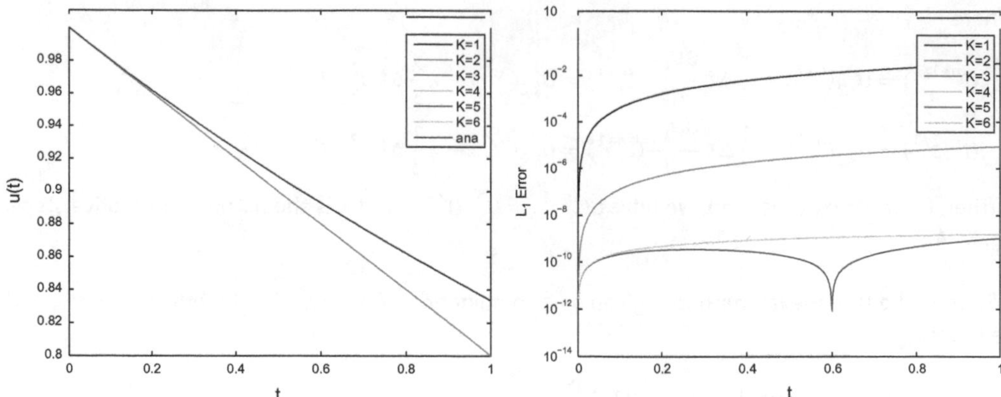

Figure 1. The two-level Picard's method compared with the analytical solution.

Table 1. Comparative results of the two-level method with k iterative steps and the analytical solution for $\Delta t = 10^{-2}$ and $T = 1.0$

Number of iterations	Δt	Err_{L1}	Comp. time
$K = 1$	10^{-1}	3.33×10^{-2}	2.3611×10^{-4}
$K = 2$	10^{-1}	9.0912×10^{-4}	3.1310×10^{-4}
$K = 3$	10^{-1}	1.1806×10^{-5}	3.7982×10^{-4}
$K = 4$	10^{-1}	1.5768×10^{-5}	4.3628×10^{-4}
$K = 5$	10^{-1}	1.4943×10^{-5}	4.8248×10^{-4}
$K = 6$	10^{-1}	1.4968×10^{-5}	5.2867×10^{-4}
$K = 1$	10^{-2}	3.33×10^{-2}	2.3×10^{-3}
$K = 2$	10^{-2}	8.7588×10^{-5}	3.1×10^{-3}
$K = 3$	10^{-2}	1.0439×10^{-7}	3.8×10^{-3}
$K = 4$	10^{-2}	1.5461×10^{-7}	4.5×10^{-3}
$K = 5$	10^{-2}	1.5385×10^{-7}	4.8×10^{-3}
$K = 6$	10^{-2}	1.5385×10^{-7}	5.3×10^{-3}
$K = 1$	10^{-3}	3.33×10^{-2}	2.3611×10^{-4}
$K = 2$	10^{-3}	8.7256×10^{-6}	3.1310×10^{-4}
$K = 3$	10^{-3}	1.0303×10^{-9}	3.7982×10^{-4}
$K = 4$	10^{-3}	1.5435×10^{-9}	4.3628×10^{-4}
$K = 5$	10^{-3}	1.5427×10^{-9}	4.8248×10^{-4}
$K = 6$	10^{-3}	1.5427×10^{-9}	5.2867×10^{-4}

4.2. Second numerical example: Bernoulli's equation

As a nonlinear differential example, we chose the Bernoulli's equation:

$$\frac{\partial u(t)}{\partial t} = (\lambda_1 + \lambda_3)u(t) + (\lambda_2 + \lambda_4)(u(t))^p, \ t \in [0, T], \ \text{with} \ u(0) = 1, \tag{54}$$

where the analytical solution can be derived as (see also Geiser, 2008):

$$u(t) = \exp((\lambda_1 + \lambda_3)t)\left[-\frac{\lambda_2 + \lambda_4}{\lambda_1 + \lambda_3}\exp((\lambda_1 + \lambda_3)(p - 1)t) + c\right]^{1/(1-p)}.$$

Using $u(0) = 1$, we find that $c = 1 + \frac{\lambda_2 + \lambda_4}{\lambda_1 + \lambda_3}$, so

$$u(t) = \exp((\lambda_1 + \lambda_3)t)\left\{1 + \frac{\lambda_2 + \lambda_4}{\lambda_1 + \lambda_3}[1 - \exp((\lambda_1 + \lambda_3)(p - 1)t)]\right\}^{1/(1-p)}.$$

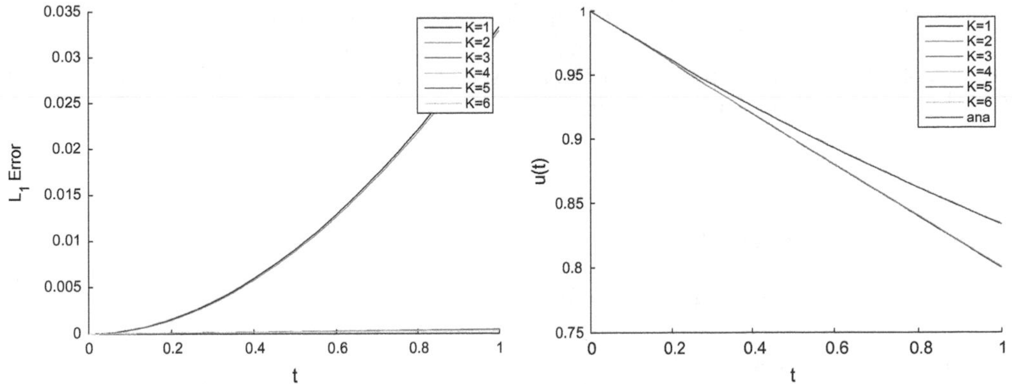

Figure 2. The three-level Picard's method compared with the analytical solution.

TAble 2. Comparative results of the three-level method with k iterative steps and the analytical solution for $\Delta t = 10^{-2}$ and $T = 1.0$

Three-Level method	Accuracy err_{L_1}	CPU-time (sec)
$k = 1$	10^{-4}	8×10^{-2}
$k = 2$	10^{-1}	3×10^{-2}
$k = 3$	10^{-1}	3×10^{-2}
$k = 4$	10^{-1}	3×10^{-2}
$k = 5$	10^{-1}	3×10^{-2}

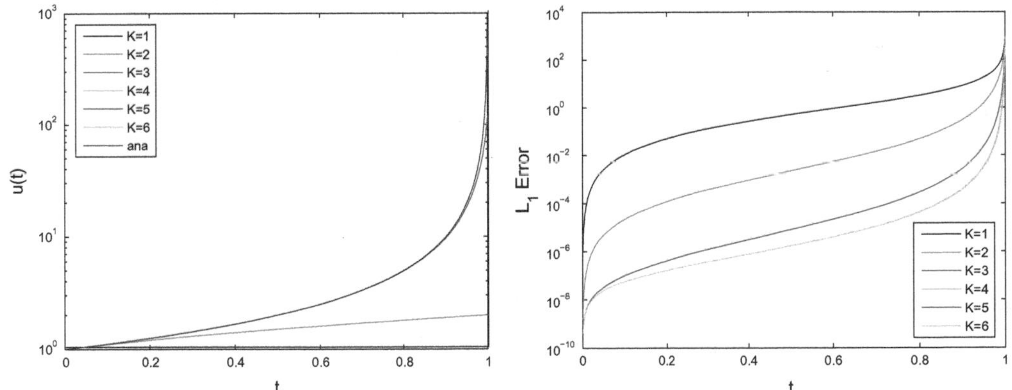

Figure 3. The two-level Picard's method compared with the analytical solution.

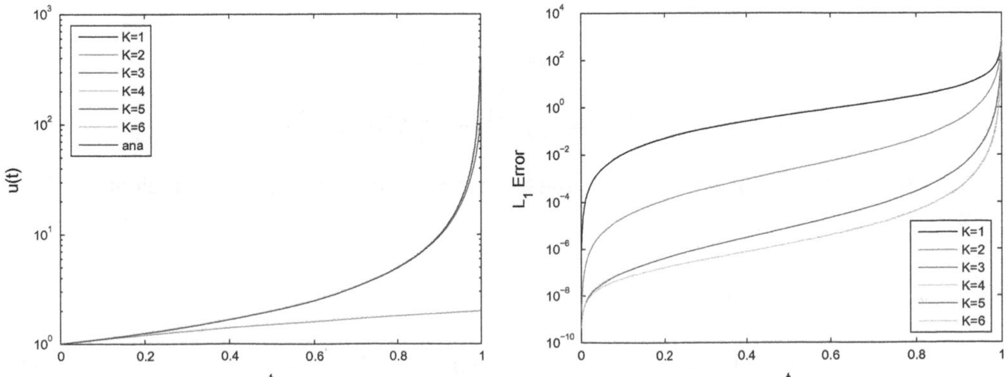

Figure 4. The three-level Picard's method compared with the analytical solution.

We choose $p = 2, \lambda_1 = -1, \lambda_2 = -0.5, \lambda_3 = -1, \lambda_4 = -1, T = 0.2$, and $u(0) = 1$.

We apply the Picard's iterative method based on approximating in each sub-interval $[t^n, t^{n+1}]$, $n = 0, 1, \dots, N$, which is given as:

- Two-level method is given as:

$$u_{k+1}(t^{n+1}) = \exp((\lambda_1 + \lambda_3)\Delta t)u(t^n) + \int_{t^n}^{t^{n+1}} \exp((\lambda_1 + \lambda_3)(t^{n+1} - s))\big((\lambda_2 + \lambda_4)u_k^p(s)\big)\, ds, \quad k = 1, 2, \dots, \tag{55}$$

where for $k = 0$; we have:

$$u_1(t^{n+1}) = \exp((\lambda_1 + \lambda_3)\Delta t)u(t^n) + \int_{t^n}^{t^{n+1}} \exp((\lambda_1 + \lambda_3)(t^{n+1} - s))(\lambda_2 + \lambda_4)u_0^p(s)\, ds, \tag{56}$$

with the time step $\Delta t = t^{n+1} - t^n$. We apply the following numerical integration rules:

- Simpson's rule: We apply the Simpson's rule for the two-level method:

$$\begin{aligned} u_{k+1,2}(t^{n+1}) &= \exp((\lambda_1 + \lambda_3)\Delta t)\, u_0(t^n) \\ &\quad + (\lambda_2 + \lambda_4)\, \frac{\Delta t}{6}\Big(u_{k,2}^p(t^{n+1}) + 4\exp((\lambda_1 + \lambda_3)\Delta t/2)\, u_{k,2}^p(t^{n+1/2}) \\ &\quad + \exp((\lambda_1 + \lambda_3)\Delta t)\, u_{k,2}^p(t^n)\Big), \quad k = 0, 1, 2, \dots, \end{aligned} \tag{57}$$

where we have the intermediate solution as:

$$\begin{aligned} u_{k,2}(t^{n+1/2}) &= u_{k,2}(t^{n+1}) - \frac{1}{2}\Delta t\, \frac{du_{k,2}}{dt}(t^{n+1}) \\ &= u_{k,2}(t^{n+1}) - \frac{1}{2}\Delta t\big((\lambda_1 + \lambda_3)u_{k,2}(t^{n+1}) + (\lambda_2 + \lambda_4)(u_{k,2}(^{n+1}))^p\big). \end{aligned} \tag{58}$$

- 3/8 Simpson's rule: We apply the 3/8-Simpson's rule for the two-level method:

$$\begin{aligned} u_{k+1,2}(t^{n+1}) &= \exp((\lambda_1 + \lambda_3)\Delta t)\, u_0(t^n) \\ &\quad + (\lambda_2 + \lambda_4)\, \frac{\Delta t}{8}\Big(u_{k,2}^p(t^{n+1}) + 3\exp((\lambda_1 + \lambda_3)\Delta t/3)\, u_{k,2}^p(t^{n+2/3}) \\ &\quad + 3\exp((\lambda_1 + \lambda_3)2\Delta t/3)\, u_{k,2}^p(t^{n+1/3}) + \exp((\lambda_1 + \lambda_3)\Delta t)\, u_{k,2}^p(t^n)\Big), \quad k = 0, 1, 2, \dots, \end{aligned} \tag{59}$$

where

$$u_{k,2}(t^{n+1/3}) = u_{k,2}(^{n+1}) - \frac{2}{3}\Delta t\, \frac{du_{k,2}}{dt}(t^{n+1}) = u_{k,2}(^{n+1}) - \lambda\frac{2}{3}\Delta t\, u_{k,2}^2(t^{n+1}),$$

$$u_{k,2}(t^{n+2/3}) = u_{k,2}(^{n+1}) - \frac{1}{3}\Delta t\, \frac{du_{k,2}}{dt}(t^{n+1}) = u_{k,2}(^{n+1}) - \lambda\frac{1}{3}\Delta t\, u_{k,2}^2(t^{n+1}).$$

- Three-level method is given as: Further, the three-level method is given as a combination of two two-level methods' means; we have:

$$\begin{aligned} \Delta u_{k+1,k,3}(t^{n+1}) &= u_{k+1,3}(t^{n+1}) - u_{k,3}(t^{n+1}) \\ &= (\lambda_2 + \lambda_4)\int_{t^n}^{t^{n+1}} \exp((\lambda_1 + \lambda_3)(t - s))\big(u_k^p(s) - u_{k-1}^p(s)\big)\, ds, \end{aligned} \tag{60}$$

$$u_{k+1,3}(t^{n+1}) = u_{k,3}(t^{n+1}) + \Delta u_{k+1,k,3}(t^{n+1}), \quad k = 0, 1, 2, \dots, \tag{61}$$

where $u_0(t) = u(t^n)$ and for $u_{1,3}(t) = u_{1,2}(t)$, we have the two-level method. Further, we apply the trapezoidal or Simpson's rule for the numerical integration of the integrals. Further, for the next time step, we have $u(t^{n+1}) = u_{K,3}(t^{n+1})$ and K is the number of iterative steps, e.g. $K = 4$.

$$u_{k+1}(t^{n+1}) = u_k(t^{n+1}) + \int_{t^n}^{t^{n+1}} \exp((\lambda_1 + \lambda_3)(t - s))\big((\lambda_2 + \lambda_4)u_k^p(s) - (\lambda_2 + \lambda_4)u_{k-1}^p(s)\big)\, ds, \quad k = 0, 1, 2, \ldots, \tag{62}$$

where $u_0(t) = u(t^n)$ and

We apply the trapezoidal or Simpson's rule for the numerical integration of the integrals.

$$u_1(t) = \exp((\lambda_1 + \lambda_3)(t - t^n))u(t^n) + \int_{t^n}^{t} \exp((\lambda_1 + \lambda_3)(t - s))(\lambda_2 + \lambda_4)u_0^p(s)\, ds$$

Remark 8 If we apply only the trapezoidal rule, we obtain less accurate results while we skip the term $\exp((\lambda_1 + \lambda_3)(\Delta t))(\lambda_2 + \lambda_4)(u_k^p(t^n) - u_{k-1}^p(t^n)) = 0$, such that it is important to have with $k = 2$ a third-order integration method, e.g. the Simpson's rule.

Based on the reference solution, we deal with the following errors:

$$E_{L_1,k,\Delta t,[0,T]} = |u_{method,k,\Delta t}(t) - u_{ref,\Delta t}(t)| = \sum_{n=0}^{N} \Delta t |u_{method,k,\Delta t}(t^n) - u_{ref,\Delta t}(t^n)|, \tag{63}$$

where *method* is the different Picard's methods and k is the number of iterative steps.

Further, we deal with different time steps $\Delta t = \Delta t_{CFL}, \Delta t_{CFL}/2, \Delta t_{CFL}/4$.

A restriction based on the CFL condition is not necessary, while we have a numerical integration, but for more accurate results, this can help. The CFL condition is given as:

$$\Delta t \leq \frac{1}{|(\lambda_1 + \lambda_2) + (\lambda_3 + \lambda_4)u_0^{p-1}|} \tag{64}$$

where we assume $u^{n+1}(t) \leq u^n$, $\forall t \in [0, T]$ and u_0 is the initial solution.

For another accurate solution, we can integrate:

We have given:

$$u_{k+1}(t^{n+1}) = u_k(t^{n+1}) + \int_{t^n}^{t^{n+1}} \exp((\lambda_1 + \lambda_3)(t - s))\big((\lambda_2 + \lambda_4)u_k^p(s) - (\lambda_2 + \lambda_4)u_{k-1}^p(s)\big)\, ds \tag{65}$$

$$= u_k(t^{n+1}) + \Delta t \phi_1(\Delta t((\lambda_1 + \lambda_3)))g(t^n) + \frac{\Delta t^2}{2}\phi_2(\Delta t((\lambda_1 + \lambda_3)))g'(t^n) + O(\Delta t^3), \tag{66}$$

where

$$g(t^n) = (\lambda_2 + \lambda_4)\big(u_k^p(t^n) - u_{k-1}^p(t^n)\big)$$

$$g'(t^n) = (\lambda_2 + \lambda_4)\, p\, \big((\lambda_1 + \lambda_3)(u_k^p(t^n) - u_{k-1}^p(t^n)) \tag{67}$$

$$+ (\lambda_2 + \lambda_4)(u_k^{p-1}(t^n)u_{k-1}^p(t^n) - u_{k-1}^p(t^n)u_{k-2}^p(t^n))\big). \tag{68}$$

Further, we have:

$$\phi_1(hA) = \frac{1}{h}\int_0^h \exp((h-\tau)A)\, d\tau, \tag{69}$$

$$\phi_i(hA) = \frac{1}{h^i}\int_0^h \exp((h-\tau)A)\frac{\tau^{i-1}}{(i-1)!}\, d\tau,\ i \geq 1, \tag{70}$$

and we apply to our formulas:

$$\phi_1(\Delta t(\lambda_1+\lambda_3)) = \frac{1}{\Delta t(\lambda_1+\lambda_3)}(\exp(\Delta t(\lambda_1+\lambda_3))-1), \tag{71}$$

$$\phi_2(\Delta t(\lambda_1+\lambda_3)) = \frac{1}{(\Delta t(\lambda_1+\lambda_3))^2}(\exp(\Delta t(\lambda_1+\lambda_3))-1) - \frac{1}{\Delta t(\lambda_1+\lambda_3)}. \tag{72}$$

We apply the Picard's iteration and compared to the analytical solutions, we obtain the following result in Figure 5.

Remark 9 We saw marginal difference between the two- and three-level Picard's methods with different iterative steps $k = 1, 2, 5, 10$. At least, it is important to deal with more than $k \approx 3$ iterative steps. Further, it is necessary to deal with accurate numerical integration methods to save the accuracy of the multilevel nonlinear methods, meaning we applied at minimum third- or fourth-order accurate methods.

4.3. Multicomponent diffusion model

We deal with the plasma model that we presented in Section 2. The multicomponent diffusion model, which is nonlinear in the diffusion operator, solves with the multilevel Picard's method.

In the following, we present a three-step method, which is based on the idea of only updating the fix-point results.

To circumvent the exponential functions, we could apply for our nonlinear method a Picard's iterative method, which is based on the following idea.

The nonlinear equation is given as:

$$\partial_t \xi_i = -\nabla N_i(\xi_i), \qquad \xi_i(0) = \xi_{i,0}, \tag{73}$$

we apply the three-step method given as:

$$\xi_{i,k}(t) = \xi_{i,k-1}(t) + \int_0^t \left(-\nabla N_i(\xi_{i,k-1}(s)) + -\nabla N_i(\xi_{i,k-2}(s))\right) ds \tag{74}$$

where the initialization is given with the initial function $\xi_{i,0}(t) = \xi_i(0)$ and the first iteration is done previously.

The explicit form of the algorithm with an explicit Euler time discretization is given as:

Algorithm 3 *1.) Initialisation $k = 0$ with an explicit time-step (CFL condition is given):*

$$\begin{pmatrix} N_1^0 \\ N_2^0 \end{pmatrix} = \begin{pmatrix} \tilde{A} \ \tilde{B} \\ \tilde{C} \ \tilde{D} \end{pmatrix} \begin{pmatrix} -D_- \xi_1^0 \\ -D_- \xi_2^0 \end{pmatrix}, \tag{75}$$

where $\xi_1^0 = (\xi_{1,0}^0, \ldots, \xi_{1,J}^0)^T$, $\xi_2^0 = (\xi_{2,0}^0, \ldots, \xi_{2,J}^0)^T$ *and* $\xi_{1,j}^0 = \xi_1^{in}(j\Delta x)$, $\xi_{2,j}^0 = \xi_2^{in}(j\Delta x)$, $j = 0, \ldots, J$ *and given as for the different intializations, we have:*

1. Uphill example

$$\xi_1^{in}(x) = \begin{cases} 0.8 & if\, 0 \leq x < 0.25, \\ 1.6(0.75 - x) & if\, 0.25 \leq x < 0.75, \\ 0.0 & if\, 0.75 \leq x \leq 1.0, \end{cases} \tag{76}$$

$$\xi_2^{in}(x) = 0.2, \ for\ all\, x \in \Omega = [0, 1], \tag{77}$$

2. Diffusion example (Asymptotic behavior)

$$\xi_1^{in}(x) = \begin{cases} 0.8 \ if\, 0 \leq x \in 0.5, \\ 0.0 \ else, \end{cases}, \tag{78}$$

$$\xi_2^{in}(x) = 0.2, \ for\ all\, x \in \Omega = [0, 1], \tag{79}$$

The inverse matrices are given as:

$$\tilde{A}, \tilde{B}, \tilde{C}, \tilde{D} \in \mathbb{R}^{J+1} \times \mathbb{R}^{J+1}, \tag{80}$$

$$\tilde{A}_{j,j} = \gamma_j\left(\frac{1}{D_{23}} + \beta \xi_{1,j}^0\right), \ j = 0 \ldots, J, \tag{81}$$

$$B_{j,j} = \gamma_j \ \alpha \xi_{1,j}^0, \ j = 0 \ldots, J, \tag{82}$$

$$C_{j,j} = \gamma_j \ \beta \xi_{2,j}^0, \ j = 0 \ldots, J, \tag{83}$$

$$D_{j,j} = \gamma_j\left(\frac{1}{D_{13}} + \alpha \xi_{2,j}^0\right), \ j = 0 \ldots, J, \tag{84}$$

$$\gamma_j = \frac{D_{13} D_{23}}{1 + \alpha D_{13} \xi_{2,j}^0 + \beta D_{23} \xi_{1,j}^0}, \ j = 0 \ldots, J, \tag{85}$$

$$\tilde{A}_{i,j} = \tilde{B}_{i,j} = \tilde{C}_{i,j} = \tilde{D}_{i,j} = 0, \ i, j = 0 \ldots, J, \ i \neq J. \tag{86}$$

Further the values of the first and the last grid points of N are zero, means $N_{1,0}^0 = N_{1,J}^0 = N_{2,0}^0 = N_{2,J}^0 = 0$ (boundary condition).

2.) Next timesteps (till $n = N_{end}$) (iterative scheme restricted via the CFL condition based on the previous iterative solutions in the matrices):

2.1) Computation of $k = 1$ *with* $\xi_1^{n+1,1}$ *and* $\xi_2^{n+1,1}$*:*

$$\xi_1^{n+1,1} = \xi_1^n - \Delta t \; D_+ N_1^{n+1,0}, \tag{87}$$

$$\xi_2^{n+1,1} = \xi_2^n - \Delta t \; D_+ N_2^{n+1,0}, \tag{88}$$

2.2) Computation of $N_1^{n+1,1}$ *and* $N_2^{n+1,1}$*:*

$$\begin{pmatrix} N_1^{n+1,1} \\ N_2^{n+1,1} \end{pmatrix} = \begin{pmatrix} \tilde{A}^{n+1,1} & \tilde{B}^{n+1,1} \\ \tilde{C}^{n+1,1} & \tilde{D}^{n+1,1} \end{pmatrix} \begin{pmatrix} -D_- \xi_1^{n+1,1} \\ -D_- \xi_2^{n+1,1} \end{pmatrix}, \tag{89}$$

where $\xi_1^n = (\xi_{1,0}^n, \ldots, \xi_{1,J}^n)^T$, $\xi_2^n = (\xi_{2,0}^n, \ldots, \xi_{2,J}^n)^T$,

2.3.1) 2 step-Method: Computation of $k = 1$ *(with* $k = 0$*)* *(*$\xi_1^{n+1,k}$, $\xi_2^{n+1,k}$ *from* $\xi_1^{n+1,k-1}$, $\xi_2^{n+1,k-1}$*):*

$$\xi_1^{n+1,k} = \xi_1^n - \Delta t \; D_+ N_1^{n+1,k-1}, \tag{90}$$

$$\xi_2^{n+1,k} = \xi_2^n - \Delta t \; D_+ N_2^{n+1,k-1}, \tag{91}$$

2.3.2) 3 step-Method: Computation of $k = 2$ *(with* $k = 1$, $k = 0$*)* *(*$\xi_1^{n+1,k}$, $\xi_2^{n+1,k}$ *from* $\xi_1^{n+1,k-1}$, $\xi_2^{n+1,k-1}$, $\xi_1^{n+1,k-2}$, $\xi_2^{n+1,k-2}$*):*

$$\xi_1^{n+1,k} = \xi_1^{n+1,k-1} - \Delta t \; D_+ N_1^{n+1,k-1} + \Delta t \; D_+ N_1^{n+1,k-2}, \tag{92}$$

$$\xi_2^{n+1,k} = \xi_2^{n+1,k-1} - \Delta t \; D_+ N_2^{n+1,k-1} + \Delta t \; D_+ N_2^{n+1,k-2}, \tag{93}$$

2.3.3) 4 step-Method: Computation of $k \geq 3$ *(with* $k-1$, $k-2$, $k-3$*)* *(*$\xi_1^{n+1,k}$, $\xi_2^{n+1,k}$
from $\xi_1^{n+1,k-1}$, $\xi_2^{n+1,k-1}$, $\xi_1^{n+1,k-2}$, $\xi_2^{n+1,k-2}$, $\xi_1^{n+1,k-3}$, $\xi_2^{n+1,k-3}$*):*

$$\begin{aligned} \xi_1^{n+1,k} &= 2\xi_1^{n+1,k-1} - \xi_1^{n+1,k-2} \\ &-\Delta t \; D_+ N_1^{n+1,k-1} + 2\Delta t \; D_+ N_1^{n+1,k-2} - \Delta t \; D_+ N_1^{n+1,k-3}, \end{aligned} \tag{94}$$

$$\begin{aligned} \xi_2^{n+1,k} &= 2\xi_2^{n+1,k-1} - \xi_2^{n+1,k-2} \\ &-\Delta t \; D_+ N_2^{n+1,k-1} + 2\Delta t \; D_+ N_2^{n+1,k-2} - \Delta t \; D_+ N_2^{n+1,k-3}, \end{aligned} \tag{95}$$

2.4) Computation of $N_1^{n+1,k}$ *and* $N_2^{n+1,k}$*:*

$$\begin{pmatrix} N_1^{n+1,k} \\ N_2^{n+1,k} \end{pmatrix} = \begin{pmatrix} \tilde{A}^{n+1,k} & \tilde{B}^{n+1,k} \\ \tilde{C}^{n+1,k} & \tilde{D}^{n+1,k} \end{pmatrix} \begin{pmatrix} -D_- \xi_1^{n+1,k} \\ -D_- \xi_2^{n+1,k} \end{pmatrix}, \tag{96}$$

where $\xi_1^n = (\xi_{1,0}^n, \ldots, \xi_{1,J}^n)^T$, $\xi_2^n = (\xi_{2,0}^n, \ldots, \xi_{2,J}^n)^T$.

Further the values of the first and the last grid points of N *are zero, means* $N_{1,0}^{n+1} = N_{1,J}^{n+1} = N_{2,0}^{n+1} = N_{2,J}^{n+1} = 0$ *(boundary condition).*

Further $\xi_1^{n+1,0} = (\xi_{1,0}^n, \ldots, \xi_{1,J}^n)^T$, $\xi_2^{n+1,0} = (\xi_{2,0}^n, \ldots, \xi_{2,J}^n)^T$ *and* $I_J \in \mathbb{R}^{J+1} \times \mathbb{R}^{J+1}$ *is the start solution given with the solution at* $t = t^n$.

Repeat 2.3.) and 2.4.) with $k = 2, 3, \ldots, K$ *and* K *is the maximal iteration index.*

3.) Do $n = n + 1$ *and goto 2.)*

The computation of the inverse matrices is given as:

$$\tilde{A}^{n+1,k-1}, \tilde{B}^{n+1,k-1}, \tilde{C}^{n+1,k-1}, \tilde{D}^{n+1,k-1} \in \mathbb{R}^{J+1} \times \mathbb{R}^{J+1}, \tag{97}$$

$$\tilde{A}_{j,j}^{n+1,k-1} = \gamma_j \left(\frac{1}{D_{23}} + \beta \xi_{1,j}^{n+1,k-1} \right), j = 0 \ldots, J, \tag{98}$$

$$B_{j,j}^{n+1,k-1} = \gamma_j \, \alpha \xi_{1,j}^{n+1,k-1}, j = 0 \ldots, J, \tag{99}$$

$$C_{j,j}^{n+1,k-1} = \gamma_j \, \beta \xi_{2,j}^{n+1,k-1}, j = 0 \ldots, J, \tag{100}$$

$$D_{j,j}^{n+1,k-1} = \gamma_j \left(\frac{1}{D_{13}} + \alpha \xi_{2,j}^{n+1,k-1} \right), j = 0 \ldots, J, \tag{101}$$

$$\gamma_j = \frac{D_{13} D_{23}}{1 + \alpha D_{13} \xi_{2,j}^{n+1,k-1} + \beta D_{23} \xi_{1,j}^{n+1,k-1}}, j = 0 \ldots, J, \tag{102}$$

$$\tilde{A}_{i,j}^{n+1,k-1} = \tilde{B}_{i,j}^{n+1,k-1} = \tilde{C}_{i,j}^{n+1,k-1} = \tilde{D}_{i,j}^{n+1,k-1} = 0, \tag{103}$$

$i,j = 0 \ldots, J,\ i \neq J.$

The numerical errors of the different schemes are given in Figure 6.

The numerical solution of the three- and four-level Picard's fix-point schemes with different iterative steps is given in Figure 7.

The numerical errors of the four-level Picard's fix-point schemes compared with a reference solution of fine time steps of a three-level method (see Figure 8).

The solutions of the numerical experiments are given in Figure 9.

Remark 10 We tested all different level properties, meaning three-level and four-level Picard's fix-point schemes, with different iterative step sizes, meaning $k = 3, 4, 5, 10$. For all applications, we saw only marginal differences, such that a three-level method is sufficient to resolve the nonlinear problem.

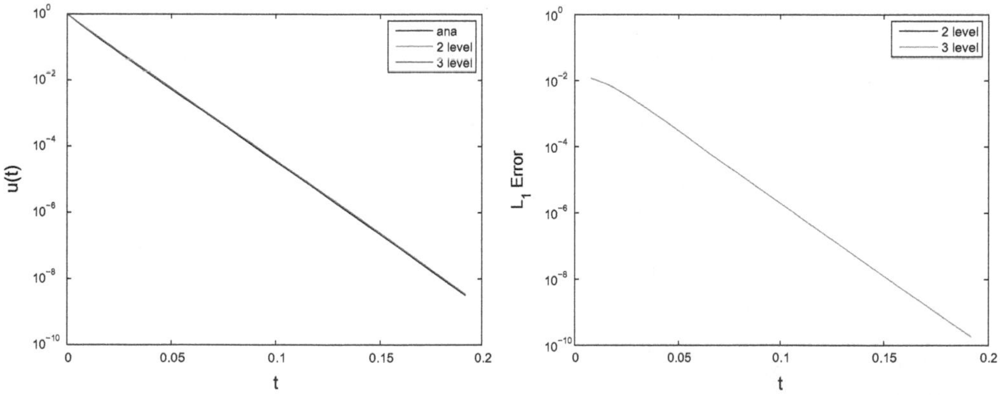

Figure 5. The three-level Picard's method compared with the analytical solution (left side: solutions, right side: errors between analytical and numerical methods).

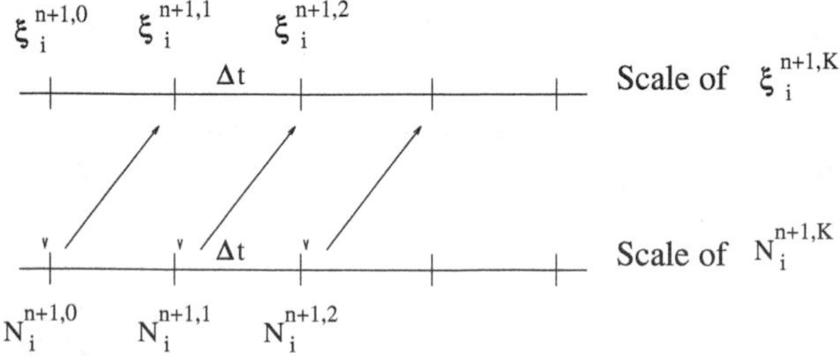

Figure 6. The numerical error of general Picard's fix-point method and a reference solution (fourth-order RK-method with fine time steps).

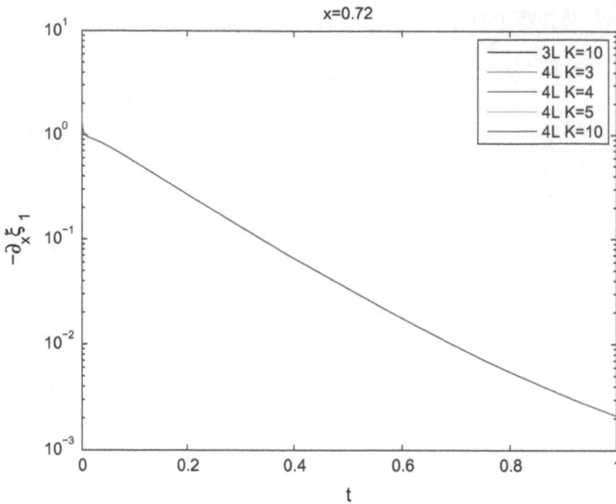

Figure 7. The numerical solution of the three- and four-level Picard's fix-point methods with different iterative steps.

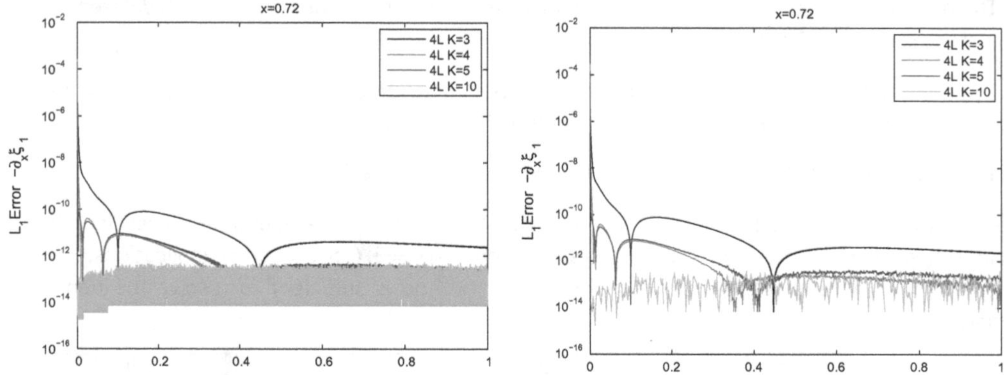

Figure 8. The figures present the numerical L_1-error between the four-level Picard's fix-point scheme and the reference solution for different iterative steps (left figure: high resolution of the error, right figure: low resolution based on averaging the high oscillating errors).

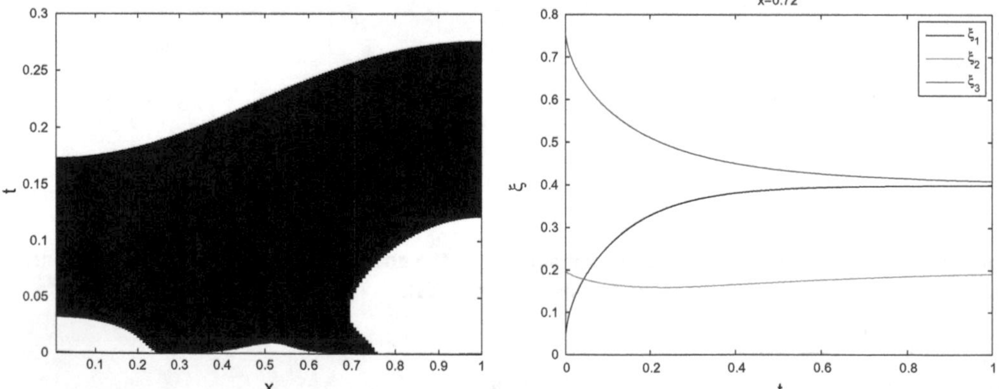

Figure 9. The figures present the numerical solutions of the uphill experiment (left side) and the concentrations at spatial point $x = 0.72$ with a four-level Picard's fix-point scheme computed (right side).

5. Conclusions and discussion

We present the coupled model for a multicomponent transport model, which can be applied for solving nonlinear diffusion equations, e.g. for multicomponent plasma transport models. The Picard's methods are flexible and we could derive multilevel methods, with exponential treatments. The benefit of the methods is to resolve the nonlinearity with more accuracy based on the intermediate levels. Here, we present the improvements based on the blow-up problems and the Bernoulli's equation. In multidiffusion applications, we saw the benefit of such methods in a fast resolution of the nonlinear diffusion and a decomposition on slow and fast time scales. Overall, the flexibility of such Picard's methods is important to solve multicomponent problems.

Funding
The author received no direct funding for this research.

Author details
Juergen Geiser[1]
E-mail: juergen.geiser@ruhr-uni-bochum.de
ORCID ID: http://orcid.org/0000-0003-1093-0001
[1] Department of Electrical Engineering and Information Technology, Ruhr University of Bochum, Universitätsstrasse 150, D-44801 Bochum, Germany.

References

Geiser, J. (2008). Iterative operator-splitting methods with higher order time-integration methods and applications for parabolic partial differential equations. *Journal of Computational and Applied Mathematics, 217*, 227–242.

Hirschfelder, J. O., Curtiss, Ch. F., & Bird, R. B. (1966). *Molecular theory of gases and liquids (Chemistry)* (1st ed.). New York, NY: Wiley.

Igitkhanov, Y. (2011). *Modelling of multi-component plasma for TOKES* (KIT Scientific Report Nr. 7564). Karlsruhe: Institut für Hochleistungsimpuls- und Mikrowellentechnik (IHM), KIT Scientific Publishing.

Ramos, J. I. (2008). On the variational iteration method and other iterative techniques for nonlinear differential equations. *Applied Mathematics and Computation, 199*, 39–69.

Ramos, J. I. (2009). Picard's iterative method for nonlinear advection reaction-diffusion equations. *Applied Mathematics and Computation, 215*, 1526–1536.

Senega, T. K., & Brinkmann, R. P. (2006). A multi-component transport model for non-equilibrium low-temperature low-pressure plasmas. *Journal of Physics D: Applied Physics, 39*, 1606–1618.

Tanaka, Y. (2004). Two-temperature chemically non-equilibrium modelling of high-power Ar-N2 inductively coupled plasmas at atmospheric pressure. *Journal of Physics D: Applied Physics, 37*, 1190–1205.

10

Dynamics of a plant–herbivore model with differential–difference equations

S. Kartal[1]*

*Corresponding author: S. Kartal, Faculty of Education, Department of Mathematics, Nevsehir Haci Bektas Veli University, Nevsehir 50300, Turkey

E-mail: senol.kartal@nevsehir.edu.tr

Reviewing editor: Amar Debbouche, Guelma University, Algeria

Abstract: This paper studies the behavior of a plant–herbivore model including both differential and difference equations. To analyze global behavior of the model, we consider the solution of the system in a certain subinterval which gives to system of difference equations. The boundedness characters, the periodic nature, both local and global stability conditions of the plant–herbivore system are investigated. Numerical studies indicate that the system exhibits Neimark–Sacker bifurcation for different parameter values in certain regions.

Subjects: Applied Mathematics; Dynamical Systems; Mathematical Modeling; Mathematics & Statistics; Science

Keywords: plant–herbivore system; difference equation; stability; Neimark–Sacker bifurcation

Mathematics subject classifications: 37N25; 39A28; 39A30

1. Introduction

Classical approaches to modeling plant–herbivore interactions are based on predator–prey system (Caughley & Lawton, 1981; May, 2001). This interaction has been described in much research using discrete and continuous model (Agiza, ELabbasy, EL-Metwally, & Elsadany, 2009; Chattopadhayay, Sarkar, Fritzsche-Hoballah, Turlings, & Bersier, 2001; Danca, Codreanu, & Bakó, 1997; Das & Sarkar, 2001; Edelstein-Keshet, 1986; Feng, Qiu, Liu, & DeAngelis, 2011; Lebon, Mailleret, Dumont, & Grognard, 2014; Li, 2011; Liu, Feng, Zhu, & DeAngelis, 2008; Mukherjee, Das, & Kesh, 2011; Ortega-Cejas, Fort, & Méndez, 2004; Owen-Smith, 2002; Saha & Bandyopadhyay, 2005; Sui, Fan, Loladze, & Kuang, 2007; Sun, Chakraborty, Liu, Jin, & Anderson, 2014; Zhao, Feng, Zheng, & Cen, 2015). The model of Li (2011) is a system of differential equations with Holling type II functional response where the plant toxin's influence in herbivores is considered. In study, Mukherjee et al. (2011) have used discrete time model with Holling type II functional response for describing the plant–herbivore interaction.

ABOUT THE AUTHOR

Senol Kartal is an associate professor in Nevsehir Haci Bektas Veli University, Turkey. His PhD thesis is about the population dynamics (Modeling of Tumor Immune System Dynamics Using System of Difference Equations and Its Stability Analysis). His research interests are related to population dynamics, discrete and continuous dynamical system, and bifurcation theory. He has published his research contributions in some internationally renowned journals whose publishers are Elsevier, Taylor & Francis, Wiley, and other journals.

PUBLIC INTEREST STATEMENT

In this study, we consider a plant–herbivore model which consists of ordinary differential equations. Our aim is to build a better understanding of how both discrete and continuous times affect the dynamic behavior of plant–herbivore interactions. Therefore, we add discrete time to this model and obtain a system of differential equations with piecewise constant arguments which gives system of difference equations. The boundedness characters, the periodic nature, both local and global stability conditions of the system are investigated.

It is well known that discrete time models governed by difference equations are more appropriate than the continuous time models when the populations have non-overlapping generations. So, a significant number of the study on the mathematical models of plant–herbivore interactions are described by the system of difference equations (Agiza et al., 2009; Danca et al., 1997; Mukherjee et al., 2011; Sui et al., 2007). In addition, working with difference equations instead of differential equations allows us to some advantages. Discrete dynamical models can bring about easier computational methods for the persistence, periodic solutions, boundedness, local and global properties of the dynamical system.

In plant–herbivore interactions, delay differential equations may widely occur due to herbivore damage and deployment of inducible defenses (Das & Sarkar, 2001; Ortega-Cejas et al., 2004; Sun et al., 2014). From this point of view, Sun et al. (2014) and et all have constructed a reaction-diffusion model with delay governed by system of partial differential equations where the effect of time delay on the herbivore cycles is investigated. In addition, the properties of delay differential equations are very close to differential equation with piecewise constant arguments. In Cooke and Györi study (1994), it was pointed out that these equations can be used to get approximate solutions to delay differential equations that include discrete delays. In such biological situations, dynamics of growth and death of populations can be described by differential equations otherwise, difference equations may reflect the interaction of two populations such as competition or predation phenomena (Gurcan, Kartal, Ozturk, & Bozkurt, 2014; Kartal & Gurcan, 2015). In the literature, various types of biological model consisting of differential equations with piecewise constant arguments have been analyzed using the method of reduction to discrete equations (Busenberg & Cooke, 1982; Gopalsamy & Liu, 1998; Gurcan et al., 2014; Kartal & Gurcan, 2015; Liu & Gopalsamy, 1999; Öztürk, Bozkurt, & Gurcan, 2012).

In the present paper, our aim is to build a better understanding of how both discrete and continuous times affect the dynamic behavior of plant–herbivore interactions. So we will reconsider the model (see Chattopadhayay et al., 2001)

$$\begin{cases} \frac{dx}{dt} = rx(t)\left(1 - \frac{x(t)}{K}\right) - \alpha x(t)y(t), \\ \frac{dy}{dt} = -sy(t) + \beta x(t)y(t), \end{cases} \tag{1.1}$$

as a system of differential equations with piecewise constant arguments such as

$$\begin{cases} \frac{dx}{dt} = rx(t)\left(1 - \frac{x(t)}{K}\right) - \alpha x(t)y([\![t]\!]), \\ \frac{dy}{dt} = -sy(t) + \beta x([\![t]\!])y(t), \end{cases} \tag{1.2}$$

which include both differential and difference equations. In this model, $x(t)$ and $y(t)$ represent the density of plant and herbivore population, respectively, $[\![t]\!]$ denotes the integer part of $t \in [0, \infty)$ and all these parameters are positive. The parameter r, K, and α is the intrinsic growth rate, environmental carrying capacity, and specific predation rate of plant species, respectively. s represents the death rate of herbivores and β is the conversion factor of herbivores (Chattopadhayay et al., 2001). The logistic term $rx(t)\left(1 - \frac{x(t)}{K}\right)$ and the term $sy(t)$ include only a continuous time for the growth of plant and for the death of herbivore, respectively. The predational form $\alpha x(t)y([\![t]\!])$ represent the loss of plant population and $\beta x([\![t]\!])y(t)$ is conversion factor of herbivores which include both discrete and continuous time for a each populations. So the plant–herbivore interaction is considered in a certain subinterval and is modeled using a system of differential equations with piecewise constant arguments.

2. Local and global stability analysis

System (1.2) can be written an interval $t \in [n, n+1)$ as follows:

$$\begin{cases} \frac{dx}{dt} - x(t)(r - \alpha y(n)) = -rkx^2(t), \\ \frac{dy}{y(t)} = ((\beta x(n) - s))dt, \end{cases} \tag{2.1}$$

where $\frac{1}{K} = k$.

By solving each equations of the system (2.1) and letting $t \to n+1$, we obtain a system of difference equations

$$\begin{cases} x(n+1) = \frac{x(n)(r-\alpha y(n))}{(r-\alpha y(n)-rkx(n))e^{-(r-\alpha y(n))}+rkx(n)}, \\ y(n+1) = y(n)e^{\beta x(n)-s}, \end{cases} \tag{2.2}$$

System (2.2) reflects the dynamical behavior of the system of differential equations with piecewise constant arguments. So we will consider the system of difference equation to analyze the global behavior of system (1.2).

The equilibrium points of system (2.2) can be obtained as

$$E_0 = (0,0), \quad E_1 = \left(\frac{1}{k}, 0\right), \quad E_* = \left(\frac{s}{\beta}, \frac{r}{\alpha}\left(1 - \frac{ks}{\beta}\right)\right).$$

We note that the positive equilibrium of the system exists if $\beta > ks$. Now, we will find Jacobian matrix of the system to investigate the dynamic behavior of the model.

THEOREM 2.1. *The equilibrium points E_0 and E_1 are saddle point.*

Proof At the equilibrium point E_0, the Jacobian matrix is the form

$$J_0 = \begin{pmatrix} e^r & 0 \\ 0 & e^{-s} \end{pmatrix}.$$

The matrix J_0 has eigenvalues $\lambda_1 = e^r, \lambda_2 = e^{-s}$. Hence $\lambda_1 > 1$ and $\lambda_2 < 1$ and consequently E_0 is saddle point. On the other hand, the Jacobian matrix J_1 at the point E_1 is

$$J_1 = \begin{pmatrix} e^{-r} & -\frac{\alpha - e^{-r}\alpha}{kr} \\ 0 & e^{-s+\frac{\beta}{k}} \end{pmatrix}$$

which gives eigenvalues $\lambda_1 = e^{-r}$ and $\lambda_2 = e^{-s+\frac{\beta}{k}}$. Considering the condition $\beta > ks$, we can say that E_1 is saddle point.

On the other hand, the Jacobian matrix J_* at the positive equilibrium point E_* is

$$J_* = \begin{pmatrix} e^{-kr\bar{x}} & \frac{(-1+e^{-kr\bar{x}})\alpha}{kr} \\ \beta\bar{y} & 1 \end{pmatrix}$$

which yields the following characteristic equation

$$p(\lambda) = \lambda^2 + \lambda\left(-1 - e^{-kr\bar{x}}\right) + e^{-kr\bar{x}} + \frac{(1 - e^{-kr\bar{x}})\alpha\beta\bar{y}}{kr} = 0.$$

Now, we can apply Schur-Cohn criterion to determine stability conditions of the system with characteristic equation $p(\lambda)$.

THEOREM 2.2 *The positive equilibrium point E_* of system (2.2) is local asymptotically stable if and only if*

$ks < \beta < k + ks$.

Proof From the Schur-Cohn criterion, E_* is local asymptotically stable if and only if

(a) $p(1) = 1 + e^{-kr\bar{x}} + \frac{(1 - e^{-kr\bar{x}})\alpha\beta\bar{y}}{kr} - 1 - e^{-kr\bar{x}} < 0,$

(b) $p(-1) = 1 + e^{-kr\bar{x}} + \dfrac{(1 - e^{-kr\bar{x}})\alpha\beta\bar{y}}{kr} + 1 + e^{-kr\bar{x}} < 0,$

(c) $D_1^+ = 1 + e^{-kr\bar{x}} + \dfrac{\left(1 - e^{-kr\bar{x}}\right)\alpha\beta\bar{y}}{kr} < 0,$

(d) $D_1^- = 1 - e^{-kr\bar{x}} - \dfrac{\left(1 - e^{-kr\bar{x}}\right)\alpha\beta\bar{y}}{kr} < 0.$

The condition (a), (b), and (c) gives the inequalities

$$p(1) = \frac{\left(1 - e^{-kr\bar{x}}\right)\alpha\beta\bar{y}}{kr} < 0, \tag{2.3}$$

$$p(-1) = 2 + 2e^{-kr\bar{x}} + \frac{\left(1 - e^{-kr\bar{x}}\right)\alpha\beta\bar{y}}{kr} < 0 \tag{2.4}$$

and

$$D_1^+ = 1 + e^{-kr\bar{x}} + \frac{\left(1 - e^{-kr\bar{x}}\right)\alpha\beta\bar{y}}{kr} < 0 \tag{2.5}$$

which always hold under the condition $\beta > ks$. From (d), we get

$$e^{-kr\bar{x}} + \frac{\left(1 - e^{-kr\bar{x}}\right)\alpha\beta\bar{y}}{kr} < 1$$

which reveal

$\beta < k + ks.$

This completes the proof. □

For the parameter values $r = 0.2, \alpha = 0.6, K = 5, \beta = 0.01, s = 0.02$ and using initial conditions $x(1) = 0.2, y(1) = 0.22$, it can be seen that the positive equilibrium point $(\bar{x}, \bar{y}) = (2, 0.2)$ is local asymptotically stable, where blue and red graphs represent population density of plant and herbivore population, respectively.

THEOREM 2.3 *Let* $\{x(n), y(n)\}_{n=-1}^{\infty}$ *be a positive solution of system (2.2); then*

$$x(n) \le \frac{e^r}{k(e^r - 1)}.$$

In addition, if $y(n) < x(n)$, then $y(n) \le \frac{e^r}{k(e^r-1)} e^{\frac{\beta e^r}{k(e^r-1)} - s}$.

Proof It can be easily seen that

$$x(n+1) = \frac{x(n)[r - \alpha y(n)]e^{r-\alpha y(n)}}{r - \alpha y(n) + rkx(n)(e^{r-\alpha y(n)} - 1)} \le \frac{[r - \alpha y(n)]e^{r-\alpha y(n)}}{rk(e^{r-\alpha y(n)} - 1)} \le \frac{e^r}{k(e^r - 1)}.$$

Also, it can be shown that $y(n+1) \le \frac{e^r}{k(e^r-1)} e^{\frac{\beta e^r}{k(e^r-1)} - s}$ under the condition $y(n) < x(n)$.

THEOREM 2.4 *The system has no prime period-two solutions.*

Proof On the contrary, suppose that the system (2.2) has a distinctive prime period-two solutions

$$\ldots, (w_1, q_1), (w_2, q_2), (w_1, q_1), (w_2, q_2) \ldots.$$

where $w_1 \neq w_2$, and $q_1 \neq q_2$, and w_i, q_i are positive real numbers for $i \in \{1, 2\}$. Then, from system (2.2) one has

$$\begin{cases} w_1 = \frac{w_2[r-\alpha q_2]}{[r-\alpha q_2-rkw_2]e^{-[r-\alpha q_2]}+rkw_2}, \\ q_1 = q_2 e^{\beta w_2 - s} \\ w_2 = \frac{w_1[r-\alpha q_1]}{[r-\alpha q_1-rkw_1]e^{-[r-\alpha q_1]}+rkw_1}, \\ q_2 = q_1 e^{\beta w_1 - s}, \end{cases}$$

Since $q_1 \neq q_2$, we have $\beta w_2 - s \neq 0$ and $\beta w_1 - s \neq 0$. From the second and last equation in the system, we have

$$q_1^2 = q_2^2 e^{\beta w_2 - s - \beta w_1 + s}.$$

If q_2 is written the above equation, we hold

$$q_1^2 = q_1^2 e^{\beta w_1 - s + \beta w_2 - s}.$$

This equation must satisfy

$$\beta w_1 - s + \beta w_2 - s = 0$$

which is a contradiction $\beta w_2 - s \neq 0$ and $\beta w_1 - s \neq 0$.

THEOREM 2.5 *Let $A_1 = r - \alpha y(n)$ and $A_2 = \beta x(n) - s$. Suppose that the conditions of Theorem 2.1 hold and*

(i) Let $y(n) < \frac{r}{\alpha}$ and $\bar{x} < \frac{A_1(1+e^{-A_1})}{2rke^{-A_1}}$ for $x(n) \in \left(0, \frac{2\bar{x}e^{-A_1}}{1+e^{-A_1}}\right)$,

(ii) Let $y(n) > \frac{r}{\alpha}$ and $\bar{x} > \frac{-A_1(1+e^{-A_1})}{2rk(e^{-A_1}-1)}$ for $x(n) \in \left(\frac{2\bar{x}e^{-A_1}}{1+e^{-A_1}}, 2\bar{x}\right)$,

(iii) Let $y(n) > \frac{r}{\alpha}$ for $x(n) \in (2\bar{x}, \infty)$,

(iv) Let $A_2 > 0$ for $y(n) \in \left(0, \frac{2\bar{y}}{1+e^{A_2}}\right)$,

(v) Let $A_2 < 0$ for $y(n) \in \left(\frac{2\bar{y}}{1+e^{A_2}}, \infty\right)$.

Then the positive equilibrium point of system (2.2) is global asymptotically stable.

Proof We define a Lyapunov function as

$$V(n) = [q(n) - \bar{q}]^2, \quad n = 0, 1, 2 \ldots$$

where $\bar{q} = (\bar{x}, \bar{y})$ is positive equilibrium point of system (2.2).

The change along the solutions of the system is

$$\Delta V(n) = V(n+1) - V(n) = \{q(n+1) - q(n)\}\{q(n+1) + q(n) - 2\bar{q}\}.$$

From the first equation in (2.2), we hold;

$$\begin{aligned}\Delta V_1(n) &= [x(n+1) - x(n)][x(n+1) + x(n) - 2\bar{x}] \\ &= x(n)[(A_1 - rkx(n))(1 - e^{-A_1})][A_1(x(n) + x(n)e^{-A_1} - 2\bar{x}e^{-A_1}) \\ &\quad + rkx(n)(x(n) - 2\bar{x})(1 - e^{-A_1})].\end{aligned}$$

By considering (i), (ii), and (iii), we have $\Delta V_1(n) < 0$. These imply that $\lim_{n\to\infty} x(n) = \bar{x}$. Additionally, we can show that $\Delta V_2(n) < 0$ which gives $\lim_{n\to\infty} y(n) = \bar{y}$.

3. Bifurcation analysis

In this section, we investigate existence of stationary bifurcation (fold, transcritical, and pitchfork bifurcation), period doubling bifurcation, and Neimark-Sacker bifurcation for the system (2.2). All of these bifurcations can be analyzed under the set of algebraic conditions that is called Schur-Cohn criterion. It is well known that the system may undergo stationary bifurcation if and only if $p(1) = 0$, $p(-1) > 0$, $D_1^+ > 0$ and $D_1^- > 0$. On the other hand, inequalities $p(1) > 0$, $p(-1) = 0$, $D_1^+ > 0$ and $D_1^- > 0$ give the conditions of period doubling bifurcation. But considering (2.3) and (2.4), it is easily seen that these conditions do not hold for the system. Therefore, stationary bifurcation and period doubling bifurcation do not exist for the system.

Now, we can investigate the existence of Neimark-Sacker bifurcation for the plant–herbivore model (Hone, Irle, & Thurura, 2010). The algebraic condition of Neimark-Sacker bifurcation can be obtained from the analysis of inequalities $p(1) > 0$, $p(-1) > 0$, $D_1^+ > 0$ and $D_1^- = 0$. In local stability analysis, we have already shown that the inequalities $p(1) > 0$, $p(-1) > 0$, $D_1^+ > 0$ are always exist. Therefore, we will only analyse the equation $D_1^- = 0$ to determine Neimark-Sacker bifurcation condition.

THEOREM 3.1 *System (2.2) undergoes Neimark-Sacker bifurcation if and only if*

$$\bar{k} = \frac{\beta}{1+s}.$$

Proof This result comes from the analysis of $D_1^- = 0$.

Using the condition of Theorem 3.1 with the parameters given in Figure 1, we have the Neimark-Sacker bifurcation point as $\bar{K} = 102$ (Figure 2).

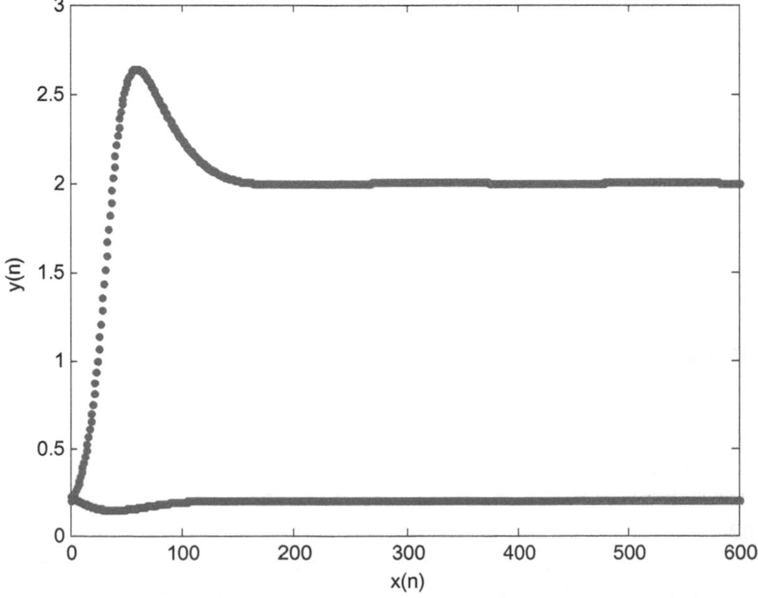

Figure 1. Stable equilibrium point of the system for $r = 0.2, \alpha = 0.6, K = 5, \beta = 0.01, = 0.02$, (1) = 0.2 and **(1) = 0.22**

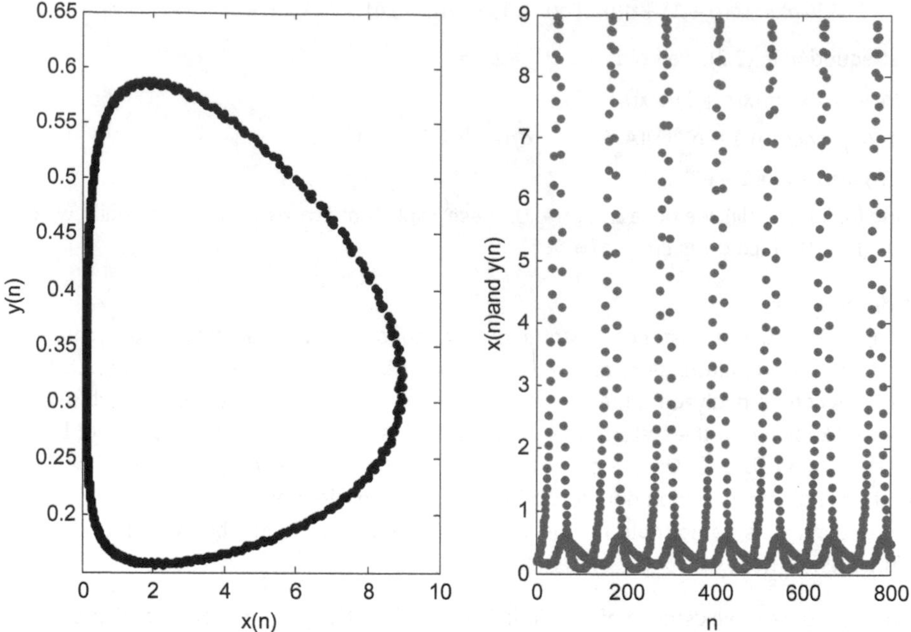

Figure 2. Stable limit cycle for $r = 0.2, \alpha = 0.6, \beta = 0.01, s = 0.02$ and $\bar{K} = 102$.

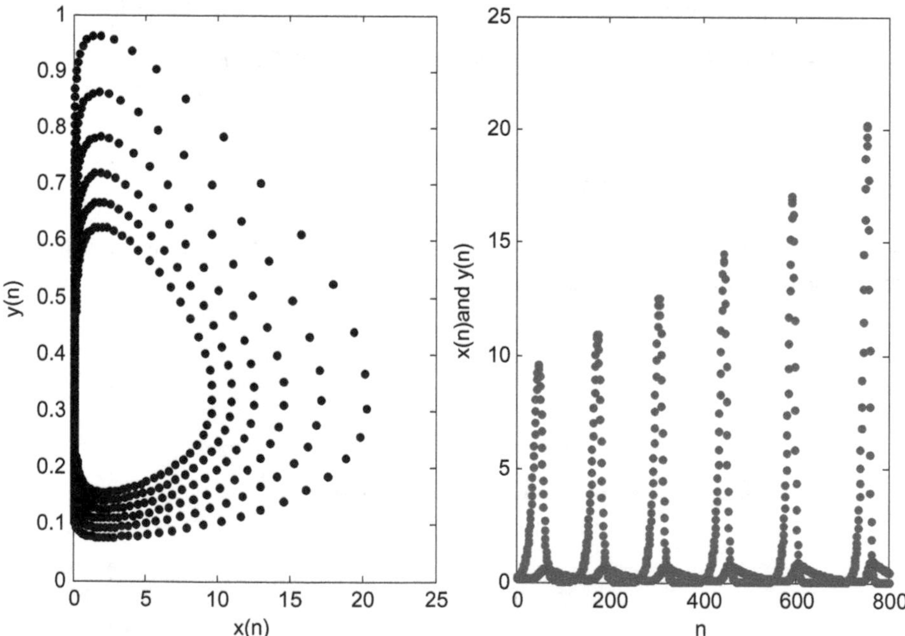

Figure 3. Graph of the iteration solution of system (2.2) for $r = 0.2$, $\alpha = 0.6$, $\beta = 0.01$, $s = 0.02$ and $K = 200$.

4. Result and discussion

In this paper, dynamics of a discrete-continuous time plant-herbivore model has been investigated. Local and global stability properties of the positive equilibrium point are analyzed. It is interesting to note that when conversion factor of herbivores becomes low then the system converges to a stable situation. On the other hand, we investigate possible bifurcation types for the system and observe that the system exhibits Neimark-Sacker bifurcation. This type of bifurcation has been observed in many plant-herbivore models (Liu et al., 2008; Saha & Bandyopadhyay, 2005; Zhao et al., 2015) and shows that periodic or quasi-periodic solutions occur as a result of a limit cycle.

In our manuscript, the parameter K (environmental carrying capacity of plant species) is determined as a bifurcation parameter. When the environmental carrying capacity of plant species reaches to $\bar{K} = 102$, the system enters a Neimark-Sacker bifurcation as a result of stable limit cycle (Figure 2). If K exceeds the $\bar{K}$, the system continues oscillatory behavior with growing amplitude (Figure 3). So we can say that the parameter K has a strong effect on the stability of the system so as to control two populations.

Funding
The author received no direct funding for this research.

Author details
S. Kartal[1]
E-mail: senol.kartal@nevsehir.edu.tr
[1] Faculty of Education, Department of Mathematics, Nevsehir Haci Bektas Veli University, Nevsehir 50300, Turkey.

References

Agiza, H. N., ELabbasy, E. M., EL-Metwally, H., & Elsadany, A. A. (2009). Chaotic dynamics of a discrete prey-predator model with Holling type II. *Nonlinear Analysis Real World Applications, 10*, 116–129. http://dx.doi.org/10.1016/j.nonrwa.2007.08.029

Busenberg, S., & Cooke, K. L. (1982). *Models of vertically transmitted diseases with sequential continuous dynamics. Nonlinear phenomena in mathematical sciences*. New York, NY: Academic Press.

Caughley, G., & Lawton, J. H. (1981). Plant-herbivore systems. In R. M. May (Ed.), *Theoretical ecology* (pp. 132–166). Sunderland: Sinauer Associates.

Chattopadhayay, J., Sarkar, R., Fritzsche-Hoballah, M. E., Turlings, T. C. J., & Bersier, L. F. (2001). Parasitoids may determine plant fitness—A mathematical model based on experimental data. *Journal of Theoretical Biology, 212*, 295–302. http://dx.doi.org/10.1006/jtbi.2001.2374

Cooke, K. L., & Győri, I. (1994). Numerical approximation of the solutions of delay-differential equations on an infinite interval using piecewise constant argument. *Computers & Mathematics with Applications, 28*, 81–92.

Danca, M., Codreanu, S., & Bakó, B. (1997). Detailed analysis of a nonlinear prey-predator model. *Journal of Biological Physics, 23*, 11–20. http://dx.doi.org/10.1023/A:1004918920121

Das, K., & Sarkar, A. K. (2001). Stability and oscillations of an autotroph-herbivore model with time delay. *International Journal of Systems Science, 32*, 585–590. http://dx.doi.org/10.1080/00207720117706

Edelstein-Keshet, L. E. (1986). Mathematical theory for plant-herbivore systems. *Journal of Mathematical Biology, 24*, 25–58. http://dx.doi.org/10.1007/BF00275719

Feng, Z., Qiu, Z., Liu, R., & DeAngelis, D. L. (2011). Dynamics of a plant–herbivore–predator system with plant-toxicity. *Mathematical Biosciences, 229*, 190–204. http://dx.doi.org/10.1016/j.mbs.2010.12.005

Gopalsamy, K., & Liu, P. (1998). Persistence and global stability in a population model. *Journal of Mathematical Analysis and Applications, 224*, 59–80. http://dx.doi.org/10.1006/jmaa.1998.5984

Gurcan, F., Kartal, S., Ozturk, I., & Bozkurt, F. (2014). Stability and bifurcation analysis of a mathematical model for tumor-immune interaction with piecewise constant arguments of delay. *Chaos Solitons & Fractals, 68*, 169–179.

Hone, A. N. W., Irle, M. V., & Thurura, G. W. (2010). On the Neimark-Sacker bifurcation in a discrete predator-prey system. *Journal of Biological Dynamics, 4*, 594–606. http://dx.doi.org/10.1080/17513750903528192

Kartal, S., & Gurcan, F. (2015). Stability and bifurcations analysis of a competition model with piecewise constant arguments. *Mathematical Methods in the Applied Sciences, 38*, 1855–1866. http://dx.doi.org/10.1002/mma.v38.9

Lebon, A., Mailleret, L., Dumont, Y., & Grognard, F. (2014). Direct and apparent compensation in plant-herbivore interactions. *Ecological Modelling, 290*, 192–203. http://dx.doi.org/10.1016/j.ecolmodel.2014.02.020

Li, Y. (2011). Toxicity impact on a plant-herbivore model with disease in herbivores. *Computers Mathematics with Applications, 62*, 2671–2680. http://dx.doi.org/10.1016/j.camwa.2011.08.012

Liu, P., & Gopalsamy, K. (1999). Global stability and chaos in a population model with piecewise constant arguments. *Applied Mathematics and Computation, 101*, 63–88. http://dx.doi.org/10.1016/S0096-3003(98)00037-X

Liu, R., Feng, Z., Zhu, H., & DeAngelis, D. L. (2008). Bifurcation analysis of a plant-herbivore model with toxin-determined functional response. *Journal of Differential Equations, 245*, 442–467. http://dx.doi.org/10.1016/j.jde.2007.10.034

May, R. M. (2001). *Stability and complexity in model ecosystems*. Princeton, NJ: Princeton University Press.

Mukherjee, D., Das, P., & Kesh, D. (2011). Dynamics of a plant-herbivore model with Holling type II functional response. *Computational and Mathematical Biology, 2*, 1–9.

Ortega-Cejas, V. O., Fort, J., & Méndez, V. (2004). The role of the delay time in the modeling of biological range expansions. *Ecology, 85*, 258–264. http://dx.doi.org/10.1890/02-0606

Owen-Smith, N. O. (2002). A metaphysiological modelling approach to stability in herbivore–vegetation systems. *Ecological Modelling, 149*, 153–178. http://dx.doi.org/10.1016/S0304-3800(01)00521-X

Öztürk, I., Bozkurt, F., & Gurcan, F. (2012). Stability analysis of a mathematical model in a microcosm with piecewise constant arguments. *Mathematical Biosciences, 240*, 85–91. http://dx.doi.org/10.1016/j.mbs.2012.08.003

Saha, T., & Bandyopadhyay M. (2005). Dynamical analysis of a plant-herbivore model: Bifurcation and global stability. *Journal of Applied Mathematics & Computing, 19*, 327–344.

Sui, G., Fan, M., Loladze, I., & Kuang, Y. (2007). The dynamics of a stoichiometric plant-herbivore model and its discrete analog. *Mathematical Biosciences Engineering, 4*, 29–46.

Sun, G. Q., Chakraborty, A., Liu, Q. X., Jin, Z., & Anderson, K. E. (2014). Influence of time delay and nonlinear diffusion on herbivore outbreak. *Communications in Nonlinear Science and Numerical Simulation, 19*, 1507–1518. http://dx.doi.org/10.1016/j.cnsns.2013.09.016

Zhao, Y., Feng, Z., Zheng, Y., & Cen, X. (2015). Existence of limit cycles and homoclinic bifurcation in a plant-herbivore model with toxin-determined functional response. *Journal of Differential Equations, 258*, 2847–2872. http://dx.doi.org/10.1016/j.jde.2014.12.029

11

A weak approximation for the Wiener–Hopf factorization

Amir T. Payandeh Najafabadi[1]* and Dan Z. Kucerovsky[2]

*Correspoding author: Amir T. Payandeh Najafabadi, Department of Mathematical Sciences, Shahid Beheshti University, G.C. Evin, Tehran 1983963113, Iran

E-mail: amirtpayandeh@sbu.ac.ir

Reviewing editor: Kok Lay Teo, Curtin University, Australia

Abstract: The Wiener–Hopf factorization plays a crucial role in studying various mathematical problems. Unfortunately, in many situations, the Wiener–Hopf factorization cannot provide closed form solutions and one has to employ some approximation techniques to find its solutions. This article provides several $L_p(R)$, $1 < p \leq 2$, approximation for a given Wiener-Hopf factorization problem. Application of our finding in spectral factorization and Lévy processes have been given.

Subjects: Applied Mathematics; Financial Mathematics; Mathematical Finance; Mathematics & Statistics; Science

Keywords: principal value integral; Hölder condition; Padé approximant; continued fraction; Fourier transform; Hilbert transform; Shannon sampling theorem; spectral factorization; Lévy processes

2010 Mathematics Subject classifications: 30E25; 11A55; 42A38; 60G51; 60j50; 60E10

1. Introduction

Roughly speaking, the Wiener–Hopf factorization problem is a technique to find a *single* complex-valued function Φ in which its radial limits, say $\Phi^{\pm}$, are respectively analytic and bounded separately in the upper and lower complex half planes (i.e. $C_+ := \{\lambda \in C : \Im(\lambda) \geq 0\}$ and $C_- := \{\lambda \in C : \Im(\lambda) \leq 0\}$) and satisfy $\Phi^+(\omega)\Phi^-(\omega) = g(\omega)$, where $\omega \in R$ and g is a zero index function which satisfies the Hölder condition.

The Wiener–Hopf factorization has proved remarkably useful in solving an enormous variety of model problems in a wide range of branches of physics, mathematics, and engineering. Subjects for which the problem is applicable range from neutron transport (Noble, 1988), geophysical fluid dynamics (Davis, 1987; Kaoullas & Johnson, 2010), diffraction theory (Noble, 1988), fracture mechanics (Freund, 1998; Shakib, Akhgarian, & Ghaderi, 2015), non-destructive evaluation of materials (Achenbach, 2012), a wide class of integral equations (Payandeh Najafabadi & Kucerovsky, 2009, 2014b), acoustics (Abrahams & Wickham, 1990), elasticity (Norris & Achenbach, 1984; Ogilat, 2013), electromagnetics (Daniele, 2014; Sautbekov & Nilsson, 2009), water wave phenomena (Chakrabarti & George, 1994; Kim, Schiavone, &

ABOUT THE AUTHOR

Amir T. Payandeh Najafabadi is an Associate Professor in Department of Mathematics sciences at Shahid Behashti University, Tehran, Evin (Email address: amirtpayandeh@ sbu.ac.ir). He was born on Sep 3, 1973. He received his PhD from University of New Brunswick, Canada in 2006. He has published 28 papers and was co-author of two books. His major research interests are: Statistical Decision Theory, Lévy processes, Risk theory, Riemann-Hilbert problem, & integral equations.

PUBLIC INTEREST STATEMENT

This article provides several weak approximation for a given Wiener–Hopf factorization problem. Application of our finding in spectral factorization and Lévy processes have been given.

Ru, 2011), geophysics (Davis, 1987), financial mathematics (Beheshti, Payandeh Najafabadi, & Farnoosh, 2013; Fusai, Abrahams, & Sgarra, 2006), distribution of extrema in a wide class of Lévy processes (Payandeh Najafabadi & Kucerovsky, 2011, in press), statistical decision problems (Kucerovsky, Marchand, Payandeh Najafabadi, & Strawderman, 2009), etc.

The key steps to solve a Wiener–Hopf factorization is decomposing of the kernel g into a product of two terms, g^+ and g^-, where g^+ and g^- are analytic and bounded in the upper and the lower complex half planes, respectively. Such decomposition can be expressed in terms of a Sokhotski–Plemelj integral (see Equation, 1), but this form presents some difficulties in numerical work due to slow evaluation and numerical problems caused by singularities near the integral contour (see Kucerovsky & Payandeh Najafabadi, 2009, for more details). To overcome these problems, several approximation methods have been considered (see Abrahams, (2000); Kudryavtsev & Levendorskiĭ, 2009; Kuznetsov, 2010; Rawlins, 2012 among others). But, as far as we known, (i) none of them provides any estimation bound for their approximation methods; (ii) most of them need uniform convergence, which is usually hard to achieve.

This article studies the problem of solving a Wiener–Hopf factorization problem, approximately. Then, it provides (i) an $L_p(R), 1 < p \leq 2$, approximation for a Wiener–Hopf factorization problem; (ii) estimation bounds for such approximation technique; (iii) application of our findings in spectral factorization and Lévy processes. This article has been developed as the following. Section 2 collects some useful elements which are used later. The main contribution of this article on approximating solutions of a given a Wiener-Hopf factorization problem has been given in Section 3. Application of our findings has been given in Section 4. Concluding remarks has been given in Section 5.

2. Preliminaries

Now, we collect some lemmas which are used later.

Definition 1 A function f in $L_1(R) \cap L_2(R)$ is said to be an exponential-type T function on the domain D if there are positive constants M and T such that $|f(\omega)| \leq M \exp\{T|\omega|\}$, for $\omega \in D$.

The well-known Paley–Wiener theorem states that the Fourier transform of an $L_2(R)$ function vanishes outside of an interval $[-T, T]$, if and only if the function is of exponential-type T (see Dym & McKean, 1972, p. 158, for more details). The exponential-type functions are continuous functions which are infinitely differentiable everywhere and have a Taylor series expansion over every interval (see Champeney, 1987, p. 77; Walnut, 2002, p. 81). These functions are also called band-limited functions (see Bracewell, 2000, p. 119, for more details on bandlimited functions) (which are equivalent to exponential-type functions by the above stated Paley–Wiener theorem). The index of a complex-valued function f on a smooth oriented curve Γ, such that $f(\Gamma)$ is closed and compact, is defined to be the winding number of $f(\Gamma)$ about the origin (see Payandeh Najafabadi, 2007, §1, for more technical details). Computing the index of a function is usually a *key step* to determine the existence and number of solutions of a Wiener–Hopf factorization problem. The *Sokhotski–Plemelj* integral of a function s which satisfies the Hölder condition and it is defined by a principal value integral, as follows.

$$\phi_s(\lambda) := \frac{1}{2\pi i} \int_{\mathbb{R}} \frac{s(x)}{x-\lambda} dx, \quad \text{for} \lambda \in C. \tag{1}$$

The following are some well-known properties of the Sokhotski-Plemelj integral, proofs can be found in Ablowitz and Fokas (1990, §7), Gakhov (1990, §2), and Pandey (1996, §4), among others. The radial limit of the Sokhotski-Plemelj integral of s, given by $\phi_s^{\pm}(\omega) = \lim_{\lambda \to \omega + i0^{\pm}} \phi_s(\lambda)$ can be represented as the jump formula. i.e. $\phi_s^{\pm}(\omega) = \pm s(\omega)/2 + \phi_s(\omega)$, (or $\phi_s^{\pm}(\omega) = \pm s(\omega)/2 + H_s(\omega)/(2i)$) where $H_s(\omega)$ is the *Hilbert transform* of s and $\omega \in R$.

The *Hausdorff-Young* theorem states that: If s is a function in $L_p(R)$. Then, its Fourier transform, say $\hat{s}$, is an $L_{p^*}(R)$ function that satisfies $||\hat{s}||_{p^*} \leq (2\pi/p)^{-1/p}/p^{*1/p^*}||s||_p$, where $1 < p \leq 2$ and $1/p + 1/p^* = 1$, see Pandey (1996) for more details. From the Hausdorff–Young Theorem, one can observe that if $\{s_n\}$ is a sequence of functions converging in $L_p(R), 1 < p \leq 2$, to s. Then, the Fourier transforms of s_n converge in $L_{p^*}(R)$, to the Fourier transform of s, whenever $1/p + 1/p^* = 1$. Using the Hausdorff–Young theorem, Payandeh Najafabadi and Kucerovsky (2014a) established that Hilbert transform of an $L_p(R), 1 < p \leq 2$ function s, say H_s, satisfies

$$||H_s||_p \leq ||s||_p. \tag{2}$$

Form this observation, one may conclude that "if $\{f_n\}$, $n \geq 1$, is a sequence of functions which converge in $L_p(R), 1 < p \leq 2$, to f. Then, the Hilbert transforms of f_n's also converge in $L_p(R)$ to the Hilbert transform of f".

The following, from Kucerovsky and Payandeh Najafabadi (2009), recalls some further useful properties of functions in $L_p(R)$ space.

LEMMA 1 *Suppose s and r are functions in $L_p(R)$, and suppose that $|s|$ and $|r|$ are bounded above by a. Then,*

(i) $||\sqrt{s} - \sqrt{r}||_p \leq \frac{1}{2\sqrt{a}}||s - r||_p$;

(ii) $||\ln s - \ln r||_p \leq a^{-1}||s - r||_p$, *whenever s and r are positive-valued functions;*

(iii) $||e^{-is/2} - e^{-ir/2}||_p \leq \frac{1}{2}||s - r||_p$, *whenever s and r are real-valued functions;*

(iv) $||1/s - 1/r||_p \leq a^{-2}||s - r||_p$.

The followings recall definition and some useful properties on a mixture-gamma distribution, which plays an important role for the next sections (see Bracewell, 2000 for more details).

Definition 2 (mixture-gamma family of distributions) A non-negative random variable X is said to be distributed according to a mixture-gamma distribution if its density function is given by

$$p(x) = \sum_{k=1}^{\nu}\sum_{j=1}^{n_\nu} c_{kj}\frac{\alpha_k^j x^{j-1}}{(j-1)!}e^{-\alpha_k x}, \quad x \geq 0, \tag{3}$$

where c_{k_j} and α_k are positive value which satisfy $\sum_{k=1}^{\nu}\sum_{j=1}^{n_\nu} c_{k_j} = 1$.

LEMMA 2 *The characteristic function of a distribution (or equivalently the Fourier transform of its density function), say $\hat{p}$, has the following properties:*

(i) $\hat{p}$ is a rational function if and only if the density function belongs to the mixture-gamma family given by;

(ii) $\hat{p}(0) = 1$; and the norm of $\hat{p}(\omega)$ bounded by 1.

3. Main results

Definition 3 The Wiener–Hopf factorization is the problem of finding a sectionally analytic function Φ whose upper and lower radial limits at the real line, say $\Phi^{\pm}$, satisfy

$$\Phi^{+}(\omega)\Phi^{-}(\omega) = g(\omega), \quad \text{for } w \in R, \tag{4}$$

where g is a given continuous function satisfying a Hölder condition on R. Moreover, g is assumed to have zero index, to be non-vanishing on R, and bounded above by 1.

Payandeh Najafabadi and Kucerovsky (2011) established that sectionally analytic functions $\Phi^{\pm}$ satisfying a zero index Wiener–Hopf factorization (3) can be found by

$$\Phi^{\pm}(\lambda) = \exp\{\pm\phi_{\ln g}(\lambda) \mp \phi_{\ln g}(0)\}, \quad \lambda \in C,$$

where $\phi_{\ln g}$ stands for the Sokhotski–Plemelj integration of $\ln g$. Using the jump formula, the above $\Phi^{\pm}$ can be re-stated as

$$\Phi^{\pm}(\omega) = \sqrt{g(\omega)}\exp\{\pm\frac{i}{2}(H_{\ln g}(0) - H_{\ln g}(\omega))\}, \tag{5}$$

where $H_{\ln g}$ stands for the Hilbert transform of $\ln g$.

In many situations, the Wiener–Hopf factorization problems (3) cannot be solved explicitly and has to be solved approximately (see Kucerovsky & Payandeh Najafabadi, 2009, for more details). The following develops an approximate technique to solve Equation 3.

THEOREM 1 *Suppose g in the the Wiener-Hopf factorization problem (3) is a given, bounded (above by a), zero index function, satisfies the Hölder condition and $g(0) = 1$. Moreover, suppose that there is a sequence of sectionally analytic functions $\Phi_n^{\pm}(\omega) = \sqrt{g(\omega)}\exp\{\pm\frac{i}{2}(H_{\ln g}(0) - H_{\ln g}(\omega))\}$, where g_n-s converge (in $L_p(R)$, $1 < p \leq 2$, sense) to g. Then, sectionally analytical solution of the Wiener-Hopf factorization problem (3), say $\Phi^{\pm}$, can be approximated by $\Phi_n^{\pm}$ and the error estimate satisfies*

$$||\Phi_n^{\pm} - \Phi^{\pm}||_p \leq \frac{1}{2a\sqrt{a}}||g_n - g||_p^2 + \frac{3}{2\sqrt{a}}||g_n - g||_p.$$

Proof Set $k(\omega) := -H_{\ln g}(\omega) + H_{\ln g}(0)$ and $k_n(\omega) := -H_{\ln g_n}(\omega) + H_{\ln g_n}(0)$. Now, from Equation 5, Equation 2, and Lemma 1, observe that

$$\begin{aligned}
||\Phi_n^{\pm} - \Phi^{\pm}||_p &= ||\sqrt{g_n}e^{\pm ik_n/2} - \sqrt{g}e^{\pm ik/2}||_p \\
&\leq \left[||\sqrt{g_n} - \sqrt{g}||_p + ||\sqrt{g}||_p\right]||e^{\pm ik_n/2} - e^{ik/2}||_p + |e^{\pm ik/2}|||\sqrt{g_n} - \sqrt{g}||_p \\
&\leq \frac{1}{2}\left[||\sqrt{g_n} - \sqrt{g}||_p + ||\sqrt{g}||_p\right]|| - H_{\ln g_n}(\omega) + H_{\ln g_n}(0) + H_{\ln g}(\omega) - H_{\ln g}(0)||_p \\
&\quad + |e^{\pm ik/2}|||\sqrt{g_n} - \sqrt{g}||_p \\
&\leq \left[||\sqrt{g_n} - \sqrt{g}||_p + ||\sqrt{g}||_p\right]||H_{\ln g_n} - H_{\ln g}||_p + ||\sqrt{g_n} - \sqrt{g}||_p
\end{aligned}$$

since k and k_n are real-valued functions

$$\begin{aligned}
&\leq \left[||\sqrt{g_n} - \sqrt{g}||_p + ||\sqrt{g}||_p\right]||\ln(g_n) - \ln(g)||_p + ||\sqrt{g_n} - \sqrt{g}||_p \\
&\leq \left[\frac{1}{2\sqrt{a}}||g_n - g||_p + \sqrt{a}\right]\frac{1}{a}||g_n - g||_p + \frac{1}{2\sqrt{a}}||g_n - g||_p \\
&= \frac{1}{2a\sqrt{a}}||g_n - g||_p^2 + \frac{3}{2\sqrt{a}}||g_n - g||_p.
\end{aligned}$$

□

In our belief, the most favorable situation is to approximate g by a sequence of rational functions, which are obtained from a Padé approximant or a continued fraction expansion.

Using the *Shannon sampling theorem*, the following develops an elegant scheme that allows to provide an explicit solution for a Wiener–Hopf factorization Equation 3.

THEOREM 2 *Suppose g in the Wiener-Hopf factorization problem (Equation 3) is a given, $L_1(R) \cap L_2(R)$, bounded, zero index function, satisfies the Hölder condition. Moreover, suppose that there is a sequence of $\ln(g)$ is an exponential-type T function, then unique solutions of the Wiener-Hopf factorization (Equation 3) can be explicitly determined by*

$$\Phi_{\pm}(\omega) = \exp\left\{ \pm \sum_{n=-\infty}^{\infty} \ln(g(\frac{2n}{T}))\frac{e^{\pm i\pi(T\omega-2n)}-1}{2i\pi(T\omega-2n)} \right\},$$

Proof Using the fact that $\ln(g)$ is an exponential-type T function, one can decompose $\ln(g(\omega))$ as $\ln(g(\omega)) = K_+(\omega) + K_-(\omega)$, where

$$K_{\pm}(\omega) = \pm \sum_{n=-\infty}^{\infty} \ln(g(\frac{2n}{T}))\frac{\exp\{\pm i\pi(T\omega-2n)\}-1}{2i\pi(T\omega-2n)}.$$

Sectionally analytical properties of $K_{\pm}(\omega)$ in $C_{\pm}$ has been established by Kucerovsky and Payandeh Najafabadi (2009). □

The following theorem provides the error bound for approximate solution arrives from the Shannon sampling theorem.

THEOREM 3 *Suppose g in the Wiener-Hopf factorization problem (3) is a given, $L_1(R) \cap L_2(R)$, bounded, zero index function, satisfies the Hölder condition and $\ln(g)$ is an exponential-type T function. Moreover, suppose that there is a sequence of g_m in $L_1(R) \cap L_2(R)$ and $\ln(g_m)$ are exponential-type T functions. Then, approximate solutions of the Wiener-Hopf factorization problem (3) can be determined by*

$$\mathbf{\Phi}_{\pm}^{(m)}(\omega) = \pm\exp\left\{ \pm \sum_{n=-\infty}^{\infty} \ln(g_m)(\frac{2n}{T}))\frac{e^{\pm i\pi(T\omega-2n)}-1}{2i\pi(T\omega-2n)} \right\}$$

and the error bound satisfies

$|\mathbf{\Phi}_{\pm}^{(m)} - \mathbf{\Phi}_{\pm}| \le \| \ln(g_m) - \ln(g)\|,$

where the norm is defined by $\|M\| := \sup_{ij}\left\{ \int_{-\infty}^{\infty} |M_{ij}(x)|^2\,dx \right\}^{1/2}$.

Proof The proof is straightforward by a double application of Lemma (3.2, Payandeh Najafabadi & Kucerovsky, 2014a) and Theorem 2. □

4. Applications

This section provides the application of the above results in two different contexts. The first subsection considers the problem of finding spectral factorization, whenever its corresponding spectral density function has been given. The second subsection derives density/probability functions of extrema random variables of a given Lévy processes.

4.1. Application to spectral factorization

Definition 4 Suppose $\rho(x)$ is an autocorrelation of the stochastically stationary process $X(t)$, i.e. $\rho(x) := Corr(X)(t), X(t-x))$. Then, the spectral function $S(\lambda)$ is defined by $\rho(\tau) = \int_{-\infty}^{\infty} e^{i\lambda\tau} S(\lambda)d\mu(\lambda)$, where μ is a given measure.

The spectral density function S has properties that: (1) for stochastically stationary processes, it can be understood as the Fourier transform of the autocorrelation $\rho(\cdot)$ [The Wiener-Khintchine's theorem: Reinsel (1997, p. 219)]; (2) it is a Hermitian function (Koopmans, 1995, p. 122); (3) it defines almost everywhere (a.e.) on the interval $[-\pi, -\pi]$ (Wilson, 1972); and (4) S is integrable and has a Fourier series expansion $S(\theta) = \sum_{k=-\infty}^{\infty} \gamma_k e^{ik\theta}$, where $\gamma_k = \int_{-\pi}^{\pi} \mathbf{S}(\theta)e^{-ik\theta}d\theta$ (Wilson, 1972).

The spectral factorization plays a crucial role in a wide range of scientific fields, such as communications (Magesacher & Cioffi, 2011), system theory (Janashia, Lagvilava, & Ephremidze, 2011), optimal control (Johannesson, Rantzer, & Bernhardsson, 2011), filtering theory (Anderson & Moore, 2005), network theory (Belevitch, 1968; Ivrlac & Nossek, 2014) prediction theory of stationary processes (Brockwell & Davis, 2002), deriving forward expression from a backward one in the ARIMA processes (Brockwell & Davis, 2009). In such applications, the spectral density factorization is the most difficult step. Since Wiener's seminal efforts create a computational method for spectral factorization, several authors have developed different methods to do so, but none of them have provided a method that has an essential superiority over all others (Janashia et al., 2011). On the other hand, most of these methods impose some extra restriction on the spectral density (e.g. to be real or rational function, or to be non-singular on the boundary). The following represents an explicit solution for the problem of spectral factorization for spectral densities.

Proposition 1 Suppose S is a spectral density function. Then, the right and left spectral factorizations $L_{\pm}$ are given by

$$L_{\pm}(\omega) = \exp\left\{ \pm \sum_{n=-\infty}^{\infty} \ln(S(\frac{2n}{T}))\frac{e^{\pm i\pi(T\omega-2n)}-1}{2i\pi(T\omega-2n)} \right\}.$$

Proof The exponential-type T condition of $\ln(S)$ arrives from the fact that the spectral density can be understood as an extension version of the inverse Fourier transform on the autocorrelation function which is bounded function. The rest of proof arrives from an application of Theorem 2 along the fact that S is Hermitian function which its index is zero (Voronin, 2010, 2011). □

4.2. Application to Lévy processes

Suppose X_t is a one-dimensional real-valued Lévy process started from $X_0 = 0$ and defined by a triple (μ, σ, ν): the drift $\mu \in R$, volatility $\sigma \geq 0$, and the jumps measure ν which is given by a non-negative function defined on $R \setminus \{0\}$ satisfying $\int_R \min\{1, x^2\}\nu(dx) < \infty$. Moreover, suppose that random stopping time $\tau(q)$ has either a geometric (with parameter $q \in (0,1)$) or an exponential distribution (with parameter $q > 0$) and independent of the Lévy process X_t which $\tau(0) = \infty$. The Lévy–Khintchine formula states that the characteristic exponent ψ (i.e. $\psi(\omega) = \ln(E(\exp(i\omega X_1)))$, $\omega \in R$) can be represented by

$$\psi(\omega) = i\mu\omega - \frac{1}{2}\sigma^2\omega^2 + \int_R (e^{i\omega x} - 1 - i\omega x I_{[-1,1]}(x))\nu(dx), \quad \omega \in R. \tag{6}$$

The extrema of the Lévy process X_t are given by

$$\begin{aligned} M_q &= \sup\{X_s : s \leq \tau(q)\} \\ I_q &= \inf\{X_s : s \leq \tau(q)\}. \end{aligned} \tag{7}$$

The Wiener–Hopf factorization is a well-known technique to study the characteristic functions of the extrema random variables (Bertoin, 1996). Namely, the Wiener–Hopf factorization states that: (i) random variables M_q and I_q are independent; (ii) product of their characteristic functions equal to the characteristic function of Lévy process X_t; (iii) random variable M_q (I_q) is infinitely divisible, positive (negative), and has zero drift.

In the cases that the characteristic function of Lévy process X_t is either a rational function or can be decomposed as a product of two sectionally analytic functions in the closed upper and lower half complex planes C^+ and C^-. The characteristic functions of random variables M_q and I_q can be determined explicitly. Lewis and Mordecki (2005) considered a Lévy process X_t which its negative jumps distributed according to a mixture-gamma family of distributions (given by Definition 2) and an arbitrary positive jumps measure. They established such process has the characteristic function which can decompose as a product of a rational function and an arbitrary function, which are analytic in C^+ and C^-, respectively. Moreover, they provided an analog result for a Lévy process whose its corresponding positive jumps measure follows from a mixture-gamma family of distributions while its negative jumps measure is an arbitrary one, more details can be found in Lewis and Mordecki (2008).

Unfortunately, in the most situations, the characteristic function of the process *neither* is a rational function *nor* can be decomposed as a product of two analytic functions in C^+ and C^-. Therefore, the characteristic functions of M_q and I_q should be expressed in terms of a Sokhotski-Plemelj integral (see

Equation 1). But, this form, also, presents some difficulties in numerical work due to slow evaluation and numerical problems caused by singularities near the integral contour. To overcome these difficulties, approximation methods have to be considered.

It is well known that a Lévy process X_t which its jumps distribution follows from the phase-type distribution has a rational characteristic function (Doney, 1987). Kuznetsov (2010) utilized this fact and approximated a jumps measure ν a ten-parameter family of Lévy processes (named β–family of Lévy process) by a sequence of the phase-type measures. Then, he determined the characteristic functions of random variables M_q and I_q, approximately.

The following theorem represents an estimation bound for approximated extrema's density/probability functions of a Lévy process.

THEOREM 4 *Suppose X_t is a Lévy process defined by a triple (μ, σ, ν) and its random stopping time $\tau(q)$ has been distributed according to either a geometric or an exponential distribution. Moreover suppose that there is a sequence of jumps measure ν_n which satisfies the following two conditions:*

(1) They converge in $L_p(R)$, $1 < p \leq 2$, to ν and $\int_{-1}^{1} x\nu_n(dx) = \int_{-1}^{1} x\nu(dx)$;

(2) their corresponding characteristic exponents ψ_n (arrived by the Lévy-Khintchine Formula 6) as well as the characteristic exponents ψ (correspondence with jumps measure ν) are bounded above by M.

Then, density/probability of supremum and infimum of Lévy process X_t, say respectively f_q^+ and f_q^-, can be approximated by sequence of density/probability functions $f_{q,n}^+$ and $f_{q,n}^-$ which has the following error bound.

(i) For exponentially distributed stopping time $\tau(q)$,

$$||f_q^{\pm} - f_{q,n}^{\pm}||_p \leq \frac{q^2}{2M^4(2\pi)^{1/p}}||\nu_n - \nu||_p^2 + \frac{3q}{2M^2}||\nu_n - \nu||_p;$$

(ii) For geometric stopping time $\tau(q)$,

$$||f_q^{\pm} - f_{q,n}^{\pm}||_p \leq \frac{(1-q)^2}{2M^4(2\pi)^{1/p}}||\nu_n - \nu||_p^2 + \frac{3(1-q)}{2M^2}||\nu_n - \nu||_p.$$

Proof From Bertoin (1996), one can observe that the Fourier transform of M_q and I_q density functions, say respectively Φ^+ and Φ^-, satisfy *either* the Wiener-Hopf factorization $\Phi^+(\omega)\Phi^-(\omega) = q/(q - \psi(\omega)$, where $\omega \in R$ (for exponentially distributed stopping time) *or* the Wiener-Hopf factorization $\Phi^+(\omega)\Phi^-(\omega) = (1-q)/(1-q\psi(\omega)$, where $\omega \in R$ (for geometric stopping time). Now, from the fact that expressions $q(q-\psi(\cdot))^{-1}$ and $(1-q)(1-q\psi(\cdot))^{-1}$ are the characteristic function of Lévy process X_t respectively for exponential and geometric stopping time, observe that both expressions are bounded by 1 (property of the characteristic function given by Lemma 2, part ii). For part (i), from Theorem 1 observe that

$$\begin{aligned}
||\Phi_n^{\pm} - \Phi^{\pm}||_{p^*} &\leq \frac{1}{2}||\frac{q}{q-\psi_n} - \frac{q}{q-\psi}||_{p^*}^2 + \frac{3}{2}||\frac{q}{q-\psi_n} - \frac{q}{q-\psi}||_{p^*} \\
&\leq \frac{q^2}{2M^4}||\psi_n - \psi||_{p^*}^2 + \frac{3q}{2M^2}||\psi_n - \psi||_{p^*} \\
&\leq \frac{q^2}{2M^4(2\pi)^{2/p}}||\nu_n - \nu||_p^2 + \frac{3q}{2M^2(2\pi)^{1/p}}||\nu_n - \nu||_p,
\end{aligned}$$

where $1/p^* + 1/p = 1$. The second inequality arrives from part (iv) of Lemma 1, while the third inequality obtains from Equation 6 along with conditions A_2 and an application of Hausdorff-Young theorem. The rest of proof arrives from an application of the Hausdorff-?Young theorem. Proof of part (ii) is quite similar. □

It would worthwhile to mention that, in the case of $\int_{-1}^{1} x\nu_n(dx) = :c_n \neq d: = \int_{-1}^{1} x\nu(dx)$. One may obtain sequence $\xi_n = \frac{d}{c_n}\nu_n$ which satisfy the desire condition.

The following utilizes result of the above theorem and provides a procedure to find the extrema's density/probability of a wide class of Lévy processes, approximately.

PROCEDURE 1 *Suppose X_t is a Lévy process with bounded characteristic exponents ψ. Moreover, suppose that random stopping time $\tau(q)$ has been distributed according to either a geometric or an exponential distribution. Then, by the following steps, one can approximate (in $L_p(R)$, $1 < p \leq 2$, sense) density/probability functions of the extrema random variables M_q and I_q.*

Step 1. Approximate jumps measure ν with either a phase-type or a mixture-gamma density function, say ν^*, where $\int_{-1}^{1} x\nu^*(dx) = \int_{-1}^{1} x\nu(dx)$ and $||\nu - \nu^*||_p \leq \varepsilon$;

Step 2. Decompose rational function $q/(q-\psi(\omega))$ (or $(1-q)/(1-q\psi(\omega))$) into product of two rational and sectionally analytic functions, say $g^{\pm}$, in $C^{\pm}$, respectively;

Step 3. Obtain, approximate, density/probability functions of M_q and I_q by the inverse Fourier transform of g^+ and g^-, respectively.

In many situations, it is more convenient to approximate the characteristic exponents ψ, rather than the jumps measure ν. The following extends results of Theorem 4 to such situations.

COROLLARY 1 *Suppose X_t is a Lévy process defined by a triple (μ, σ, ν) and its random stopping time $\tau(q)$ has been distributed according to either a geometric or an exponential distribution. Moreover suppose that there is a sequence of bounded characteristic exponents ψ_n (i.e. $|\psi_n| \leq M$) which converge in $L_p(R)$, $1 < p \leq 2$, to the characteristic exponent of the process ψ, and their corresponding jumps measure ν_n satisfies $\int_{-1}^{1} x\nu_n(dx) = \int_{-1}^{1} x\nu(dx)$.*

Then, density/probability function of the supremum and the infimum of Lévy process X_t, say respectively f_q^+ and f_q^-, can be approximated by sequence of density/probability functions $f_{q,n}^+$ and $f_{q,n}^-$ which has the following error bound.

(i) For exponentially distributed stopping time $\tau(q)$,

$$||f_q^{\pm} - f_{q,n}^{\pm}||_p \leq \frac{q^2(2\pi)^{1/p}}{2M^4}||\psi_n - \psi||_{p^*}^2 + \frac{3q(2\pi)^{1/p}}{2M^2}||\psi_n - \psi||_{p^*};$$

(ii) For discrete stopping time $\tau(q)$,

$$||f_q^{\pm} - f_{q,n}^{\pm}||_p \leq \frac{(1-q)^2(2\pi)^{1/p}}{2M^4}||\psi_n - \psi||_{p^*}^2 + \frac{3(1-q)(2\pi)^{1/p}}{2M^2}||\psi_n - \psi||_{p^*}.$$

Using the fact that, the characteristic exponent $\psi(i\omega)$, $\omega \in R$, is a real-valued function, Bertoin (1996). One can suggest that following procedure to generate approximation density/probability functions for M_q and I_q.

PROCEDURE 2 *Suppose X_t is a Lévy process with bounded characteristic exponents ψ. Moreover, suppose that random stopping time $\tau(q)$ has been distributed according to either a geometric or an exponential distribution. Then, by the following steps, one can approximate (in $L_p(R)$, $1 < p \leq 2$, sense) density/probability functions of the extrema random variables M_q and I_q.*

Step 1. Approximate (in $L_p(R)$, $1 < p \leq 2$, sense) the characteristic exponent $\psi(i\omega)$ by a rational function, generated by the Padé approximant or the continued fraction, say $\psi^*(i\xi)$;

Step 2. Decompose rational function $q/(q-\psi^*(\omega))$ (or $(1-q)/(1-q\psi^*(\omega))$) into product of two rational and sectionally analytic functions, say $g^{\pm}$, in $C^{\pm}$, respectively;

Step 3. Obtain, approximate, density/probability functions of M_q and I_q by the inverse Fourier transform of g^+ and g^-, respectively.

Using Lemma 2, one can readily, conclude that the above two procedures approximate density/probability functions of M_q and I_q by the mixture-gamma density functions.

Now, we provide several examples.

Example 1 Consider a 1-stable Lévy process with a jumps measure $\nu(dx) = c|x|^{-2}I_{(-\infty,\infty)}(x)dx$. One can, readily, show that the characteristic exponent of such process is $\psi(\omega) = (c_1 + c_2)|\omega|\{1 + i\beta \operatorname{sgn}\omega \frac{2}{\pi} \ln(|\omega|)\} + i\omega\eta$, where η is a normalized real-valued. The natural logarithm $\ln(|\lambda|)$ has the continued fraction

$$\ln(|\lambda|) = \frac{\lambda - 1}{1} \underset{+}{} \frac{1^2(\lambda - 1)}{2} \underset{+}{} \frac{1^2(\lambda - 1)}{3} \underset{+}{} \frac{2^2(\lambda - 1)}{4} \underset{+}{} \frac{2^2(\lambda - 1)}{5} \underset{+}{} \frac{3^2(\lambda - 1)}{7} \underset{+}{} \frac{3^2(\lambda - 1)}{9} \underset{+}{} \cdots, \quad \lambda \in C.$$

(see Jones & Thron, 1931 or Cuyt, Petersen, Verdonk, Waadeland, & Jones, 2008, among others). In Practice, in the above-continued fraction, it has to cut off somewhere and obtain a rational function for $\ln(\cdot)$. Consequently, an expression $q(q - \psi(\cdot))^{-1}$ can be approximated by rational function $P(\lambda)/Q(\lambda)$. Now, the characteristic functions for the extrema can be obtained after decomposing $P(\lambda)/Q(\lambda)$ into product of two rational, analytic, and bounded functions, in C^+ and C^-.

Example 2 Suppose X_t is a Lévy process with independent and continuous $\tau(q)$ and a jumps measure $\nu(dx) = \exp\{\alpha x\} cosech^2(x/2)dx$. The characteristic exponent for such Lévy process is given by

$$\psi(\omega) = \frac{\sigma^2\omega^2}{2} + i\rho\omega + 4\pi(\omega - i\alpha)\coth(\pi(\omega - i\alpha)) - 4\gamma,$$

where $\gamma = \pi\alpha\cot(\pi\alpha)$, $\rho = 4\pi^2\alpha + \frac{4\gamma(\gamma-1)}{\alpha} - \mu$, $\omega \in R$, and α, μ, and σ are given. The continued fraction for $\tanh(\lambda)$ is given by

$$\tanh(\lambda) = \frac{\lambda}{1} + \frac{\lambda^2}{3} \underset{+}{} \frac{\lambda^2}{5} \underset{+}{} \frac{\lambda^2}{7} \underset{+}{} \frac{\lambda^2}{9} \underset{+}{} \cdots, \quad \lambda \in C,$$

(Cuyt et al., 2008). After cutting off the above-continued fraction at the nth term, one may be obtained a rational function, say ρ_n, for $\tanh(\cdot)$. Substituting the ρ_n in an expression $q(q - \psi(\cdot))$, we may obtain a rational function, say $P(\lambda)/Q(\lambda)$. Therefore, the characteristic functions for the extrema can be found by decomposing a rational function $P(\lambda)/Q(\lambda)$ as a product of two rational, analytic, and bounded functions in C^+ and C^-. This observation verifies Kuznetsov's (2010) result.

Example 3 Metron model is a Lévy process with a jumps measure $\nu(x) = a(\delta\sqrt{2\pi})^{-1}\exp\{-(x - \mu)^2/(2\delta^2)\}$ and characteristic exponent $\psi(\omega) = i\mu\omega - \sigma^2\omega^2/2 + a\{e^{-\delta^2\omega^2/2 + i\mu\omega} - 1\}$, where a, δ, μ, and σ are given and $\omega \in R$. Metron model has tail behaviors heavier than the Gaussian but all exponential moments are finite. The continued fraction for the exponential function $\exp(\cdot)$ is given by

$$e^{\lambda} = \frac{1}{1} \underset{-}{} \frac{\lambda}{1} \underset{+}{} \frac{\lambda}{2} \underset{-}{} \frac{\lambda}{3} \underset{+}{} \frac{\lambda}{2} \underset{-}{} \frac{\lambda}{5} \underset{+}{} \frac{\lambda}{2} \underset{-}{} \cdots, \quad \lambda \in C,$$

(Cuyt et al., 2008). After cutting off the above-continued fraction somewhere, and substituting the arrived rational function in an expression $q(q - \psi(\lambda))$, we may obtain a rational function, say $P(\lambda)/Q(\lambda)$. The characteristic functions for the extrema can be found by decomposing $P(\lambda)/Q(\lambda)$ as a product of two rational, analytic, and bounded functions in C^+ and C^-.

5. Conclusion and suggestion

This paper provides two techniques to solve a Wiener-Hopf factorization problem. Application of our findings, in two different contexts, has been given. It would be worthwhile mentioning that: (1) In the situation where given function $\ln(g)$ is not of exponential-type function. We suggest to approximate such function with an exponential-type function which pointwise converges to such function (see Kucerovsky and Payandeh Najafabadi (2009) for more details); (2) Other applications our findings can be developed to other situations where the Wiener-Hopf factorization is applicable, such as finding first/last passage time and the overshoot, the last time the extrema was archived, several kind of option pricing, etc.

Acknowledgements

Thanks to an anonymous reviewer for his/her constructive comments. The author Dan Z. Kucerovsky acknowledges Natural Sciences and Engineering Research Council (NSERC).

Funding

This work was supported by Natural Sciences and Engineering Research Council (NSERC) of Canada.

Author details

Amir T. Payandeh Najafabadi[1]

E-mail: amirtpayandeh@sbu.ac.ir

Dan Z. Kucerovsky[2]

E-mail: dkucerov@unb.ca

[1] Department of Mathematical Sciences, Shahid Beheshti University, G.C. Evin, Tehran, 1983963113, Iran.

[2] Department of Mathematics and Statistics, University of New Brunswick, Fredericton, New Brunswick, Canada, E3B 5A3.

References

Ablowitz, M. J., & Fokas, A. S. (1990). *Complex variable, introduction and application*. Berlin: Springer-Verlag.

Abrahams, I. D. (2000). The application of Padé approximants to Wiener–Hopf factorization. *IMA Journal of Applied Mathematics, 65*, 257–281.

Abrahams, I. D., & Wickham, G. R. (1990). General Wiener–Hopf factorization of matrix kernels with exponential phase factors. *SIAM Journal of Applied Mathematics, 50*, 819–838.

Achenbach, J. D. (2012). *Wave propagation in elastic solids*. Amsterdam: North-Holland.

Anderson, B. D., & Moore, J. B. (2005). *Optimal filtering*. New York, NY: Dover.

Beheshti, M. H., Payandeh Najafabadi, A. T., & Farnoosh, R. (2013). An analytical approach to pricing discrete barrier options under time-dependent models. *European Journal of Scientific Research, 103*, 304–312.

Belevitch, V. (1968). *Classical network theory*. San Francisco, CA: Holden Day.

Bertoin, J. (1996). *Lévy processes*. New York, NY: Cambridge University Press.

Bracewell, R. N. (2000). *The Fourier transform and its applications* (3rd ed.). New York, NY: McGraw-Hill.

Brockwell, P. J., & Davis, R. A. (2002). *Introduction to time series and forecasting*. New York, NY: Springer-Verlag.

Brockwell, P. J., & Davis, R. A. (2009). *Time series: Theory and methods*. New York, NY: Springer-Verlag.

Chakrabarti, A., & George, A. J. (1994). Solution of a singular integral equation involving two intervals arising in the theory of water waves. *Applied Mathematics Letters, 7*, 43–47.

Champeney, D. C. (1987). *A handbook of Fourier theorems*. New York, NY: Cambridge University Press.

Cuyt, A., Petersen, V. B., Verdonk, B., Waadeland, H., & Jones, W. B. (2008). *Handbook of continued fractions for special functions*. New York, NY: Springer.

Daniele, V. G. (2014). *The Wiener-Hopf method in electromagnetics* (Technical report). New York, NY: Institution of Engineering and Technology.

Davis, A. M. J. (1987). Continental shelf wave scattering by a semi-infinite coastline. *Geophysical & Astrophysical Fluid Dynamics, 39*, 25–55.

Doney, R. A. (1987). On Wiener–Hopf factorisation and the distribution of extrema for certain stable processes. *The Annals of Probability, 15*, 1352–1362.

Dym, H., & Mckean, H. P. (1972). *Fourier series and integrals. Probability and mathematical statistics*. New York, NY: Academic Press.

Freund, L. B. (1998). *Dynamic fracture mechanics*. Cambridge: Cambridge University Press.

Fusai, G., Abrahams, I. D., & Sgarra, C. (2006). An exact analytical solution for discrete barrier options. *Finance and Stochastics, 10*, 1–26.

Gakhov, F. D. (1990). *Boundary value problem* (Translated from the Russian Reprint of the 1966 translation). New York, NY: Dover Publications.

Ivrlac, M., & Nossek, J. (2014). The multiport communication theory. *IEEE Circuits and Systems Magazine, 14*, 27–44.

Janashia, G., Lagvilava, E., & Ephremidze, L. (2011). A new method of matrix spectral factorization. *IEEE Transactions on Information Theory, 57*, 2318–2326.

Johannesson, E., Rantzer, A., & Bernhardsson, B. (2011). Optimal linear control for channels with signal-to-noise ratio constraints. In *American Control Conference*. San Francisco, CA.

Jones, W. B., & Thron, W. J. (1931). *Continued fractions: Analytic theory and applications*. New York, NY: Addison-Wesley.

Kaoullas, G., & Johnson, E. R. (2010). Fast accurate computation of shelf waves for arbitrary depth profiles. *Continental Shelf Research, 30*, 833–836.

Kim, C. I., Schiavone, P., & Ru, C. Q. (2011). The effect of surface elasticity on a Mode-III interface crack. *Archives of Mechanics, 63*, 267–286.

Koopmans, L. H. (1995). *The spectral analysis of time series*. New York, NY: Academic Press.

Kucerovsky, D., Marchand, É., Payandeh, A. T., & Strawderman, W. (2009). On the Bayesianity of maximum likelihood estimators of restricted location parameters under absolute value error loss. *Statistics & Decisions, 27*, 145–168.

Kucerovsky, D., & Payandeh Najafabadi, A. T. (2009). An approximation for a subclass of the Riemann–Hilbert problems. *IMA Journal of Applied Mathematics, 74*, 533–547.

Kudryavtsev, O., & Levendorskiĭ, S. (2009). Fast and accurate pricing of barrier options under Lévy processes. *Finance and Stochastics, 13*, 531–562.

Kuznetsov, A. (2010). Wiener–Hopf factorization and distribution of extrema for a family of Lévy processes. *The Annals of Applied Probability, 20*, 1801–1830.

Lewis, A., & Mordecki, E. (2005). *Wiener–Hopf factorization for Lévy processes having negative jumps with rational transforms* (Technicall report). Retrieved from http://www.cmat.edu.uy/ mordecki/articles/

Lewis, A., & Mordecki, E. (2008). Wiener–Hopf factorization for Lévy processes having positive jumps with rational transforms. *Journal of Applied Probability, 45*, 118–134.

Magesacher, T., & Cioffi, J. M. (2011). On minimum peak-to-average power ratio spectral factorization. In *The 8th International Workshop on Multi-Carrier Systems & Solutions*. Herrsching.

Noble, B. (1988). *Methods based on the Wiener–Hopf technique* (2nd ed.). New York, NY: Chelsea Publishing.

Norris, A. N., & Achenbach, J. D. (1984). Elastic wave diffraction by a semi infinite crack in a transversely isotropic material. *Quarterly Journal of Mechanics and Applied Mathematics, 37*, 565–580.

Ogilat, O. N. (2013). *Steady and unsteady free surface flow past a two-dimensional stern* (Doctoral dissertation). University of Southern Queensland, Queensland.

Pandey, J. (1996). *The Hilbert transform of Schwartz distributions and application*. New York, NY: Wiley.

Payandeh Najafabadi, A. T. (2007). *Riemann–Hilbert and statistical inference problems in restricted parameter spaces* (Doctoral dissertation). Department of Mathematics and Statistics, University of New Brunswick, Feredricton.

Payandeh Najafabadi, A. T., & Kucerovsky, D. (2009). A weak approximated solution for a subclass of Wiener–Hopf integral equation. *IAENG International Journal of Applied Mathematics, 39*, 247–252.

Payandeh Najafabadi, A. T., & Kucerovsky, D. (2011). On distribution of extrema for a class of Lévy processes. *Journal of Probability and Statistical Science, 9*, 127–138.

Payandeh Najafabadi, A. T., & Kucerovsky, D. (2014a). Exact solutions for a class of matrix Riemann–Hilbert problems. *IMA Journal of Applied Mathematics, 79*, 109–123.

Payandeh Najafabadi, A. T., & Kucerovsky, D. (2014b). On solutions of a system of Wiener–Hopf integral equations. *IAENG International Journal of Applied Mathematics, 44*, 1–6.

Payandeh Najafabadi, A. T., & Kucerovsky, D. (in press). *Approximate Wiener–Hopf factorization for finance problems.*

Rawlins, A. D. (2012). The method of finite-product extraction and an application to Wiener–Hopf theory. *IMA Journal of Applied Mathematics, 77*, 590–602.

Reinsel, G. C. (1997). *Elements of multivariate time series analysis*. New York, NY: Springer-Verlag.

Sautbekov, S., & Nilsson, B. (2009). Electromagnetic scattering theory for gratings based on the Wiener–Hopf method. *AIP Conference Proceedings, 1106*, 110–117.

Shakib, J. T., Akhgarian, E., & Ghaderi, A. (2015). The effect of hydraulic fracture characteristics on production rate in thermal EOR methods. *Fuel, 141*, 226–235.

Voronin, A. F. (2010). Partial indices of unitary and Hermitian matrix functions. *Siberian Mathematical Journal, 51*, 805–809.

Voronin, A. F. (2011). A method for determining the partial indices of symmetric matrix functions. *Siberian Mathematical Journal, 52*, 41–53.

Walnut, D. F. (2002). *An introduction to wavelet analysis* (2nd ed.). New York, NY: Birkhauser.

Wilson, G. T. (1972). The factorization of matricial spectral densities. *SIAM, Journal of applied mathematics, 23*, 420–426.

12

The existence of global weak solutions to the shallow water wave model with moderate amplitude

Ying Wang[1*]
Corresponding author: Ying Wang, College of Science, Sichuan University of Science and Engineering 643000, Zigong, China
*E-mail: matyingw@126.com

Reviewing editor: Lishan Liu, Qufu Normal University, China

Abstract: The existence of global weak solutions to the shallow water model with moderate amplitude, which is firstly introduced in Constantin and Lannes's work (2009), is investigated in the space $C([0,\infty)\times\mathbf{R})\bigcap L^\infty((0,\infty);H^1(\mathbf{R}))$ without the sign condition on the initial value by employing the limit technique of viscous approximation. A new one-sided lower bound and the higher integrability estimate act a key role in our analysis.

Subjects: Applied Mathematics; Dynamical System; Mathematics & Statistics; Science

Keywords: global weak solutions; viscous approximation; shallow water model

AMS subject classifications: 35D05; 35G25; 35L05; 35Q35

1. Introduction

In this paper, we consider the following model for shallow water wave with moderate amplitude

$$\begin{cases} u_t + u_x + \frac{3}{2}\sigma u u_x + \iota\sigma^2 u^2 u_x + \kappa\sigma^3 u^3 u_x + \frac{\mu}{12}(u_{xxx} - u_{txx}) \\ \qquad = -\frac{7}{24}\sigma\mu(u u_{xxx} + 2u_x u_{xx}), \\ u(0,x) = u_0. \end{cases} \tag{1}$$

system (1) is firstly found in Constantin and Lannes (2009) as a model for the evolution of the free surface u. Here the function $u = u(t,x)$, σ and ι are parameters, $\mu > 0$ is shallowness parameter, $\sigma > 0$ is amplitude parameter (see Alvarez-Samaniego & Lannes, 2008; Constantin & Lannes, 2009; Mi & Mu, 2013). It is shown in Constantin (2011) that unlike KdV and Camassa–Holm equation, system (1) does not have a bi-Hamiltonian integrable structure. However, the equation possesses solitary wave profiles that resemble those of C-H (Constantin & Escher, 2007). Recently, Constantin and Lannes (2009) established the local well-posedness of system (1) for any initial data $u_0 \in H^{s+1}(\mathbf{R})$

ABOUT THE AUTHOR

Ying Wang is a senior lecturer in the college of Science at Sichuan University of Science and Engineering, Zigong city, Sichuan province, China. She has 9 years' teaching experience. Her research interests are blow-up theory of Partial Differential Equation and exact travelling wave solutions for Partial Differential Equation. She has published some research articles in reputed international journals.

PUBLIC INTEREST STATEMENT

In this paper, we use the limit technique of viscous approximation to prove the existence of global weak solutions for a shallow water wave model with moderate amplitude. It is shown from the proof of main Theorem that the weak solutions are stable when a regularizing term vanishes. The method is so effective and can be applied to solve some control problems and economic model. Moreover, the shallow water wave model with moderate amplitude we investigate captures breaking wave, which is a major interest in shallow water wave. Overall, the results we obtain can be applied in many hydrodynamic problems.

with $s > \frac{3}{2}$ and also claimed that if the maximal existence time is finite, then blow-up occurs in form of wave breaking. In Duruk Mutlubas (2013), the local well-posedness of system (1) is proved for initial data in H^s with $s > \frac{3}{2}$ using Kato's semigroup method for quasi-linear equations. Orbital stability and existence of solitary waves for system (1) was obtained in Duruk Mutlubas and Geyer (2013), Geyer (2012). Mi and Mu (2013) investigated the local well-posedness of system (1) in Besov space using Littlewood–Paley decomposition and transport equation theory, and proposed that if initial data u_0 is analytic its solutions are analytic. Moreover, persistence properties on strong solutions were also presented (see Mi & Mu, 2013).

One of the close relatives of the first equation of problem (1) is the rod wave equation (Dai, 1998; Dai & Huo, 2000)

$$u_t - u_{txx} + 3uu_x = \gamma(2u_x u_{xx} + uu_{xxx}), \quad t > 0, \quad x \in \mathbf{R}, \tag{2}$$

where $\gamma \in \mathbf{R}$ and $u = u(t, x)$ stands for the radial stretch relative to a prestressed state in non-dimensional variables. Equation (2) is a model for finite-length and small-amplitude axial-radial deformation waves in the cylindrical compressible hyperelastic rods. Since Equation (2) was derived by Dai (1998), Dai and Huo (2000, many works have been carried out to investigate its dynamic properties. In Constantin and Strauss (2000), Constantin and Strauss studied the Cauchy problem of the rod equation on the line (nonperiodic case), where the local well-posedness and blow-up solutions were discussed. Moreover, they also proved the stability of solitary waves for the equation (see Constantin & Strauss, 2000). Later, Yin (2003,2004) and Hu and Yin (2010) discussed the smooth solitary waves and blow-up solutions. Zhou (2006), the precise blow-up scenario and several blow-up results of strong solutions to the rod equation on the circle (periodic case) were presented. For other techniques to study the problems relating to various dynamic properties of other shallow water wave equations, the reader is referred to Coclite, Holden, and Karlsen (2005), Yan, Li, and Zhang (2014), Fu, Liu, and Qu (2012), Guo and Wang (2014), Himonas, Misiolek, Ponce, and Zhou (2007), Holden and Raynaud (2009), Li and Olver (2000), Qu, Fu, and Liu (2014), Lai (2013) and the reference therein.

Xin and Zhang (2000) use the limit method of viscous approximations to analyze the existence of global weak solutions for Equation (2) with $\gamma = 1$ (Namely, Camassa–Holm equation). Motivated by the desire to extend the works (Xin & Zhang, 2000), the objective of this paper was to establish the existence of global weak solutions for the system (1) in the space $C([0, \infty) \times \mathbf{R}) \bigcap L^\infty([0, \infty); H^1(\mathbf{R}))$ under the assumption $u_0(x) \in H^1(\mathbf{R})$. Following the idea in Xin and Zhang (2000), the limit method of viscous approximations is employed to establish the existence of the global weak solution for system(1). In our analysis, a new one-sided lower bound (see Lemma 3.4) and the higher integrability estimate (see Lemma 3.3), which ensure that weak convergence of q_ε is equal to strong convergence, play a crucial role in establishing the existence of global weak solutions.

The rest of this paper is as follows. The main result is presented in Section 2. In Section 3, we state the viscous problem and give a corresponding result. Strong compactness of the derivative of viscous approximations is obtained in Section 4. Section 5 completes the proof of the main result.

2. The main results

Using the Green function $G(x) = \sqrt{\frac{3}{\mu}} e^{-2\sqrt{\frac{3}{\mu}}|x|}$, we have $(1 - \frac{\mu}{12}\partial_x^2)^{-1} f = G(x) * f$ for all $f \in L^2$, and $G * (u - \frac{\mu}{12}u_{xx}) = u$, where we denote by $*$ the convolution. Then we can rewrite system(1) as follows

$$\begin{cases} u_t - (1 + \frac{7}{2}\sigma u)u_x + \partial_x(1 - \frac{\mu}{12}\partial_x^2)^{-1}(2u + \frac{5}{2}\sigma u^2 + \frac{l}{3}\sigma^2 u^3 + \frac{k}{4}\sigma^3 u^4 \\ \qquad\qquad\qquad\qquad\qquad - \frac{7}{48}\sigma\mu u_x^2) = 0, \\ u(0, x) = u_0, \end{cases} \tag{3}$$

which is also equivalent to the elliptic-hyperbolic system

$$\begin{cases} u_t - \left(1 + \frac{7}{2}\sigma u\right)u_x + \frac{\partial P}{\partial x} = 0, \\ \frac{\partial P}{\partial x} = \partial_x \Lambda^{-2}\left[2u + \frac{5}{2}\sigma u^2 + \frac{\iota}{3}\sigma^2 u^3 + \frac{\kappa}{4}\sigma^3 u^4 - \frac{7}{48}\sigma\mu u_x^2\right], \\ u(0,x) = u_0(x), \end{cases} \tag{4}$$

where $\Lambda = (1 - \frac{\mu}{12}\partial_x^2)^{\frac{1}{2}}$.

Now we give the definition of a weak solution to the Cauchy problem (3) or (4).

Definition 2.1 A continuous function u:$[0,\infty) \times \mathbf{R} \to \mathbf{R}$ is said to be a global weak solution to the Cauchy problem (4) if

(i) $u \in C([0,\infty) \times \mathbf{R}) \cap L^\infty([0,\infty); H^1(\mathbf{R}))$;

(ii) $\| u(t,\cdot) \|_{H^1(\mathbf{R})} \leq \| u_0 \|_{H^1(\mathbf{R})}$, for every $t > 0$;

(iii) $u = u(t,x)$ satisfies (4) in the sense of distributions and takes on the initial value pointwise.

The existence of global weak solutions to the Cauchy problem (4) will be established by proving compactness of a sequence of smooth functions $\{u_\varepsilon\}_{\varepsilon>0}$ solving the following viscous problem

$$\begin{cases} \frac{\partial u_\varepsilon}{\partial t} - \left(1 + \frac{7}{2}\sigma u_\varepsilon\right)\frac{\partial u_\varepsilon}{\partial x} + \frac{\partial P_\varepsilon}{\partial x} = \varepsilon\frac{\partial^2 u_\varepsilon}{\partial x^2}, \\ \frac{\partial P_\varepsilon}{\partial x} = \partial_x \Lambda^{-2}\left[2u_\varepsilon + \frac{5}{2}\sigma u_\varepsilon^2 + \frac{\iota}{3}\sigma^2 u_\varepsilon^3 + \frac{\kappa}{4}\sigma^3 u_\varepsilon^4 - \frac{7}{48}\sigma\mu(\frac{\partial u_\varepsilon}{\partial x})^2\right], \\ u_\varepsilon(0,x) = u_{\varepsilon,0}(x). \end{cases} \tag{5}$$

The main result of present paper is collected in following theorem.

THEOREM 2.2 *Assume that $u_0(x) \in H^1(\mathbf{R})$. Then the Cauchy problem (4) has a global weak solution u(t, x) in the sense of Definition 2.1. In addition, there is a positive constant $C = C(\| u_0 \|_{H^1(\mathbf{R})})$, independent of ε, such that*

$$-\frac{4}{7\sigma t} - \sqrt{\frac{4C}{7\sigma}} \leq \frac{\partial u_\varepsilon(t,x)}{\partial x}, \text{ for } (t,x) \in [0,T) \times \mathbf{R}. \tag{6}$$

3. Viscous approximations

Defining

$$\phi(x) = \begin{cases} e^{\frac{1}{x^2-1}}, & |x| < 1, \\ 0, & |x| \geq 1, \end{cases} \tag{7}$$

and setting the mollifier $\phi_\varepsilon(x) = \varepsilon^{-\frac{1}{4}}\phi(\varepsilon^{-\frac{1}{4}}x)$ with $0 < \varepsilon < \frac{1}{4}$ and $u_{\varepsilon,0} = \phi_\varepsilon * u_0$, we know that $u_{\varepsilon,0} \in C^\infty$ for any $u_0 \in H^s, s > 0$ (see Lai & Wu, 2010).

In fact, suitably choosing the mollifier, we have

$$\| u_{\varepsilon,0} \|_{H^1(\mathbf{R})} \leq \| u_0 \|_{H^1(\mathbf{R})}, \quad \text{and} \quad u_{\varepsilon,0} \to u_0 \quad \text{in} \quad H^1(\mathbf{R}). \tag{8}$$

Differentiating the first equation of problem (5) with respect to variable x and letting $q_\varepsilon(t,x) = \frac{\partial u_\varepsilon}{\partial x}$, we have

$$\begin{aligned}
&\frac{\partial q_\varepsilon}{\partial t} - \frac{\partial q_\varepsilon}{\partial x} - \varepsilon\frac{\partial^2 q_\varepsilon}{\partial x^2} - \frac{7}{4}\sigma q_\varepsilon^2 - \frac{7}{2}\sigma u_\varepsilon \frac{\partial q_\varepsilon}{\partial x}\\
&= \frac{12}{\mu}(2u_\varepsilon + \frac{5}{2}\sigma u_\varepsilon^2 + \frac{l}{3}\sigma^2 u_\varepsilon^3 + \frac{k}{4}\sigma^3 u_\varepsilon^4)\\
&\quad - \frac{12}{\mu}\Lambda^{-2}\left[2u_\varepsilon + \frac{5}{2}\sigma u_\varepsilon^2 + \frac{l}{3}\sigma^2 u_\varepsilon^3 + \frac{k}{4}\sigma^3 u_\varepsilon^4 - \frac{7}{48}\sigma\mu(\frac{\partial u_\varepsilon}{\partial x})^2\right]\\
&= Q_\varepsilon(t,x).
\end{aligned} \tag{9}$$

The starting point of our analysis is the following well-posedness result for problem (5).

Lemma 3.1 *Assume $u_0 \in H^1(\mathbf{R})$. For any $l \geq 2$, there exists a unique solution $u_\varepsilon \in C([0,\infty); H^l(\mathbf{R}))$ to the Cauchy problem (5). Moreover, for any $t > 0$, it holds that*

$$\begin{aligned}
&\int_{\mathbf{R}} (u_\varepsilon^2 + \frac{\mu}{12}(\frac{\partial u_\varepsilon}{\partial x}))^2 dx + 2\varepsilon\int_0^t\int_{\mathbf{R}} ((\frac{\partial u_\varepsilon}{\partial x})^2 + \frac{\mu}{12}(\frac{\partial^2 u_\varepsilon}{\partial x^2})^2)(s,x)dxds\\
&= \int_{\mathbf{R}} (u_{\varepsilon,0}^2 + \frac{\mu}{12}(\frac{\partial u_{\varepsilon,0}}{\partial x}))^2 dx < (1+\frac{\mu}{12})\int_{\mathbf{R}} (u_{\varepsilon,0}^2 + (\frac{\partial u_{\varepsilon,0}}{\partial x}))^2 dx\\
&= (1+\frac{\mu}{12}) \parallel u_{\varepsilon,0} \parallel_{H^1(\mathbf{R})}^2 .
\end{aligned} \tag{10}$$

Proof For any $l \geq 2$ and $u_0 \in H^1(\mathbf{R})$, we have $u_{\varepsilon,0} \in C([0,\infty); H^l(\mathbf{R}))$. From Theorem 2.1 in Coclite et al. (2005), we infer that problem (5) has a unique solution $u_\varepsilon \in C([0,\infty); H^l(\mathbf{R}))$.

The first equation of (5) is rewritten as

$$\begin{aligned}
&\frac{\partial u_\varepsilon}{\partial t} + \frac{\partial u_\varepsilon}{\partial x} + \frac{3}{2}\sigma u_\varepsilon \frac{\partial u_\varepsilon}{\partial x} + l\sigma^2 u_\varepsilon^2 \frac{\partial u_\varepsilon}{\partial x} + k\sigma^3 u_\varepsilon^3 \frac{\partial u_\varepsilon}{\partial x} + \frac{\mu}{12}(\frac{\partial^3 u_\varepsilon}{\partial x^3} - \frac{\partial^3 u_\varepsilon}{\partial t\partial x^2})\\
&+ \frac{7}{24}\sigma\mu(u_\varepsilon \frac{\partial^3 u_\varepsilon}{\partial x^3} + 2\frac{\partial u_\varepsilon}{\partial x}\frac{\partial^2 u_\varepsilon}{\partial x^2})\\
&= \varepsilon(\frac{\partial^2 u_\varepsilon}{\partial x^2} - \frac{\mu}{12}\frac{\partial^4 u_\varepsilon}{\partial x^4}).
\end{aligned} \tag{11}$$

Multiplying (11) by u_ε, we derive that

$$\frac{1}{2}\frac{d}{dt}\int_{\mathbf{R}} (u_\varepsilon^2 + \frac{\mu}{12}(\frac{\partial u_\varepsilon}{\partial x}))^2 dx + \varepsilon\int_{\mathbf{R}} ((\frac{\partial u_\varepsilon}{\partial x})^2 + \frac{\mu}{12}(\frac{\partial^2 u_\varepsilon}{\partial x^2})^2)dx = 0, \tag{12}$$

which finishes the proof. □

For simplicity, in this paper, let c denote any positive constant which is independent of the parameter ε. From Lemma 3.1, we have

$$\parallel u_\varepsilon \parallel_{L^\infty(\mathbf{R})} \leq c \parallel u_\varepsilon \parallel_{H^1(\mathbf{R})} \leq c \parallel u_{\varepsilon,0} \parallel_{H^1(\mathbf{R})} \leq c \parallel u_0 \parallel_{H^1(\mathbf{R})} . \tag{13}$$

Lemma 3.2 *For $0 < t < T$, there exists a positive constant $C = C(\parallel u_0 \parallel_{H^1(\mathbf{R})})$, independent of ε, such that*

$$\parallel P_\varepsilon(t,\cdot) \parallel_{L^\infty(\mathbf{R})} \leq C, \tag{14}$$

$$\parallel P_\varepsilon(t,\cdot) \parallel_{L^1(\mathbf{R})} \leq C, \tag{15}$$

$$\parallel P_\varepsilon(t,\cdot) \parallel_{L^2(\mathbf{R})} \leq C \tag{16}$$

and

$$\| \frac{\partial P_\varepsilon(t,\cdot)}{\partial x} \|_{L^\infty(\mathbf{R})} \le C, \tag{17}$$

$$\| \frac{\partial P_\varepsilon(t,\cdot)}{\partial x} \|_{L^1(\mathbf{R})} \le C, \tag{18}$$

$$\| \frac{\partial P_\varepsilon(t,\cdot)}{\partial x} \|_{L^2(\mathbf{R})} \le C, \tag{19}$$

$$\| Q_\varepsilon(t,\cdot) \|_{L^\infty(\mathbf{R})} \le C, \tag{20}$$

where $u_\varepsilon = u_\varepsilon(t,x)$ is the unique solution of (5) and

$$\begin{aligned} Q_\varepsilon(t,\cdot) &= \frac{12}{\mu}(2u_\varepsilon + \frac{5}{2}\sigma u_\varepsilon^2 + \frac{l}{3}\sigma^2 u_\varepsilon^3 + \frac{k}{4}\sigma^3 u_\varepsilon^4) \\ &\quad - \frac{12}{\mu}\Lambda^{-2}\left[2u_\varepsilon + \frac{5}{2}\sigma u_\varepsilon^2 + \frac{l}{3}\sigma^2 u_\varepsilon^3 + \frac{k}{4}\sigma^3 u_\varepsilon^4 - \frac{7}{48}\sigma\mu q_\varepsilon^2\right], \\ q_\varepsilon &= \frac{\partial u_\varepsilon(t,x)}{\partial x}. \end{aligned} \tag{21}$$

Proof In the proof of this lemma, we will use the identity

$$\Lambda^{-2}g(x) = \sqrt{\frac{3}{\mu}} \int_{\mathbf{R}} e^{-\sqrt{\frac{12}{\mu}}|x-y|} g(y)dy \text{ for } \quad g(x) \in L^2(\mathbf{R}). \tag{22}$$

For simplicity, setting $u_\varepsilon(t,x) = u(t,x)$, we have

$$P_\varepsilon(t,\cdot) = \frac{12}{\mu}\Lambda^{-2}\left[2u + \frac{5}{2}\sigma u^2 + \frac{l}{3}\sigma^2 u^3 + \frac{k}{4}\sigma^3 u^4 - \frac{7}{48}\sigma\mu q^2\right] \tag{23}$$

and

$$\frac{\partial P_\varepsilon(t,\cdot)}{\partial x} = \frac{12}{\mu}\partial_x\Lambda^{-2}\left[2u + \frac{5}{2}\sigma u^2 + \frac{l}{3}\sigma^2 u^3 + \frac{k}{4}\sigma^3 u^4 - \frac{7}{48}\sigma\mu q^2\right]. \tag{24}$$

Note that $\int_{\mathbf{R}} e^{-\sqrt{\frac{12}{\mu}}|x-y|}dx = \sqrt{\frac{\mu}{3}}$ for $x \in \mathbf{R}$. Using (22), one has

$$\begin{aligned} |P_\varepsilon| &= \frac{12}{\mu}|\Lambda^{-2}\left[2u + \frac{5}{2}\sigma u^2 + \frac{l}{3}\sigma^2 u^3 + \frac{k}{4}\sigma^3 u^4 - \frac{7}{48}\sigma\mu q^2\right]| \\ &\le \frac{12}{\mu}\sqrt{\frac{3}{\mu}}|\int_{\mathbf{R}} e^{-\sqrt{\frac{12}{\mu}}|x-y|}[2u + \frac{5}{2}\sigma u^2 + \frac{l}{3}\sigma^2 u^3 + \frac{k}{4}\sigma^3 u^4]dy| \\ &\quad + \frac{7\sigma}{4}\sqrt{\frac{3}{\mu}}|\int_{\mathbf{R}} e^{-\sqrt{\frac{12}{\mu}}|x-y|}q^2 dy| \\ &\le c(\|u_0\|_{H^1(\mathbf{R})} + \|u_0\|^2_{H^1(\mathbf{R})} + \|u_0\|^3_{H^1(\mathbf{R})} + \|u_0\|^4_{H^1(\mathbf{R})}), \end{aligned} \tag{25}$$

which proves (14). □

In view of Lemma 3.1 and Tonelli theorem, one has

$$|\frac{7\sigma}{4}\Lambda^{-2}(q^2)| = \frac{7\sigma}{4}\sqrt{\frac{3}{\mu}}|\int_{\mathbf{R}} e^{-\sqrt{\frac{12}{\mu}}|x-y|}q^2 dy| \le c\|u_0\|^2_{H^1(\mathbf{R})}, \tag{26}$$

and then, we get

$$\begin{aligned} \int_{\mathbf{R}} |\frac{7\sigma}{4}\Lambda^{-2}(q^2)|dx &= \frac{7\sigma}{4}\sqrt{\frac{3}{\mu}}\int_{\mathbf{R}}|\int_{\mathbf{R}} e^{-\sqrt{\frac{12}{\mu}}|x-y|}q^2 dy|dx \\ &\le c\|u_0\|^2_{H^1(\mathbf{R})}. \end{aligned} \tag{27}$$

Using the Tonelli theorem and the Hölder inequality, it holds

$$\begin{aligned}&\int_{\mathbf{R}}\left|\frac{12}{\mu}\Lambda^{-2}\left[2u+\frac{5}{2}\sigma u^2+\frac{l}{3}\sigma^2u^3+\frac{k}{4}\sigma^3u^4\right]\right|dx\\&\leq\frac{12}{\mu}\sqrt{\frac{3}{\mu}}\int_{\mathbf{R}}\int_{\mathbf{R}}e^{-\sqrt{\frac{12}{\mu}}|x-y|}\left|2u+\frac{5}{2}\sigma u^2+\frac{l}{3}\sigma^2u^3+\frac{k}{4}\sigma^3u^4\right|dydx\\&\leq c\left(\|u_0\|^1_{H^1(\mathbf{R})}+\|u_0\|^2_{H^1(\mathbf{R})}+\|u_0\|^3_{H^1(\mathbf{R})}+\|u_0\|^4_{H^1(\mathbf{R})}\right).\end{aligned}\tag{28}$$

Making use of (27) and (28), we complete the proof of (15).

From (26)–(27) and the Hölder inequality, we have

$$\begin{aligned}\int_{\mathbf{R}}\left|\frac{7\sigma}{4}\Lambda^{-2}(q^2)\right|^2dx&\leq\left\|\frac{7\sigma}{4}\Lambda^{-2}(q^2)\right\|_{L^\infty}\left\|\frac{7\sigma}{4}\Lambda^{-2}(q^2)\right\|_{L^1}\\&\leq c\|u_0\|^4_{H^1(\mathbf{R})}.\end{aligned}\tag{29}$$

Hence,

$$\left\|\frac{7\sigma}{4}\Lambda^{-2}(q^2)\right\|_{L^2}\leq c\|u_0\|^2_{H^1(\mathbf{R})}.\tag{30}$$

By (25) and (28), one has

$$\begin{aligned}&\int_{\mathbf{R}}\left|\frac{12}{\mu}\Lambda^{-2}[2u+\frac{5}{2}\sigma u^2+\frac{l}{3}\sigma^2u^3+\frac{k}{4}\sigma^3u^4]\right|^2dx\\&\leq c\left(\|u_0\|^2_{H^1(\mathbf{R})}+\|u_0\|^4_{H^1(\mathbf{R})}+\|u_0\|^6_{H^1(\mathbf{R})}+\|u_0\|^8_{H^1(\mathbf{R})}\right).\end{aligned}\tag{31}$$

From (30) and (31), we deduce (16).

On the other hand, from (24), we derive that

$$\begin{aligned}\frac{\partial P_\varepsilon}{\partial x}&=\frac{12}{\mu}\Lambda^{-2}\left[2u+\frac{5}{2}\sigma u^2+\frac{l}{3}\sigma^2u^3+\frac{k}{4}\sigma^3u^4-\frac{7}{48}\sigma\mu q^2\right]_x\\&=\frac{12}{\mu}\sqrt{\frac{3}{\mu}}\int_{\mathbf{R}}e^{-\sqrt{\frac{12}{\mu}}|x-y|}\left[2u+\frac{5}{2}\sigma u^2+\frac{l}{3}\sigma^2u^3+\frac{k}{4}\sigma^3u^4-\frac{7}{48}\sigma\mu q^2\right]_y dy\\&=\frac{36}{\mu^2}\int_{\mathbf{R}}e^{-\sqrt{\frac{12}{\mu}}|x-y|}[2u+\frac{5}{2}\sigma u^2+\frac{l}{3}\sigma^2u^3+\frac{k}{4}\sigma^3u^4\\&\quad-\frac{7}{48}\sigma\mu q^2]sign(y-x)dy.\end{aligned}\tag{32}$$

Inequalities (17), (18), and (19) are direct consequences of (25), (27), (28), (30), and (31).

Finally, note that

$$\begin{aligned}&\left|\frac{12}{\mu}(2u+\frac{5}{2}\sigma u^2+\frac{l}{3}\sigma^2u^3+\frac{k}{4}\sigma^3u^4)\right|\\&\leq\frac{12}{\mu}\left(2|u|+\frac{5}{2}\sigma|u|^2+\left|\frac{l}{3}\right|\sigma^2|u|^3+\left|\frac{k}{4}\right|\sigma^3|u|^4\right)\\&\leq c\left(\|u_0\|_{H^1(\mathbf{R})}+\|u_0\|^2_{H^1(\mathbf{R})}+\|u_0\|^3_{H^1(\mathbf{R})}+\|u_0\|^4_{H^1(\mathbf{R})}\right).\end{aligned}\tag{33}$$

Using (25), we obtain (20).

Lemma 3.3 *Let $0 < \phi < 1, T > 0$, and $a, b \in \mathbf{R}, a < b$. Then there exists a positive constant C_1 depending only on $\|u_0\|_{H^1(\mathbf{R})}$, ϕ, T, a and b, but independent of ε, such that*

$$\int_0^T \int_a^b |\frac{\partial u_\varepsilon}{\partial x}(t,x)|^{2+\phi} dtdx \leq C_1, \tag{34}$$

where $u_\varepsilon = u_\varepsilon(t,x)$ is the unique solution of (5).

Proof The proof of Lemma 3.3 is similar to that of Lemma 4.1 in Xin and Zhang (2000). Here, we omit its proof. □

Lemma 3.4 *For an arbitrary $T > 0$, the following estimate on the first-order spatial derivative holds*

$$-\frac{4}{7\sigma t} - \sqrt{\frac{4C}{7\sigma}} \leq \frac{\partial u_\varepsilon(t,x)}{\partial x}, \quad \text{for} \quad (t,x) \in [0,T) \times \mathbf{R}. \tag{35}$$

Proof Using (9), we get

$$\frac{\partial(-q_\varepsilon)}{\partial t} - \frac{\partial(-q_\varepsilon)}{\partial x} - \varepsilon\frac{\partial^2(-q_\varepsilon)}{\partial x^2} + \frac{7}{4}\sigma(-q_\varepsilon)^2 - \frac{7}{2}\sigma u_\varepsilon \frac{\partial(-q_\varepsilon)}{\partial x} = -Q_\varepsilon(t,x) \leq C. \tag{36}$$

Let $f = f(t)$ be the solution of

$$\frac{df}{dt} + \frac{7}{4}\sigma f^2 = C, \quad t > 0, \quad f(0) = \|\frac{\partial u_{\varepsilon,0}}{\partial x}\|_{L^\infty}. \tag{37}$$

Since $f = f(t)$ is a supersolution of the parabolic equation (36) with initial value $u_{\varepsilon,0}$, due to the comparison principle for parabolic equations, we get

$$-q_\varepsilon(t,x) \leq f(t,x).$$

Consider the function $F(t) = \frac{4}{7\sigma t} + \sqrt{\frac{4C}{7\sigma}}$, observing that $\frac{dF}{dt} + \frac{7}{4}\sigma F^2 - C = \frac{2}{t}\sqrt{\frac{4C}{7\sigma}} > 0$ for any $t > 0$ and using the comparison principle for ordinary differential equations, we have $f(t) \leq F(t)$ for all $t > 0$. It completes the proof. □

Lemma 3.5 *There exists a sequence $\{\varepsilon_j\}_{j\in N}$ tending to zero and a function $u \in L^\infty([0,\infty); H^1(\mathbf{R})) \cap H^1([0,T] \times \mathbf{R})$, for each $T \geq 0$, such that*

$$u_{\varepsilon_j} \rightharpoonup u \quad \text{in} \quad H^1([0,T] \times \mathbf{R}), \quad \text{for each} \quad T \geq 0, \tag{38}$$

$$u_{\varepsilon_j} \to u \quad \text{in} \quad L^\infty_{loc}([0,T] \times \mathbf{R}), \tag{39}$$

where $u_\varepsilon = u_\varepsilon(t,x)$ is the unique solution of (5).

Proof For fixed $T > 0$, using Lemmas 3.1 and 3.3, and

$$\frac{\partial u_\varepsilon}{\partial t} - \left(1 + \frac{7}{2}\sigma u_\varepsilon\right)\frac{\partial u_\varepsilon}{\partial x} + \frac{\partial P_\varepsilon}{\partial x} = \varepsilon\frac{\partial^2 u_\varepsilon}{\partial x^2},$$

we obtain

$$\|\frac{\partial u_\varepsilon}{\partial t}\|_{L^2([0,T]\times\mathbf{R})} \le \sqrt{\frac{12T}{\mu}}\|u_0\|_{H^1(\mathbf{R})} + \sqrt{\frac{42\sigma T}{\mu}}\|u_0\|^2_{H^1(\mathbf{R})} + \sqrt{C}T + \sqrt{\frac{6\varepsilon}{\mu}}\|u_0\|_{H^1(\mathbf{R})}. \tag{40}$$

Hence $\{u_\varepsilon\}$ is uniformly bounded in $L^\infty([0,\infty); H^1(\mathbf{R})) \cup H^1([0,T]\times\mathbf{R})$ and (38) follows.

Observe that, for each $0 \le s,t \le T$,

$$\begin{aligned}\|u_\varepsilon(t,\cdot) - u_\varepsilon(s,\cdot)\|^2_{L^2(\mathbf{R})} &= \int_{\mathbf{R}}(\int_s^t \frac{\partial u_\varepsilon}{\partial t}(\iota,x)d\iota)^2 dx \\ &\le \sqrt{|t-s|}\int_0^T\int_{\mathbf{R}}(\frac{\partial u_\varepsilon}{\partial t}(\iota,x))^2 d\iota dx.\end{aligned} \tag{41}$$

Moreover, $\{u_\varepsilon\}$ is uniformly bounded in $L^\infty([0,\infty); H^1(\mathbf{R}))$ and $H^1(\mathbf{R}) \subset L^\infty_{loc}(\mathbf{R}) \subset L^2_{loc}(\mathbf{R})$. Using the results in Coclite et al. (2005), we know that (39) holds. □

LEMMA 3.6 *There exists a sequence $\{\varepsilon_j\}_{j\in N}$ tending to zero and a function $P \in L^\infty([0,\infty)\times\mathbf{R})$ such that for each $0 < p < \infty$,*

$$P_{\varepsilon_j} \to P \quad \text{strongly in} \quad L^p_{loc}([0,\infty)\times\mathbf{R}). \tag{42}$$

Proof Using Lemma 3.2, we have the existence of pointwise convergence subsequence P_{ε_j} which is uniformly bounded in $L^\infty([0,\infty)\times\mathbf{R})$. Inequalities (15) and (16) derive that (42) holds. □

Throughout this paper we use overbars to denote weak limits.

LEMMA 3.7 *There exists a sequence $\{\varepsilon_j\}_{j\in N}$ tending to zero and two function $q \in L^p_{loc}([0,\infty)\cap\mathbf{R})$, $q^2 \in L^r_{loc}([0,\infty)\cap\mathbf{R})$ such that*

$$q_{\varepsilon_j} \rightharpoonup q \quad \text{in} \quad L^p_{loc}([0,\infty)\cap\mathbf{R}), \quad q_{\varepsilon_j} \overset{\star}{\rightharpoonup} q \quad \text{in} \quad L^\infty_{loc}([0,\infty);L^2(\mathbf{R})), \tag{43}$$

$$q^2_{\varepsilon_j} \rightharpoonup \overline{q^2} \quad \text{in} \quad L^\infty_{loc}([0,T]\times\mathbf{R}), \tag{44}$$

for each $1 < p < 3$ and $1 < r < \frac{3}{2}$. Moreover,

$$q^2(t,x) \le \overline{q^2}(t,x) \quad \text{for almost every} \quad (t,x)\in[0,T]\times\mathbf{R} \tag{45}$$

and

$$\frac{\partial u}{\partial x} = q \quad \text{in the sense of distributions on} \quad [0,T]\times\mathbf{R}. \tag{46}$$

Proof Equations (43) and (44) are direct consequences of Lemmas 3.1. and 3.3. Inequality (45) is valid because of the weak convergence in (44). Finally, (46) is a consequence of definition of q_ε, Lemma 3.5. and (43).

In the following, for notational convenience, we replace the sequence $\{u_{\varepsilon_j}\}_{j\in N}$, $\{q_{\varepsilon_j}\}_{j\in N}$, and $\{Q_{\varepsilon_j}\}_{j\in N}$ by $\{u_\varepsilon\}_{\varepsilon>0}$, $\{q_\varepsilon\}_{\varepsilon>0}$, and $\{Q_\varepsilon\}_{\varepsilon>0}$, separately.

Using (43), we conclude that for any convex function $\eta \in C^1(\mathbf{R})$ with η' bounded, Lipschitz continuous on $\mathbf{R}$ and any $1 < p < 3$ we get

$$\eta(q_\varepsilon) \rightharpoonup \overline{\eta(q)} \quad \text{in} \quad L^p_{loc}([0,\infty)\times\mathbf{R}) \tag{47}$$

and

$$\eta(q_\varepsilon) \rightharpoonup \overline{\eta(q)} \quad \text{in} \quad L^\infty_{loc}([0,\infty); L^2(\mathbf{R})). \tag{48}$$

Multiplying Equation (9) by $\eta'(q_\varepsilon)$ yields □

$$\begin{aligned}&\frac{\partial}{\partial t}\eta(q_\varepsilon) - \frac{\partial}{\partial x}\eta(q_\varepsilon) - \varepsilon\frac{\partial^2}{\partial x^2}\eta(q_\varepsilon) + \varepsilon\eta''(q_\varepsilon)(\frac{\partial}{\partial x}q_\varepsilon)^2 - \frac{\partial}{\partial x}(\frac{7\sigma}{2}u_\varepsilon\eta(q_\varepsilon))\\ &+\frac{7\sigma}{2}q_\varepsilon\eta(q_\varepsilon) = \frac{7\sigma}{4}q_\varepsilon^2\eta'(q_\varepsilon) + Q_\varepsilon\eta'(q_\varepsilon).\end{aligned} \tag{49}$$

LEMMA 3.8 *For any convex $\eta \in C^1(\mathbf{R})$ with η' bounded, Lipschitz continuous on $\mathbf{R}$, it holds that*

$$\begin{aligned}&\frac{\partial}{\partial t}\overline{\eta(q)} - \frac{\partial}{\partial x}\overline{\eta(q)} - \frac{\partial}{\partial x}(\frac{7\sigma}{2}\overline{u\eta(q)}) + \frac{7\sigma}{2}\overline{q\eta(q)}\\ &\le \frac{7\sigma}{4}\overline{q^2\eta'(q)} + Q\overline{\eta'(q)},\end{aligned} \tag{50}$$

in the sense of distributions on $[0,T]\times\mathbf{R}$. Here $\overline{q\eta(q)}$ and $\overline{q^2\eta'(q)}$ denote the weak limits of $q_\varepsilon\eta(q_\varepsilon)$ and $q_\varepsilon^2\eta'(q_\varepsilon)$ in $L^r_{loc}([0,\infty)\times\mathbf{R})$, $1<r<2$, respectively.

Proof In (49), by the convexity of η, Lemmas 3.5.–3.7, sending $\varepsilon\to 0$, gives rise to the desired result. □

Remark 3.9 We know that

$$q = q_+ + q_- = \overline{q_+} + \overline{q_-}, \quad q^2 = (q_+)^2 + (q_-)^2, \quad \overline{q^2} = \overline{(q_+)^2} + \overline{(q\_)^2} \tag{51}$$

almost everywhere in $([0,\infty)\times\mathbf{R})$, where $\xi_+ := \xi_{\chi[0,+\infty)}(\xi)$, $\xi_- := \xi_{\chi(-\infty,0]}(\xi)$ for $\xi\in\mathbf{R}$.

LEMMA 3.10 *In the sense of distributions on $([0,\infty)\times\mathbf{R})$, it holds that*

$$\frac{\partial q}{\partial t} - \frac{\partial q}{\partial x} - \frac{7\sigma}{2}\frac{\partial}{\partial x}(uq) = -\frac{7\sigma}{4}\overline{q^2} + Q. \tag{52}$$

Proof Using Lemmas 3.5–3.8, (52) holds by sending $\varepsilon\to 0$ in (9). □

LEMMA 3.11 *For any $\eta\in C^1(\mathbf{R})$ with $\eta'\in L^\infty(\mathbf{R})$, it has*

$$\begin{aligned}&\frac{\partial\eta(q)}{\partial t} - \frac{7\sigma}{2}\frac{\partial}{\partial x}(u\eta(q))\\ &= -\frac{7\sigma}{2}q\eta(q) + \frac{7\sigma}{2}\eta'(q)(q^2 - \frac{1}{2}\overline{q^2}) + \frac{\partial\eta(q)}{\partial x} + Q\eta'(q)\end{aligned} \tag{53}$$

in the sense of distributions on $([0,\infty)\times\mathbf{R})$.

Proof Let $\{\omega_\delta\}_\delta$ be a family of mollifiers defined on $\mathbf{R}$. Defined $q_\delta(t,x) := (q(t,\cdot)*\omega_\delta)(x)$. The notation $*$ is the convolution with respect to the x variable. Multiplying (52) by $\eta'(q_\delta)$, it has

$$\begin{aligned}\frac{\partial\eta(q_\delta)}{\partial t} &= \eta'(q_\delta)\frac{\partial q_\delta}{\partial t} = \eta'(q_\delta)\frac{\partial q}{\partial t}*\omega_\delta\\ &= \eta'(q_\delta)\left(\frac{7\sigma}{2}\frac{\partial}{\partial x}(uq)*\omega_\delta + \frac{\partial}{\partial x}q*\omega_\delta - \frac{7\sigma}{4}\overline{q^2}*\omega_\delta + Q*\omega_\delta\right)\\ &= \eta'(q_\delta)\left(\frac{7\sigma}{2}uq_x*\omega_\delta + \frac{7\sigma}{2}q^2*\omega_\delta\right)\\ &\quad+\eta'(q_\delta)\left(q_x*\omega_\delta - \frac{7\sigma}{4}\overline{q^2}*\omega_\delta + Q*\omega_\delta\right)\end{aligned} \tag{54}$$

and

$$-\frac{7\sigma}{2}\frac{\partial}{\partial x}(u\eta(q_\delta)) = -\frac{7\sigma}{2}q\eta(q_\delta) - \frac{7\sigma}{2}u\eta'(q_\delta)(q_x * \omega_\delta). \tag{55}$$

Using the boundedness of η, η' and letting $\delta \to 0$ in the above two equations, we obtain (53).

Following the ideas in Xin and Zhang (2000), in next section we hope to improve the weak convergence of q_ε in (43) to strong convergence, and then we have an existence result for problem (4). Since the measure $(\overline{q^2} - q^2) \geq 0$, we will prove that if the measure is zero initially, then it will continue to be zero at all times $t > 0$. □

4. Strong convergence of q_ε

LEMMA 4.1 (see Coclite et al., 2005) *Assume $u_0 \in H^1(\mathbf{R})$. It holds that*

$$\lim_{t\to 0}\int_{\mathbf{R}} q^2(t,x)dx = \lim_{t\to 0}\int_{\mathbf{R}} \overline{q^2}(t,x)dx = \int_{\mathbf{R}}(\frac{\partial u_0}{\partial x})^2 dx. \tag{56}$$

LEMMA 4.2 (see Coclite et al., 2005) *If $u_0 \in H^1(\mathbf{R})$, for each $M > 0$, it has*

$$\lim_{t\to 0}\int_{\mathbf{R}}(\overline{\eta_M^{\pm}(q)}(t,x) - \eta_M^{\pm}(q(t,x)))dx = 0, \tag{57}$$

where

$$\eta_M(\xi) := \begin{cases} \frac{1}{2}\xi^2, & \text{if} |\xi| \leq M, \\ M|\xi| - \frac{1}{2}M^2, & \text{if} |\xi| > M, \end{cases} \tag{58}$$

and $\eta_M^+(\xi) := \eta_M(\xi)\chi_{[0,+\infty)}(\xi)$, $\eta_M^-(\xi) := \eta_M(\xi)\chi_{(-\infty,0]}(\xi)$ for $\xi \in \mathbf{R}$.

LEMMA 4.3 (see Coclite et al., 2005) *Let $M > 0$. Then for each $\xi \in \mathbf{R}$*

$$\begin{cases} \eta_M(\xi) = \frac{1}{2}\xi^2 - \frac{1}{2}(M - |\xi|)^2\chi_{(\infty,-M)\cap(M,\infty)}(\xi), \\ \eta_M'(\xi) = \xi + (M - |\xi|)sign(\xi)\chi_{(\infty,-M)\cap(M,\infty)}(\xi), \\ \eta_M^+(\xi) = \frac{1}{2}(\xi_+)^2 - \frac{1}{2}(M - \xi)^2\chi_{(M,\infty)}(\xi), \\ (\eta_M^+)'(\xi) = \xi_+ + (M - \xi)\chi_{(M,\infty)}(\xi), \\ \eta_M^-(\xi) = \frac{1}{2}(\xi_-)^2 - \frac{1}{2}(M + \xi)^2\chi_{(-\infty,-M)}(\xi), \\ (\eta_M^-)'(\xi) = \xi_- - (M + \xi)\chi_{(-\infty,-M)}(\xi). \end{cases} \tag{59}$$

LEMMA 4.4 *For almost all $t > 0$, it holds that*

$$\begin{aligned} \int_{\mathbf{R}}[\overline{\eta_M^+(q)} - \eta_M^+(q))]dx \leq & -\frac{7\sigma M}{4}\int_0^t\int_{\mathbf{R}} q(M-q)\chi_{(M,\infty)}(q)dxdt \\ & + \frac{7\sigma M}{4}\int_0^t\int_{\mathbf{R}} \overline{q(M-q)\chi_{(M,\infty)}(q)}dxdt \\ & + \int_0^t\int_{\mathbf{R}} \frac{7\sigma M}{4}(\overline{q^2} - q^2)dxds \\ & + \int_0^t\int_{\mathbf{R}} Q(s,x)(\overline{(\eta_M^+)'(q)} - (\eta_M^+)'(q))dxds. \end{aligned} \tag{60}$$

Proof For an arbitrary $T > 0\,(0 < t < T)$. Using (50) minus (53), and the entropy η_M^+ results in

$$\begin{aligned} & \frac{\partial}{\partial t}(\overline{\eta_M^+(q)} - \eta_M^+(q)) - \frac{7\sigma}{2}\frac{\partial}{\partial x}(u(\overline{\eta_M^+(q)} - \eta_M^+(q))) \\ & \leq -\frac{7\sigma}{2}(\overline{q\eta_M^+(q)} - q\eta_M^+(q)) + \frac{7\sigma}{4}(\overline{q^2(\eta_M^+)'(q)} - q^2(\eta_M^+)'(q)) \\ & \quad + \frac{\partial}{\partial x}(\overline{\eta_M^+(q)} - \eta_M^+(q)) + Q(\overline{(\eta_M^+)'(q)} - (\eta_M^+)'(q)) \\ & \quad - \frac{7\sigma}{4}(q^2 - \overline{q^2})(\eta_M^+)'(q). \end{aligned} \tag{61}$$

Since η_M^+ is increasing and $(\eta_M^+)' \leq M$, from (45), we have

$$0 \leq -\frac{7\sigma}{4}(q^2 - \overline{q^2})(\eta_M^+)'(q) \leq -\frac{7\sigma}{4}(q^2 - \overline{q^2})M. \tag{62}$$

It follows from Lemma 4.3 that

$$\begin{aligned} &q\eta_M^+(q) - \frac{1}{2}q^2(\eta_M^+)'(q) = -\frac{M}{2}q(M-q)\chi_{(M,\infty)}(q), \\ &\overline{q\eta_M^+(q)} - \frac{1}{2}\overline{q^2(\eta_M^+)'(q)} = -\frac{M}{2}\overline{q(M-q)\chi_{(M,\infty)}(q)}. \end{aligned} \tag{63}$$

From (61)–(63), we obtain the following result

$$\begin{aligned} &\frac{\partial}{\partial t}(\overline{\eta_M^+(q)} - \eta_M^+(q)) - \frac{7\sigma}{2}\frac{\partial}{\partial x}(u(\overline{\eta_M^+(q)} - \eta_M^+(q))) \\ &\leq -\frac{7\sigma M}{4}q(M-q)\chi_{(M,\infty)}(q) + \frac{7\sigma M}{4}\overline{q(M-q)\chi_{(M,\infty)}(q)} \\ &\quad + \frac{\partial}{\partial x}(\overline{\eta_M^+(q)} - \eta_M^+(q)) + Q(\overline{(\eta_M^+)'(q)} - (\eta_M^+)'(q)) \\ &\quad + \frac{7\sigma M}{4}(\overline{q^2} - q^2). \end{aligned} \tag{64}$$

Integrating the resultant inequality over $(0,t) \times \mathbf{R}$ yields

$$\begin{aligned} \int_{\mathbf{R}} [\overline{\eta_M^+(q)} - \eta_M^+(q)]dx \leq \lim_{t\to 0}&\int_{\mathbf{R}} [\overline{\eta_M^+(q)} - \eta_M^+(q)]dx \\ &- \frac{7\sigma M}{4}\int_0^t\int_{\mathbf{R}} q(M-q)\chi_{(M,\infty)}(q)dxdt \\ &+ \frac{7\sigma M}{4}\int_0^t\int_{\mathbf{R}} \overline{q(M-q)\chi_{(M,\infty)}(q)}dxdt \\ &+ \int_0^t\int_{\mathbf{R}} \frac{7\sigma M}{4}(\overline{q^2} - q^2)dxds \\ &+ \int_0^t\int_{\mathbf{R}} Q(s,x)(\overline{(\eta_M^+)'(q)} - (\eta_M^+)'(q))dxds. \end{aligned} \tag{65}$$

Using Lemma 4.2, we complete the proof. □

LEMMA 4.5 *For almost all $t > 0$, it holds that*

$$\int_{\mathbf{R}} \frac{1}{2}(\overline{q_-^2} - q_-^2)dx \leq \int_0^t\int_{\mathbf{R}} Q(\overline{q_-} - q_-)dxds. \tag{66}$$

Proof Let $M > 0$. Subtracting (53) from (50) and using entropy η_M^-, we deduce

$$\begin{aligned} &\frac{\partial}{\partial t}(\overline{\eta_M^-(q)} - \eta_M^-(q)) - \frac{7\sigma}{2}\frac{\partial}{\partial x}(u(\overline{\eta_M^-(q)} - \eta_M^-(q))) \\ &\leq -\frac{7\sigma}{2}(\overline{q\eta_M^-(q)} - q\eta_M^-(q)) + \frac{7\sigma}{4}(\overline{q^2(\eta_M^-)'(q)} - q^2(\eta_M^-)'(q)) \\ &\quad + \frac{\partial}{\partial x}(\overline{\eta_M^-(q)} - \eta_M^-(q)) + Q(\overline{(\eta_M^-)'(q)} - (\eta_M^-)'(q)) \\ &\quad - \frac{7\sigma}{4}(q^2 - \overline{q^2})(\eta_M^-)'(q). \end{aligned} \tag{67}$$

Since $-M \leq (\eta_M^-)' \leq 0$, we get

$$\frac{7\sigma M}{4}(q^2 - \overline{q^2}) \leq -\frac{7\sigma}{4}(q^2 - \overline{q^2})(\eta_M^-)'(q) \leq 0. \tag{68}$$

It follows from Lemma 4.3 that

$$q\eta_M^-(q) - \frac{1}{2}q^2(\eta_M^-)'(q) = -\frac{M}{2}q(M+q)\chi_{(-\infty,-M)}(q),$$
$$\overline{q\eta_M^-(q)} - \frac{1}{2}\overline{q^2(\eta_M^-)'(q)} = -\frac{M}{2}\overline{q(M+q)\chi_{(-\infty,-M)}(q)}. \tag{69}$$

Using (35), we can find sufficiently large $M > 0$ such that $q \geq -M$. Let $\Omega_M = (\frac{4}{7\sigma(M-\sqrt{\frac{4C}{7\sigma}})}, \infty) \times \mathbf{R}$. Applying Lemma 4.2 gives rise to

$$q\eta_M^-(q) - \frac{1}{2}q^2(\eta_M^-)'(q) = \overline{q\eta_M^-(q)} - \frac{1}{2}\overline{q^2(\eta_M^-)'(q)} = 0, \quad \text{in} \quad \Omega_M. \tag{70}$$

In Ω_M, it has

$$\eta_M^-(q) = \frac{1}{2}(q_-)^2, \qquad (\eta_M^-)'(q) = q_-,$$
$$\overline{\eta_M^-(q)} = \frac{1}{2}\overline{(q_-)^2}, \qquad \overline{(\eta_M^-)'(q)} = \overline{q_-}. \tag{71}$$

Substituting (68) and (69) into (67) gives

$$\frac{\partial}{\partial t}(\overline{\eta_M^-(q)} - \eta_M^-(q)) - \frac{7\sigma}{2}\frac{\partial}{\partial x}(u(\overline{\eta_M^-(q)} - \eta_M^-(q)))$$
$$\leq \frac{\partial}{\partial x}(\overline{\eta_M^-(q)} - \eta_M^-(q)) + Q(\overline{(\eta_M^-)'(q)} - (\eta_M^-)'(q)). \tag{72}$$

Integrating the above inequality over $(0,t) \times \mathbf{R}$, by (71), we obtain/ □

$$\int_{\mathbf{R}} \frac{1}{2}(\overline{q_-^2} - q_-^2)dx \leq \int_0^t \int_{\mathbf{R}} Q(\overline{q_-} - q_-)dxds. \tag{73}$$

LEMMA 4.6 *It holds that*

$$\overline{q^2} = q^2$$

almost everywhere in $[0,t) \times (-\infty,\infty)$.

Proof It follows from Lemma 4.3 that

$$\overline{\eta_M^+(q)} - \eta_M^+(q) = \frac{1}{2}(\overline{(q_+)^2} - (q_+)^2) + \frac{1}{2}(M-q)^2\chi_{(-\infty,-M)}(q)$$
$$- \frac{1}{2}\overline{(M-q)^2\chi_{(-\infty,-M)}(q)}. \tag{74}$$

From (60) and (74) , we have

$$\int_{\mathbf{R}}(\overline{\eta_M^+(q)} - \eta_M^+(q))dx \leq -\frac{7\sigma M^2}{4}\int_0^t\int_{\mathbf{R}}(M-q)\chi_{(M,\infty)}(q)dxds$$
$$+ \frac{7\sigma M^2}{4}\int_0^t\int_{\mathbf{R}}\overline{(M+q)\chi_{(M,\infty)}(q)}dxds$$
$$+ \frac{7\sigma M}{2}\int_0^t\int_{\mathbf{R}}(\overline{\eta_M^+(q)} - \eta_M^+(q))dxds$$
$$+ \frac{7\sigma M}{2}\int_0^t\int_{\mathbf{R}}\frac{1}{2}(\overline{q_-^2} - q_-^2)dxds$$
$$+ \int_0^t\int_{\mathbf{R}} Q(\overline{(\eta_M^+)'(q)} - (\eta_M^+)'(q))dxds, \tag{75}$$

where we used the identity $M(M-q)^2 + Mq(M-q) = M^2(M-q)$.

Combining (66) with (75) gets

$$\begin{aligned}
&\int_{\mathbf{R}}(\frac{1}{2}[\overline{(q_-)^2}-(q_-)^2]+[\overline{\eta_M^+(q)}-\eta_M^+(q)])dx\\
&\leq \frac{7\sigma M^2}{4}\int_0^t\int_{\mathbf{R}}\overline{(M-q)\chi_{(M,\infty)}(q)}dxds\\
&\quad-\frac{7\sigma M^2}{4}\int_0^t\int_{\mathbf{R}}(M-q)\chi_{(M,\infty)}(q)dxds\\
&\quad+\frac{7\sigma M}{2}\int_0^t\int_{\mathbf{R}}[\overline{\eta_M^+(q)}-\eta_M^+(q)]dxds\\
&\quad+\frac{7\sigma M}{2}\int_0^t\int_{\mathbf{R}}\frac{1}{2}(\overline{(q_-)^2}-q_-^2)dxds\\
&\quad+\int_0^t\int_{\mathbf{R}}Q(s,x)([\overline{q_-}-q_-]+[\overline{(\eta_M^+)'(q)}-(\eta_M^+)'(q)])dxds.
\end{aligned} \tag{76}$$

In fact, for $0<t<T$, there exists a constant $L>0$, depending only on $\|u_0\|_{H^1(\mathbf{R})}$ and T such that

$$\|Q(t,x)\|_{L^\infty([0,T)\times\mathbf{R})}\leq L.$$

From Lemma 4.3, it has

$$\begin{aligned}
&q_-+(\eta_M^+)'(q)=q+(M-q)\chi_{(M,\infty)}(q),\\
&\overline{q_-}+\overline{(\eta_M^+)'(q)}=q+\overline{(M-q)\chi_{(M,\infty)}(q)}.
\end{aligned} \tag{77}$$

Since the map $\xi\to\xi_-+(\eta_M^+)'(\xi)$ is convex and concave, we get

$$\begin{aligned}
&[\overline{q_-}-q_-]+[\overline{(\eta_M^+)'(q)}-(\eta_M^+)'(q)]\\
&=-(M-q)\chi_{(M,\infty)}(q)+\overline{(M-q)\chi_{(M,\infty)}(q)}\leq 0.
\end{aligned} \tag{78}$$

Therefore,

$$\begin{aligned}
&Q(s,x)([\overline{q_-}-q_-]+[\overline{(\eta_M^+)'(q)}-(\eta_M^+)'(q)])\\
&\leq -L\left(\overline{(M-q)\chi_{(M,\infty)}(q)}-(M-q)\chi_{(M,\infty)}(q)\right).
\end{aligned} \tag{79}$$

Choosing M large enough,

$$\begin{aligned}
&\frac{7\sigma M^2}{4}\overline{(M-q)\chi_{(M,\infty)}(q)}-\frac{7\sigma M^2}{4}(M-q)\chi_{(M,\infty)}(q)\\
&\qquad\qquad+Q(s,x)([\overline{q_-}-q_-]+[\overline{(\eta_M^+)'(q)}-(\eta_M^+)'(q)])\\
&\leq(-L+\frac{7\sigma M^2}{4})\left(\overline{(M-q)\chi_{(M,\infty)}(q)}-(M-q)\chi_{(M,\infty)}(q)\right)\\
&\leq 0.
\end{aligned} \tag{80}$$

Hence, from (76) and (80), we obtain

$$\begin{aligned}
&0\leq\int_{\mathbf{R}}(\frac{1}{2}[\overline{(q_-)^2}-(q_-)^2]+[\overline{\eta_M^+(q)}-\eta_M^+(q)])dx\\
&\leq\frac{7\sigma M}{2}\int_0^t\int_{\mathbf{R}}(\frac{1}{2}[\overline{(q_-)^2}-(q_-)^2]+[\overline{\eta_M^+(q)}-\eta_M^+(q)])dxds.
\end{aligned} \tag{81}$$

For $t>0$, we conclude from Gronwall'inequality and Lemma 4.1 and 4.2 that

$$0\leq\int_{\mathbf{R}}(\frac{1}{2}[\overline{(q_-)^2}-(q_-)^2]+[\overline{\eta_M^+(q)}-\eta_M^+(q)])dx=0. \tag{82}$$

By the Fatou lemma, sending $M \to \infty$, we obtain

$$0 \leq \int_{\mathbf{R}} (\overline{q^2} - q^2) dx \leq 0, \tag{83}$$

which completes the proof. □

5. Proof of main theorem

Proof of Theorem 2.2. From (8), (10), and Lemma 3.5, we know that the conditions (i) and (ii) in definition 2.1 are satisfied. We have to verify (iii). Due to Lemma 4.6, we have

$$q_\varepsilon \longrightarrow q \quad \text{in} \quad L^2_{loc}([0, \infty) \times \mathbf{R}). \tag{84}$$

Using (84) and Lemmas 3.5 and 3.6, we know that u is a distributional solution to problem (1). In addition, Inequality (6) is deduced from Lemma 3.4. Then the proof of Theorem 2.2 is finished.

Acknowledgements
The author thanks the referees for their valuable comments and suggestions.

Funding
This work was supported by Sichuan Province University, Key Laboratory of Bridge Non-destruction Detecting and Engineering Computing [grant number 2013QZJ02], [grant number 2014QYJ03]; Scientific Research Foundation of the Education Department of Sichuan Province Project [grant number 16ZA0265]; SUSER [grant number 2014RC03].

Author details
Ying Wang[1]
E-mail: matyingw@126.com
[1] College of Science, Sichuan University of Science and Engineering, 643000 Zigong, China.

References

Alvarez-Samaniego, B., & Lannes, D. (2008). Large time existence for 3D water waves and asymptotics. *Inventiones Mathematicae, 171*, 485–541.

Coclite, G. M., Holden, H., & Karlsen, K. H. (2005). Global weak solutions to a generalized hyperelastic-rod wave equation. *SIAM Journal on Mathematical Analysis, 37*, 1044–1069.

Constantin, A. (2011). Nonlinear water waves with applications to wave-current interactions and Tsunamis. In *CBMS-NSF Regional Conference Series in Applied Mathematics* (Vol. 81). Philadephia, PA: SIAM.

Constantin, A., & Escher, J. (2007). Particle trajectories in solitary water waves. *Bulletin of the American Mathematical Society, 44*, 423–431.

Constantin, A., & Lannes, D. (2009). The hydrodynamical relevance of the Camassa-Holm and Degasperis–Procesi equations. *Archive for Rational Mechanics and Analysis, 192*, 165–186.

Constantin, A., & Strauss, W. A. (2000). Stability of a class of solitary waves in compressible elastic rods. *Physics Letters A, 270*, 140–148.

Dai, H. H. (1998). Model equations for nonlinear dispersive waves in a compressible Mooney--Rivlin rod. *Acta Mechanica, 127*, 193–207.

Dai, H. H., & Huo, Y. (2000). Solitary shock waves and other travelling waves in a general com-pressible hyperelastic rod. *Royal Society A: Mathematical, Physical and Engineering, Science, 456*, 331–363.

Duruk Mutlubas, N. (2013). On the Cauchy problem for a model equation for shallow water waves of moderate amplitude. *Nonlinear Analysis: Real World Applications, 14*, 2022–2026.

Duruk Mutlubas, N., & Geyer, A. (2013). Orbital stability of solitary waves of moderate amplitude in shallow water. *Journal of Differential Equations, 255*, 254–263.

Fu, Y., Liu, Y., & Qu, C. (2012). On the blow-up structure for the generalized periodic Camassa-Holm and Degasperis-Procesi equations. *Journal of Functional Analysis, 262*, 3125–3158.

Geyer, A. (2012). Solitary traveling waves of moderate amplitude. *Journal of Nonlinear Mathematical Physics, 19*, 1240010.

Guo, Y. X., & Wang, Y. (2014). Wave-breaking criterion and global solution for a generalized periodic coupled Camassa-Holm system. *Boundary value problems, 2014*(155), 1–21.

Himonas, A., Misiolek, G., Ponce, G., & Zhou, Y. (2007). Persistence properties and unique continuation of solutions of the Camassa--Holm equation. *Communications in Mathematical Physics, 271*, 511–512.

Holden, H., & Raynaud, X. (2009). Dissipative solutions for the Camassa-Holm equation. *Discrete and Continuous Dynamical Systems, 24*, 1047–1112.

Hu, Q., & Yin, Z. (2010). Blowup phenomena for a new periodic nonlinearly dispersive wave equation. *Mathematische Nachrichten, 283*, 1613–1628.

Lai, S. Y., & Wu, Y. H. (2010). The local well-posedness and existence of weak solutions for a generalized Camassa-Holm equation. *Journal of Differential Equations, 248*, 2038–2063.

Lai, S. Y. (2013). Global weak solutions to the Novikov equation. *Journal of Functional Analysis, 265*, 520–544.

Li, Y. A., & Olver, P. J. (2000). Well-posedness and blow-up solutions for an integrable nonlinearly dispersivemodel wave equation. *Journal of Differential Equations, 162*, 27–63.

Mi, Y., & Mu, C. (2013). On the solutions of a model equation for shallow water waves of moderate amplitude. *Journal of Differential Equations, 255*, 2101–2129.

Qu, C. Z., Fu, Y., & Liu, Y. (2014). Well-posedness, wave breaking and peakons for a modified μ-Camassa-Holm equation. *Journal of Functional Analysis, 266*, 433–477.

Xin, Z., & Zhang, P. (2000). On the weak solutions to a shallow water equation. *Communications on Pure and Applied Mathematics, 53*, 1411–1433.

Yan, W., Li, Y., & Zhang, Y. (2014). The Cauchy problem for the generalized Camassa-Holm equation in Besov space. *Journal of Differential Equations, 256*, 2876–2901.

Yin, Z. (2004). Well-posedness, global solutions and blowup phenomena for a nonlinearly dispersive equation. *Journal of Evolution Equations, 4*, 391–491.

Yin, Z. (2003). On the blow-up of solutions of a periodic nonlinear dispersive wave equation in compressible elastic rods. *Journal of Mathematical Analysis and Applications, 288*, 232–245.

Zhou, Y. (2006). Blow-up phenomena for a periodic rod equation. *Physics Letters A, 353*, 479–486.

13

Dynamics of the interaction of plankton and planktivorous fish with delay

Samares Pal[1]* and Anal Chatterjee[2]

*Corresponding author: Samares Pal, Department of Mathematics, University of Kalyani, Kalyani 741235, India
E-mail: samaresp@yahoo.co.in

Reviewing editor: Ryan Loxton, Curtin University, Australia

Abstract: This paper is devoted to the study of a plankton–fish ecosystem model. The model represents the interaction between phytoplankton, zooplankton, and fish with Holling II functional response consisting of carrying capacity and constant intrinsic growth rate of phytoplankton. It is observed that if the carrying capacity of phytoplankton population crosses a certain critical value, the system enters into Hopf bifurcation. We have introduced discrete time delay due to gestation in the functional response term involved with the growth equation of planktivorous fish. We have studied the effect of time delay on the stability behavior. In addition, we have obtained an estimate for the length of time delay to preserve the stability of the model system. Existence of Hopf bifurcating small amplitude periodic solutions is derived by considering time delay as a bifurcation parameter. It is observed that constant intrinsic growth rate of phytoplankton and mortality rate of planktivorous fish play an important role in changing one steady state to another steady state and oscillatory behavior of the system. Computer simulations illustrate the results.

Subjects: Applied Mathematics; Mathematical Biology; Mathematical Modeling; Mathematics & Statistics; Non-Linear Systems; Science

Keywords: plankton; fish; stability; equilibria; delay

AMS subject classifications: 92D40; 92D25

ABOUT THE AUTHORS

Our research group is headed by Samares Pal and includes Anal Chatterjee as a post-doctoral researcher. Research in the group addresses a wide range of questions broadly concerning the dynamics of plankton ecosystems under environmental conditions. The main emphasis of our work includes mathematical modeling of marine ecosystems affected by fish induced, overfishing of zooplankton. The research reported in this paper will help in studying the interspecies between phytoplankton and zooplankton in presence of fish and the subsequent changes the stability behavior.

PUBLIC INTEREST STATEMENT

Economically important fish species have long been regarded in isolation from each other and their habitat. In order to comprehensively assess the impacts of fisheries, the entire habitat must be considered. Only then will a sustainable and economic fishery system be possible. Predation by fish determines the abundance of herbivorous zooplankton which in turn regulates the level of phytoplankton. Also, changes in the abundance of planktivorous fish do affect both the phytoplankton and zooplankton. The relationship between phytoplankton, zooplankton, and fish culture is of paramount importance in determining the water quality, on one hand, and the natural productivity and the fish production, on the other. It is observed that constant intrinsic growth rate of phytoplankton and mortality rate of planktivorous fish play an important role in changing one steady state to another steady state and oscillatory behavior of the system.

1. Introduction

It is well known that plankton plays an integral role in marine ecosystem. Phytoplankton and zooplankton are the main two types of plankton. Many models have already been built to simulate zooplankton–phytoplankton interactions. In marine ecosystem, much attention has been paid to the effect of fish and plankton biomass. Top-down and bottom-up effects on plankton–fish dynamics have been discussed by Scheffer (1991) and Pal and Chatterjee (2012). The authors have discussed that at high fish densities, zooplankton is controlled by fish predation and algal biomass is light or nutrient limited, whereas at low fish densities, zooplankton is food limited and phytoplankton density is controlled by zooplankton grazing. Scheffer, Rinaldi, and Kuznetsov (2000) have discussed about isocline analysis and simulations to show how an increase in fish predation may affect plankton dynamics in the model and to explain which type of bifurcations may arise. Role of gestation delay in a plankton–fish model under stochastic fluctuations has been discussed by Mukhopadhyay and Bhattacharyya (2008). Realistic patterns of patchiness in plankton–fish dynamics have been studied by Upadhyay, Kumari, and Rai (2009). The authors also studied both one-dimensional and two-dimensional reaction–diffusion model of nutrient–phytoplankton–zooplankton–fish interaction. As studied by Biktashev, Brindley, and Horwood (2003), the food source for fish larvae depends on the zooplankton dynamics, which in turn is coupled to larval populations through the zooplankton mortality term. A conceptual mathematical model of fish and zooplankton populations inhabiting the lakes Naroch and Myastro has been developed and studied by Medvinsky et al. (2009).

In general, delay differential equations exhibit much more complicated dynamics than ordinary differential equations since a time delay can cause destabilization of equilibria and can induce various oscillations and periodic solutions. As observed by Bischi (1992), the effects of the time delay involved nutrient recycling on resilience, that is, the rate at which a system returns to a stable steady state following a perturbation. The researchers have suggested that characterizing a system by oscillatory behavior, an increase in the distributed time delay, can have a stabilizing effect. But it has been found that the introduction of time delays is a destabilizing process, in the sense that increasing the time delay could cause a stable equilibrium to become unstable and/or cause the populations to fluctuate (Cushing, 1977; Freedman & Rao, 1983). Chemostat-type models incorporating discrete delays have been investigated by Freedman, So, and Waltman (1989). Also, delayed models in population biology have been discussed in Gopalsamy (1984), Kuang (1993), Khare, Misra, Singh, and Dhar (2011) and Liu, Beretta, and Breda (2010). A discrete time delay term to the model of Beretta, Bischi, and Solimano (1991) was introduced in Ruan (1995). This discrete time delay term may be considered as delay due to gestation. The author also allowed the washout rates for nutrient and plankton to be different. Recently, effect of delay on nutrient cycling in phytoplankton–zooplankton interaction model has been studied by Das and Ray (2008). Rehim and Imran (2012) have induced a discrete time delay to both the consume response function and distribution of toxic substance terms in phytoplankton–zooplankton model. They have found out a range of gestation delays which initially impart stability, then induce instability and ultimately lead to periodic behavior. Due to gestation of prey, a discrete time delay toxin producing phytoplankton–zooplankton system shows rich dynamic behavior including chaos and limit cycles (Singh & Gakkhar, 2012). In particular, Gazi and Bandyopadhyay (2006) have introduced discrete time delay due to recycling of dead organic matters and gestation of nutrients to the growth equations of various trophic levels. They have studied the effect of time delay on the stability behavior and obtained an estimate for the length of time delay to preserve the stability of the model system. Effects of time delay on a harvested predator–prey model have discussed in Gazi and Bandyopadhyay (2008), Maity, Patra, and Samanta (2007) and Yongzhen, Min, and Changguo (2011).

Here, an open system is considered with three interacting components consisting of phytoplankton (P), zooplankton (Z), and planktivorous fish (F). In this paper, a plankton–fish interaction model is described. The stability of equilibrium point is described. We have derived the conditions for instability of the system around the interior equilibrium and Hopf bifurcation in presence of delay and non-delay models. Direction of Hopf bifurcation is discussed. Numerical simulations under a set of parameter values have been performed to support our analytical results.

2. The mathematical model

Let $P(t)$ be the concentration of the phytoplankton at time t with carrying capacity K and constant intrinsic growth rate r. Let $Z(t)$ be the concentration zooplankton biomass and $F(t)$ be the concentration of planktivorous fish biomass present at any instant of time t.

Let α_1 be the maximal zooplankton ingestion rate and α_2 maximal zooplankton conversion rate for the growth of zooplankton, respectively ($\alpha_2 \leq \alpha_1$). Let γ_1 be the maximal planktivorous fish ingestion rate and $\gamma_2(\gamma_2 \leq \gamma_1)$ be the maximal planktivorous fish conversion rate due to grazing of herbivorous zooplankton. Further, μ_1, μ_2, μ_3 denote the mortality rates of the phytoplankton, zooplankton, and planktivorous fish biomass, respectively. We choose Holling type II functional form to describe the grazing phenomena with K_1 and K_2 as half saturation constant.

The basic equations with all of the parameters are:

$$\left.\begin{aligned} \frac{dP}{dt} &= rP\left(1-\frac{P}{K}\right) - \frac{\alpha_1 PZ}{K_1+P} - \mu_1 P \equiv J_1(P,Z,F) \\ \frac{dZ}{dt} &= \frac{\alpha_2 PZ}{K_1+P} - \frac{\gamma_1 ZF}{K_2+Z} - \mu_2 Z \equiv J_2(P,Z,F) \\ \frac{dF}{dt} &= \frac{\gamma_2 ZF}{K_2+Z} - \mu_3 F \equiv J_3(P,Z,F). \end{aligned}\right\} \tag{1}$$

Set $X = (P,\ Z,\ F)^T \in \mathbf{R}_+^{\,3}$ and $J(X) = \left[J_1(X),\ J_2(X),\ J_3(X)\right]^T$, with $J\colon \mathbf{R}_+^{\,3} \to \mathbf{R}^3$. The system (Equation 1) can be written in compact form as $\dot{X} = J(X)$. Its Jacobian is

$$V = \begin{bmatrix} r - \frac{2rP}{K} - \frac{\alpha_1 K_1 Z}{(K_1+P)^2} - \mu_1 & -\frac{\alpha_1 P}{K_1+P} & 0 \\ \frac{\alpha_2 K_1 Z}{(K_1+P)^2} & \frac{\alpha_2 P}{K_1+P} - \frac{\gamma_1 K_2 F}{(K_2+Z)^2} - \mu_2 & -\frac{\gamma_1 Z}{K_2+Z} \\ 0 & \frac{\gamma_2 K_2 F}{(K_2+Z)^2} & \frac{\gamma_2 Z}{K_2+Z} - \mu_3 \end{bmatrix}. \tag{2}$$

3. Some preliminary results

3.1. Positive invariance

The system (Equation 1) is not homogeneous. However, it is easy to check that whenever choosing $X(0) \in \mathbf{R}_+^{\,3}$ with $X_i = 0$, for $i = 1,\ 2,\ 3$, then $J_i(X)\,|_{X_i=0} \geq 0$. This ensures that the solution remains within the positive orthant, ensuring the biological well posedness of the system.

3.2. Boundedness of the system

THEOREM 1 *All the solutions of (Equation 1) are ultimately bounded.*

Proof We define a function

$w = P + Z + F.$

We have $\frac{dP}{dt} \leq rP(1-\frac{r}{K})$ which gives $P \leq \frac{c_1 K}{c_1+e^{-rt}} \to K$ as $t \to \infty$, where c_1 is constant.

For the large value of t, we get $\frac{dw}{dt} \leq rP - D_0 w$, where $D_0 = \min\{\mu_1, \mu_2, \mu_3\}$.

Let $x(t)$ be the solution of $\frac{dx}{dt} + xD_0 = rK$, satisfying $x(0) = w(0)$.

Then, $x(t) = \frac{rK}{D_0} + (w(0) - \frac{rK}{D_0})e^{-D_0 t} \to \frac{rK}{D_0}$ as $t \to \infty$. By comparison, it follows that lim $\sup_{t\to\infty}[P(t) + Z(t) + F(t)] \leq \frac{rK}{D_0}$, proving the theorem.

3.3. Equilibria

The system (Equation 1) possesses the following four equilibria: the plankton free equilibrium $E_0 = (0,0,0)$, the zooplankton free equilibrium $E_{01} = \left(\frac{K(r-\mu_1)}{r}, 0, 0\right)$, the planktivorous fish free equilibrium $E_1 = (P_1, Z_1, 0)$, and possibly the coexistence of the three populations $E^* = (P^*, Z^*, F^*)$.

3.3.1. Plankton free equilibrium

$E_0(0,0,0)$ is always feasible; the Jacobian (Equation 2) evaluated at this equilibrium has the eigenvalues $-\mu_2 < 0, -\mu_3 < 0$ and $r - \mu_1 > 0$. Clearly, E_0 is always unstable.

3.3.2. Zooplankton free equilibrium

At E_{01}, the Jacobian (Equation 2) factorizes to give three explicit eigenvalues $-(r-\mu_1) < 0, -\mu_3 < 0$ and $\mu_2(R_0 - 1)$, where R_0 Is defined below. Thus, E_{01} is locally asymptotically stable if and only if

$$R_0 = \frac{\alpha_2 K(r-\mu_1)}{\mu_2[r(K_1+K) - K\mu_1]} < 1. \qquad (3)$$

3.3.3. Planktivorous fish free equilibrium

At E_1, the population levels are

$$P_1 = \frac{\mu_2 K_1}{\alpha_2 - \mu_2}, \; Z_1 = \frac{\alpha_2 K_1[K(\alpha_2-\mu_2)(r-\mu_1) - r\mu_2 K_1]}{\alpha_1 K(\alpha_2-\mu_2)^2}.$$

This equilibrium is feasible if it is satisfying

$$\alpha_2 < \mu_2, \quad K < \frac{r\mu_2 K_1}{(\alpha_2-\mu_2)(r-\mu_1)}. \qquad (4)$$

At E_1, the Jacobian (Equation 2) factorizes to give one explicit eigenvalue $\gamma_2 Z_1(K_2+Z_1)^{-1} - \mu_3$ and the quadratic $\lambda^2 + \left(\frac{rP_1}{K} - \frac{P_1\alpha_1 Z_1}{(K_1+P_1)^2}\right)\lambda + \frac{\alpha_1\alpha_2 K_1 P_1 Z_1}{(K_1+P_1)^3} = 0$. The Routh–Hurwitz conditions for the latter are easily seen to hold, but only when (Equation 4) is satisfied. Thus, stability of E_1 is then ensured by

$$R_1 = \frac{\alpha_2 K_1[K(\alpha_2-\mu_2)(r-\mu_1) - rK_1\mu_2](\gamma_2-\mu_3)}{\mu_3 K_2(\alpha_2-\mu_2)^2} < 1. \qquad (5)$$

3.3.4. The coexistence equilibrium

The coexistence equilibrium $E^* = (P^*, Z^*, F^*)$ cannot be found explicitly since $Z^* = \frac{\mu_3 K_2}{\gamma_2-\mu_3}$, $F^* = \frac{[(\alpha_2-\mu_2)P^* - \mu_2 K_1]K_2\gamma_2}{\gamma_1(\gamma_2-\mu_3)(K_1+P^*)}$ and $P^* = \frac{-B+\sqrt{B^2+4AC}}{2A}$, where $A = r(\gamma_2-\mu_3)$,

$B = -[(K-K_1)r - K\mu_1](\gamma_2-\mu_3)$, $C = -K[(r-\mu_1)K_1 - \alpha_1\mu_3K_2]$.

Thus, the condition for the existence of the interior equilibrium point $E^*(P^*, Z^*, F^*)$ is given by,

$P^* > 0,\ Z^* > 0,\ F^* > 0$. For feasibility, we certainly need

$$\gamma_2 > \mu_3 \text{ and } 0 < r < \min\left\{\frac{K\mu_1}{K-K_1}, \frac{\alpha_1\mu_3K_2 + \mu_1K_1}{K_1}\right\}.$$

Stability analysis of the positive interior equilibrium of the system (Equation 1)

The variational matrix of system (Equation 1) around the positive equilibrium $E^* = (P^*, Z^*, F^*)$ is

$$V^* = \begin{bmatrix} n_{11} & n_{12} & 0 \\ n_{21} & n_{22} & n_{23} \\ 0 & n_{32} & 0 \end{bmatrix},$$

where $n_{11} = \frac{\alpha_1 P^* Z^*}{(K_1+P^*)^2} - \frac{rP^*}{K}$, $n_{12} = -\frac{\alpha_1 P^*}{K_1+P^*} < 0, n_{21} = \frac{K_1 \alpha_2 Z^*}{(K_1+P^*)^2} > 0, n_{22} = \frac{\gamma_1 Z^* F^*}{(K_2+Z^*)^2} > 0, n_{23} = -\frac{\gamma_1 Z^*}{K_2+Z^*} < 0,$ $n_{32} = \frac{K_2 \gamma_2 F^*}{(K_2+Z^*)^2} > 0.$

The characteristic equation is $Q^3 + A_1 Q^2 + A_2 Q + A_3 = 0,$

where $A_1 = -(n_{11} + n_{22}), A_2 = n_{11}n_{22} - n_{12}n_{21} - n_{32}n_{23}, A_3 = n_{11}n_{32}n_{23}.$

Case1: If $n_{11} > 0$, it implies that $A_3 < 0$. Thus, E^* is unstable.
Case2: If $n_{11} < 0$, it implies that $A_3 > 0$. Here, $A_1 > 0$, if $\frac{rP^*}{K} > \frac{\alpha_1 P^* Z^*}{(K_1+P^*)^2} + \frac{\gamma_1 Z^* F^*}{(K_2+Z)^2}$.

Also, $A_2 > 0$ if $-n_{12}n_{21} - n_{32}n_{23} > -n_{11}n_{22}$ where $n_{23}n_{32} < 0, n_{12}n_{21} < 0$ and $n_{11}n_{22}$ is negative. So, $A_1A_2 - A_3 > 0$ if $A_1A_2 > A_3$. Therefore, according the Routh–Hurwitz criteria, all roots of the above cubic equation have negative real parts satisfying $A_1 > 0, A_3 > 0$, and $A_1A_2 - A_3 > 0$. Then, the system becomes locally asymptotically stable around E^*. Thus, depending upon system parameters, the system may exhibit stable or unstable behavior in this case.

The analytical results are summarized in Table 1.

3.4. Hopf bifurcation at coexistence

THEOREM 2 *When the carrying capacity K crosses a critical value, say K^*, the system (Equation 1) enters into a Hopf bifurcation around the positive equilibrium, which induces oscillations of the populations.*

Proof The necessary and sufficient conditions for the existence of the Hopf bifurcation for $K = K^*$, if it exists, are (i) $A_i(K^*) > 0$, $i = 1, 2, 3$ (ii) $A_1(K^*)A_2(K^*) - A_3(K^*) = 0$ and (iii) $A_1'(K^*)A_2(K^*) + A_1(K^*)A_2'(K^*) - A_3'(K^*) \neq 0$. The transversality condition, the third (iii), is obtained observing that the eigenvalues of the characteristic equation, of the form $\chi_i = u_i + \mathbf{i}v_i$, must satisfy the transversality condition $\left.\frac{du_i}{dK}\right|_{K=K^*} \neq 0$.

We will now verify the Hopf bifurcation condition (iii), putting $\chi = u + \mathbf{i}v$ in the characteristic equation, we get

$$(u + \mathbf{i}v)^3 + A_1(u + \mathbf{i}v)^2 + A_2(u + \mathbf{i}v) + A_3 = 0, \tag{6}$$

On separating the real and imaginary parts and eliminating v, we get

$$8u^3 + 8A_1u^2 + 2u(A_1^2 + A_2) + A_1A_2 - A_3 = 0. \tag{7}$$

It is clear from the above that $u(K^*) = 0$ iff $A_1(K^*)A_2(K^*) - A_3(K^*) = 0$. Further, at $K = K^*$, $u(K^*)$ is the only root since the discriminate $8u^2 + 8A_1u + 2(A_1^2 + A_2) = 0$ if $64A_1^2 - 64(A_1^2 + A_2) < 0$.

Further, differentiating (Equation 7) with respect to K, we have $24u^2\frac{du}{dK} + 16A_1u\frac{du}{dK} + 2(A_1^2 + A_2)\frac{du}{dK} + 2u[2A_1\frac{dA_1}{dK} + \frac{dA_2}{dK}] + \frac{dS}{dK} = 0$, where $S = A_1A_2 - A_3$.

Table 1. The table representing thresholds and stability of steady states

Thresholds (R_0, R_1)	$E_0(0,0,0)$	$E_{01}\left(\frac{K(r-\mu_1)}{K},0,0\right)$	$E_1(P_1,Z_1,0)$	$E^*(P^*,Z^*,F^*)$
$R_0 < 1$	Unstable	Asymptotically stable	Not feasible	Not feasible
$R_0 > 1$, $R_1 < 1$	Unstable	Unstable	Asymptotically stable	Not feasible
$R_1 > 1$	Unstable	Unstable	Unstable	Asymptotically stable

At $K = K^*$, we have $u(K^*) = 0$, so that above equation becomes $\left[\frac{du}{dK}\right]_{K=K^*} = \frac{-\frac{dS}{dK}}{2(A_1^2+A_2)} \neq 0$, providing the third condition (iii).

This ensures that the above system has a Hopf bifurcation around the positive interior equilibrium E^*.

Next, we perform detailed analysis of the bifurcation solutions to study the nature of Hopf bifurcation. We start by rewriting system (Equation 1) in the form $\dot{X} = MX + \mathbf{F}(X) + O\|X\|^4$ where

$$X = \begin{pmatrix} p \\ z \\ f \end{pmatrix} : p = P - P^*, z = Z - Z^*, f = F - F^* \text{ and } M = V^*.$$

Here, $\mathbf{F} = \begin{bmatrix} F_1 \\ F_2 \\ F_3 \end{bmatrix}$,

where $F_1 = -\frac{\alpha_1 Z^*}{(K_1+P^*)^3}p^3 + \frac{\alpha_1 K_1}{(K_1+P^*)^3}p^2 z + \left(\frac{\alpha_1 K_1 Z^*}{(K_1+P^*)^3} - \frac{r}{K}\right)p^2 - \frac{\alpha_1 K_1}{(K_1+P^*)^2}pz,$

$$F_2 = \frac{\alpha_2 Z^*}{(K_1+P^*)^3}p^3 - \frac{\gamma_1 F^*}{(K_2+Z^*)^3}z^3 - \frac{\alpha_2 K_1}{(K_1+P^*)^3}p^2 z + \frac{\gamma_1 K_2}{(K_2+Z^*)^3}z^2 f - \frac{\alpha_2 K_1 Z^*}{(K_1+P^*)^3}p^2 + \frac{\gamma_1 K_2 F^*}{(K_2+Z^*)^3}z^2 + \frac{\alpha_2 K_1}{(K_1+P^*)^2}pz - \frac{\gamma_1 K_2}{(K_2+Z^*)^2}zf,$$

$$F_3 = \frac{\gamma_2 F^*}{(K_2+Z^*)^3}z^3 - \frac{\gamma_2 K_2}{(K_2+Z^*)^3}z^2 f - \frac{\gamma_2 K_2 F^*}{(K_2+Z^*)^3}z^2 + \frac{\gamma_2 K_2}{(K_2+Z^*)^2}zf.$$

Now, let the normalized eigenvectors of M and M^T corresponding to the eigenvalues $\mathbf{i}\omega$ and $-\mathbf{i}\omega$ be q and p, respectively, where

$\mathbf{q} = \frac{1}{|q|}\begin{bmatrix} -\frac{n_{12}}{n_{11}-\mathbf{i}\omega} \\ 1 \\ \frac{n_{32}}{\mathbf{i}\omega} \end{bmatrix}$, $\mathbf{p} = \frac{1}{|p|}\begin{bmatrix} \frac{-n_{21}}{n_{11}+\mathbf{i}\omega} \\ 1 \\ -\frac{n_{23}}{\mathbf{i}\omega} \end{bmatrix}$. Let us define the following multi-linear functions for the three vectors $x = (x_1, x_2, x_3) \in R^3$, $y = (y_1, y_2, y_3) \in R^3$, and $w = (w_1, w_2, w_3) \in R^3$.

$$B_i(x,y) = \sum_{j,k=1}^{3}\left[\frac{\delta^2 F_i}{\delta\xi_j \delta\xi_k}\right]_{\xi=0} x_j y_k$$

$$C_i(x,y,w) = \sum_{j,k,l=1}^{3}\left[\frac{\delta^3 F_i}{\delta\xi_j \delta\xi_k \delta\xi_l}\right]_{\xi=0} x_j y_k w_l.$$

where $B(x,y) = \begin{bmatrix} 2\left(\frac{\alpha_1 K_1 Z^*}{(K_1+P^*)^3} - \frac{r}{K}\right)x_1 y_1 - \frac{\alpha_1 K_1}{(K_1+P^*)^2}x_1 y_2 \\ -\frac{2\alpha_2 K_1 Z^*}{(K_1+P^*)^3}x_1 y_1 + \frac{2\gamma_1 K_2 F^*}{(K_2+Z^*)^3}x_2 y_2 + \frac{\alpha_2 K_1}{(K_1+P^*)^2}x_1 y_2 - \frac{\gamma_1 K_2}{(K_2+Z^*)^2}x_2 y_3 \\ -\frac{2\gamma_2 K_2 F^*}{(K_2+Z^*)^3}x_2 y_2 + \frac{\gamma_2 K_2}{(K_2+Z^*)^2}x_2 y_3 \end{bmatrix}$,

$$C(x,y,w) = \begin{bmatrix} -\frac{6\alpha_1 Z_1^*}{(K_1+P^*)^3}x_1 y_1 w_1 + \frac{2\alpha_1 K_1}{(K_1+P^*)^3}x_1 y_1 w_2 \\ \frac{6\alpha_2 Z^*}{(K_1+P^*)^3}x_1 y_1 w_1 - \frac{6\gamma_1 F^*}{(K_2+Z^*)^3}x_2 y_2 w_2 - \frac{2\alpha_2 K_1}{(K_1+P^*)^3}x_1 y_1 w_2 + \frac{2\gamma_1 K_2}{(K_2+Z^*)^3}x_2 y_2 w_3 \\ \frac{6\gamma_2 F^*}{(K_2+Z^*)^3}x_2 y_2 w_2 - \frac{2\gamma_2 K_2}{(K_2+Z^*)^3}x_2 y_2 w_3 \end{bmatrix}.$$

With these, $\mathbf{F}(X)$ will be of the form $\mathbf{F}(X) = \frac{1}{2}B(X,X) + \frac{1}{2}C(X,X,X)$.

Solving the corresponding linear system, we get

$$M^{-1}B(\mathbf{q},\bar{\mathbf{q}}) = \begin{bmatrix} s_1 \\ s_2 \\ s_3 \end{bmatrix},$$

where

$$s_1 = -\frac{1}{n_{11}n_{32}n_{23}}\left[-n_{23}n_{32}\left[2\left(\frac{\alpha_1 K_1 Z^*}{(K_1+P^*)^3} - \frac{r}{K}\right)\frac{n_{12}^2}{n_{11}^2+\omega^2} + \frac{\alpha_1 K_1}{(K_1+P^*)^2}\frac{n_{12}(n_{11}+\mathbf{i}\omega)}{n_{11}^2+\omega^2}\right]\right.$$
$$\left.+n_{12}n_{23}\left(\frac{\gamma_2 K_2}{(K_2+Z^*)^2}\frac{\mathbf{i}n_{32}}{\omega} - \frac{2\gamma_2 K_2 F^*}{(K_2+Z^*)^3}\right)\right],$$

$$s_2 = -\frac{1}{n_{11}n_{32}n_{23}}\left[-n_{11}n_{23}\left(\frac{\gamma_2 K_2}{(K_2+Z^*)^2}\frac{\mathbf{i}n_{32}}{\omega} - \frac{2\gamma_2 K_2 F^*}{(K_2+Z^*)^3}\right)\right],$$

$$s_3 = -\frac{1}{n_{11}n_{32}n_{23}}\left[n_{21}n_{32}\left[2\left(\frac{\alpha_1 K_1 Z^*}{(K_1+P^*)^3} - \frac{r}{K}\right)\frac{n_{12}^2}{n_{11}^2+\omega^2} + \frac{\alpha_1 K_1}{(K_1+P^*)^2}\frac{n_{12}(n_{11}+\mathbf{i}\omega)}{n_{11}^2+\omega^2}\right]\right.$$
$$-n_{11}n_{32}\left[\frac{-2\gamma_1 K_1 Z^*}{(K_1P^*)^3}\frac{n_{12}^2}{n_{11}^2+\omega^2} + \frac{2\gamma_1 K_2 F^*}{(K_2+Z^*)^3} - \frac{\alpha_2 K_1}{(K_1+P^*)^2}\frac{n_{12}(n_{11}+\mathbf{i}\omega)}{n_{11}^2+\omega^2} - \frac{\gamma_1 K_2}{(K_2+Z^*)^2}\frac{\mathbf{i}n_{32}}{\omega}\right]$$
$$\left.+(n_{11}n_{22}-n_{12}n_{21})\left(\frac{\gamma_2 K_2}{(K_2+Z^*)^2}\frac{\mathbf{i}n_{32}}{\omega} - \frac{2\gamma_2 K_2 F^*}{(K_2+Z^*)^3}\right)\right].$$

Again,

$$(2i\omega I_3 - M)^{-1} = \frac{1}{|2\mathbf{i}\omega I_3 - M|}\begin{pmatrix} -4\omega^2 - 2n_{22}\mathbf{i}\omega - n_{23}n_{32} & 2\mathbf{i}\omega n_{12} & n_{12}n_{23} \\ 2\mathbf{i}\omega n_{21} & -4\omega^2 - 2n_{11}\mathbf{i}\omega & 2n_{23}\mathbf{i}\omega - n_{11}n_{23} \\ n_{21}n_{32} & n_{32}(2\mathbf{i}\omega - n_{11}) & \mathbf{E} \end{pmatrix},$$

where $\mathbf{E} = (2\mathbf{i}\omega - n_{11})(2\mathbf{i}\omega - n_{12}) - n_{12}n_{21}$.

Therefore,

$$(2\mathbf{i}\omega I_3 - M)^{-1}\begin{bmatrix} b_{11}q_1^2 + b_{12}q_1q_2 \\ b_{21}q_1^2 + b_{22}q_2^2 + b_{23}q_1q_2 + b_{24}q_2q_3 \\ b_{31}q_2q_3 + b_{32}q_2^2 \end{bmatrix}$$

$$= \frac{1}{|2\mathbf{i}\omega I_3 - M|}\begin{bmatrix} (-4\omega^2 - 2n_{22}\mathbf{i}\omega - n_{23}n_{32})\tau_1 + 2\mathbf{i}\omega n_{12}\tau_2 + n_{12}n_{23}\tau_3 \\ 2\mathbf{i}\omega n_{21}\tau_1 - (4\omega^2 + 2n_{11}\mathbf{i}\omega)\tau_2 + (2n_{23}\mathbf{i}\omega - n_{11}n_{23})\tau_3 \\ n_{21}n_{32}\tau_1 + n_{32}(2\mathbf{i}\omega - n_{11})\tau_2 + \mathbf{E}\tau_3 \end{bmatrix},$$

where $\tau_1 = b_{11}q_1^2 + b_{12}q_1q_2$; $\tau_2 = b_{21}q_1^2 + b_{22}q_2^2 + b_{23}q_1q_2 + b_{24}q_2q_3$; $\tau_3 = b_{31}q_2q_3 + b_{32}q_2^2$.

Using the invariant expression for the first lyapunov coefficient $l_1(0)$, we get the first lyapunov coefficients as Mukhopadhyay and Bhattacharyya (2011). $l_1(0) = \frac{1}{2\omega}Re[< \mathbf{p}, \mathbf{c}(\mathbf{q},\mathbf{q},\bar{\mathbf{q}}) - 2 < \mathbf{p}, B(\mathbf{q}, M^{-1}, B(\mathbf{q},\bar{\mathbf{q}}))) > + < \mathbf{p}, B(\bar{\mathbf{q}}, (2\mathbf{i}\omega I - M)^{-1}B(\mathbf{q},\mathbf{q})) >]$.

Now, if $l_1(0) < 0$, the Hopf bifurcation will be of supercritical nature and there will be emergence of stable limit cycles.

4. The mathematical model with time delay

We have already discussed the usefulness of time delay in realistic modeling of ecosystems. Here, we consider the following modification of the model (Equation 1) incorporating discrete time delay in it. In this paper, we have introduced a discrete time delay term to the above model; this term may be considered as delay due to gestation. Here, τ is the discrete time delay.

$$\left.\begin{aligned} \frac{dP}{dt} &= rP\left(1-\frac{P}{K}\right)-\frac{\alpha_1 PZ}{K_1+P}-\mu_1 P \\ \frac{dZ}{dt} &= \frac{\alpha_2 PZ}{K_1+P}-\frac{\gamma_1 ZF}{K_2+Z}-\mu_2 Z \\ \frac{dF}{dt} &= \frac{\gamma_2 Z(t-\tau)F(t-\tau)}{K_2+Z(t-\tau)}-\mu_3 F \end{aligned}\right\}. \tag{8}$$

The system (Equation 8) has to be analyzed with the following initial conditions,

$$P(0)>0, \quad Z(0)>0, \quad F(0)>0, \quad t\in[-\tau,0]. \tag{9}$$

4.1. Qualitative analysis of the model with time delay

$$\overline{V}=\begin{bmatrix} r-\frac{2r\overline{P}}{K}-\frac{\alpha_1 K_1\overline{Z}}{(K_1+\overline{P})^2}-\mu_1 & -\frac{\alpha_1\overline{P}}{K_1+\overline{P}} & 0 \\ \frac{\alpha_2 K_1\overline{Z}}{(K_1+\overline{P})^2} & \frac{\alpha_2\overline{P}}{K_1+\overline{P}}-\frac{\gamma_1 K_2\overline{F}}{(K_2+\overline{Z})^2}-\mu_2 & -\frac{\gamma_1\overline{Z}}{K_2+\overline{Z}} \\ 0 & \frac{\gamma_2 K_2\overline{F}}{(K_2+\overline{Z})^2}e^{-\lambda\tau} & \frac{\gamma_2\overline{Z}}{K_2+Z}e^{-\lambda\tau}-\mu_3 \end{bmatrix}.$$

Remark The characteristic equation for the variational matrix V_0 about the plankton free steady states E_0 and E_{01} remains the same as obtained for the non-delayed system (Equation 8). Thus, in our model system, the delay has no effect on the stability nature of the system about E_0 and E_{01}.

To find conditions for the other local asymptotic stability of system (Equation 8), we use the following theorem from Gopalsamy (1992).

THEOREM 3 *If $D_E(\lambda,\tau)=0$ denotes the characteristic equation, then a set of necessary and sufficient conditions for the equilibrium to be asymptotically stable for all $\tau\geq 0$ are*

(i) The real parts of all the roots of $D_E(\lambda,0)=0$ are negative.
(ii) For all real ω and any $\tau\geq 0$, $D_E(\mathbf{i}\omega,\tau)\neq 0$, where $\mathbf{i}=\sqrt{-1}$.

The characteristic equation for the variational matrix V_1 about E_1 takes the following form

$$D_{E_1}(\lambda,\tau)=\lambda^3+l\lambda^2+m\lambda+n-e^{-\lambda\tau}(f\lambda^2+g\lambda+h)=0, \tag{10}$$

where $l=\frac{rP_1}{K}-\frac{\alpha_1P_1Z_1}{(K_1+P_1)^2}+\mu_3$, $m=\frac{\alpha_1\alpha_2K_1P_1Z_1}{(K_1+P_1)^3}+\mu_3\left(\frac{rP_1}{K}-\frac{\alpha_1P_1Z_1}{(K_1+P_1)^2}\right)$, $n=\frac{\alpha_1\alpha_2\mu_3K_1P_1Z_1}{(K_1+P_1)^3}$, $f=\frac{\gamma_2Z_1}{K_2+Z_1}$, $g=\frac{\gamma_2Z_1}{K_2+Z_1}\left(\frac{rP_1}{K}-\frac{\alpha_1P_1Z_1}{(K_1+P_1)^2}\right)$, $h=\frac{\alpha_1\alpha_2\gamma_2K_1P_1Z_1^2}{(K_1+P_1)^3(K_2+Z_1)}$.

Here, $D_{E_1}(\lambda,0)=0$ has roots with negative real parts, provided system (Equation 8) is locally asymptotically stable about equilibrium E_1 (for the condition, see Section 3.3.3).

For $\omega=0$, $D_{E_1}(0,\tau)=n-h\neq 0$.

Now, for $\omega\neq 0$,

$$D_{E_1}(\mathbf{i}\omega,\tau)=-\mathbf{i}\omega^3-l\omega^2+m\mathbf{i}\omega+n-e^{-\mathbf{i}\omega\tau}(-f\omega^2+g\mathbf{i}\omega+h).$$

Let $D_{E_1}(\mathbf{i}\omega,\tau)=0$ and separating the real and imaginary parts, we get
$-l\omega^2+n=-f\omega^2\cos\omega\tau+h\cos\omega\tau+g\omega\cos\omega\tau,$
$-\omega^3+m\omega=f\omega^2\sin\omega\tau-h\sin\omega\tau+g\omega\cos\omega\tau.$

Squaring and adding the above two equations, we get $\omega^6++A_{11}\omega^4+A_{22}\omega^2+A_{33}=0$, where $A_{11}=l^2-f^2-2m, A_{22}=m^2-2ln+2fh-g^2, A_{33}=n^2-h^2$.

Sufficient conditions for the non-existence of a real number ω satisfying $D_{E_1}(\mathbf{i}\omega,\tau)=0$ can be written as $\omega^6+A_{11}\omega^4+A_{22}\omega^2+A_{33}>0$,

which can be transformed to $\omega^6+A_{11}\left[\omega^2+\frac{A_{22}}{2A_{11}}\right]^2+A_{33}-\frac{A_{22}^2}{4A_{11}}>0.$

Therefore, a sufficient condition for E_1 to be stable is (i) $A_{11}>0$ and (ii) $A_{33}>\frac{A_{22}^2}{4A_{11}}$.

4.2. Estimation of the length of delay to preserve stability at E_1

In this section, we assume that in the absence of delays, E_1 is locally asymptotically stable. Let $u_1(t)$, $v_1(t)$, and $w_1(t)$ be the respective linearized variables of this model. The system (Equation 8) becomes

$$\left.\begin{aligned}\frac{du_1}{dt}&=\widehat{A_1}u_1(t)+\widehat{A_2}v_1(t)\\ \frac{dv_1}{dt}&=\widehat{B_1}u_1(t)+\widehat{B_2}w_1(t)\\ \frac{dw_1}{dt}&=\widehat{C_1}w_1(t)+\widehat{C_2}w_1(t-\tau)\end{aligned}\right\}, \tag{11}$$

where $\widehat{A_1}=\frac{\alpha_1P_1Z_1}{(K_1+P_1)^2}-\frac{rP_1}{K},\widehat{A_2}=-\frac{\alpha_1P_1}{K_1+P_1},\widehat{B_1}=\frac{\alpha_2K_1Z_1}{(K_1+P_1)^2},\widehat{B_2}=-\frac{\gamma_1Z_1}{K_2+Z_1},\widehat{C_1}=-\mu_3,\widehat{C_2}=\frac{\gamma_2Z_1}{K_2+Z_1}.$

Let $\hat{u}_1(L)$, $\hat{v}_1(L)$, and $\hat{w}_1(L)$ be the Laplace transform of $u_1(t)$, $v_1(t)$, and $w_1(t)$, respectively. Taking the Laplace transformation of system (Equation 11), we have

$$\begin{aligned}(L-\widehat{A_1})\hat{u}_1(L)&=\widehat{A_2}\hat{v}_1(L)+u_1(0)\\ L\hat{v}_1(L)&=\widehat{B_1}\hat{u}_1(L)+\widehat{B_2}\hat{w}_1(L)+v_1(0)\\ (L-\widehat{C_1})\hat{w}_1(L)&=\widehat{C_2}M_1(L)e^{-L\tau}+\widehat{C_2}e^{-L\tau}\hat{w}_1(L),\end{aligned}$$

where $M_1(L)=\int_{-\tau}^{0}e^{-Lt}w_1(t)dt$.

The inverse Laplace transform of $\hat{u}_1(L)$, $\hat{v}_1(L)$, and $\hat{w}_1(L)$ will have terms which exponentially increase with time, if $\hat{u}_1(L)$, $\hat{v}_1(L)$, and $\hat{w}_1(L)$ have pole with positive real parts. Since E_1 needs to be locally asymptotically stable, it is necessary and sufficient that all poles of $\hat{u}_1(L)$, $\hat{v}_1(L)$, and $\hat{w}_1(L)$ have negative real parts. We shall employ the Nyquist criterion which states that if L is the arc length of a curve encircling the right half-plane, the curve of $\hat{u}_1(L)$, $\hat{v}_1(L)$, and $\hat{w}_1(L)$ will encircle the origin a number of times equal to the difference between the number of poles and the number of zeroes of $\hat{u}_1(L)$, $\hat{v}_1(L)$, and $\hat{w}_1(L)$ in the right half-plane.

Let $G(L)=L^3+lL^2+mL+n-e^{-L\tau}(fL^2+gL+h)$ from (Equation 10). Then, E_1 is locally asymptotically stable if

$$ReG(\mathbf{i}v_2)=0, \tag{12}$$

$$ImG(\mathbf{i}v_2)>0, \tag{13}$$

where v_2 is the smallest positive root of Equation 12 (Freedman, Erbe, & Rao, 1986).

Now, (Equation 12) and (Equation 13) become

$$-lv_2^2 + n = -fv_2^2 \cos v_2\tau + h\cos v_2\tau + gv_2 \sin v_2\tau,$$
$$-v_2^3 + mv_2 < fv_2^2 \sin v_2\tau - h\sin v_2\tau + gv_2 \cos v_2\tau.$$

To get an estimation on the length of delay, we utilize the following conditions,

$$-lv^2 + n = -fv^2 \cos v\tau + h\cos v\tau + gv \sin v\tau, \tag{14}$$

$$-v^3 + mv > fv^2 \sin v\tau - h\sin v\tau + gv \cos v\tau. \tag{15}$$

Therefore, E_1 will be stable if the above inequality holds at $v = v_2$, where v_2 is the first positive root of Equation 14. We shall now estimate an upper bound v_{++} of v_2, which would be independent of τ. Then, we estimate τ so that (Equation 15) holds for all $0 \le v \le v_{++}$, and hence in particular at $v = v_2$.

Maximizing $-fv^2 \cos v\tau + h\cos v\tau + gv \sin v\tau$, subject to $|\sin v\tau| \le 1, |\cos v\tau| \le 1$, we obtain

$$-lv^2 + n \le fv^2 + h + gv. \tag{16}$$

Thus, the unique positive solution of $(f+l)v^2 + gv - (n-h) = 0$ is denoted by v_{++}. Hence, if $v_{++} = \frac{-g+\sqrt{g^2+4(f+l)(n-h)}}{2(f+l)}$, then from (Equation 16) we have $v_2 \le v_{++}$ which ensures that v_{++} is independent of τ. Then, we estimate τ so that (Equation 15) holds for all values $0 \le v \le v_{++}$; now, rearranging (Equation 15) by $|\sin v\tau| \le \tau v$ and $|1 - \cos v\tau| \le \frac{1}{2}\tau^2 v^2$, we have

$$\frac{g}{2}\tau^2 v^2 + (h - f\,v^2)\tau + (m - g - v^2) > 0. \tag{17}$$

Thus, (Equation 15) will be satisfied if $A_{\theta_1}\tau^2 + B_{\theta_1}\tau + C_{\theta_1} > 0$, where $A_{\theta_1} = \frac{g}{2}v_{++}^2$, $B_{\theta_1} = h - fv_{++}^2$, $C_{\theta_1} = m - g - v_{++}^2$.

Then, the Nyquist criterion holds for $0 \le \tau \le \tau_1$, where $\tau_1 = \frac{1}{2A_{\theta_1}}(-B_{\theta_1} + \sqrt{B_{\theta_1}^2 + 4A_{\theta_1}C_{\theta_1}})$ and τ_1 gives an estimate for the length of delay for which stability is preserved.

THEOREM 4 *If there exist τ in $0 \le \tau \le \tau_1$ such that $A_\theta\tau^2 + B_\theta\tau + C_\theta > 0$, then τ_1 is the maximum value (length of delay) of τ for which E_1 is asymptotically stable.*

4.3. Qualitative analysis of the model at E^* with gestation time delay ($\tau \ne 0$)

The characteristic equation for the variational matrix V^* about E^* takes the following form:

$$D_E^*(\lambda_1, \tau) = \lambda_1^3 + \bar{l}\lambda_1^2 + \bar{m}\lambda_1 + \bar{n} - e^{-\lambda_1\tau}(\bar{f}\lambda_1^2 + \bar{g}\lambda_1 + \bar{h}) = 0, \tag{18}$$

where $\bar{l} = \frac{rP^*}{K} - \frac{\alpha_1 P^* Z^*}{(K_1 + P^*)^2} - \frac{\gamma_1 Z^* F^*}{(K_2 + Z^*)^2} + \mu_3,$

$$\bar{m} = \left(\frac{\alpha_1 P^* Z^*}{(K_2 + Z^*)^2} - \frac{rP^*}{K}\right)\left(\frac{\gamma_1 Z^* F^*}{(K_2 + Z^*)^2} - \mu_3\right) + \frac{\alpha_1\alpha_2 K_1 P^* Z^*}{(K_1 + P^*)^3} - \frac{\gamma_1\mu_3 Z^* F^*}{(K_2 + Z^*)^3},$$

$$\bar{n} = \left(\frac{\alpha_1 P^* Z^*}{(K_1 + P^*)^2} - \frac{rP^*}{K}\right)\frac{\gamma_1\mu_3 Z^* F^*}{(K_2 + Z^*)^3} + \frac{\alpha_1\alpha_2 K_1\mu_3 P^* Z^*}{(K_1 + P^*)^3},$$

$$\bar{f} = \frac{\gamma_2 Z^*}{K_2 + Z^*},\ \bar{g} = \left(\frac{rP^*}{K} - \frac{\alpha_1 P^* Z^*}{(K_1 + P^*)^2}\right)\frac{\gamma_2 Z^*}{K_2 + Z^*} - \frac{\gamma_1\gamma_2 Z^{*2}F^*}{(K_2 + Z^*)^3} - \frac{\gamma_1\gamma_2 K_2 Z^* F^*}{(K_2 + Z^*)^3},$$

$$\bar{h} = \left(\frac{\alpha_1 P^* Z^*}{(K_1 + P^*)^2} - \frac{rP^*}{K}\right)\left(\frac{\gamma_1\gamma_2 Z^{*2}F^*}{(K_2 + Z^*)^3} + \frac{\gamma_1\gamma_2 K_2 Z^* F^*}{(K_2 + Z^*)^3}\right) - \frac{\alpha_1\alpha_2\gamma_2 K_1 P^* Z^{*2}}{(K_1 + P^*)^3 (K_2 + Z^*)}.$$

Now,

$D_E{}^*(\mathbf{i}\bar{\omega}, \tau) = -\mathbf{i}\bar{\omega}^3 - \bar{l}\bar{\omega}^2 + \bar{m}\mathbf{i}\bar{\omega} + \bar{n} - e^{-\bar{\omega}\lambda\tau}(-\bar{f}\bar{\omega}^2 + \bar{g}\mathbf{i}\bar{\omega} + \bar{h})$ (Using Theorem 3).

Let $D_E{}^*(\mathbf{i}\bar{\omega}, \tau) = 0$ and separating the real and imaginary parts, we get

$$-\bar{l}\bar{\omega}^2 + \bar{n} = -\bar{f}\bar{\omega}^2 \cos\bar{\omega}\tau + \bar{h}\cos\bar{\omega}\tau + \bar{g}\bar{\omega}\sin\bar{\omega}\tau,$$

$$-\bar{\omega}^3 + \bar{m}\bar{\omega} = \bar{f}\bar{\omega}^2 \sin\bar{\omega}\tau - \bar{h}\sin\bar{\omega}\tau + \bar{g}\bar{\omega}\cos\bar{\omega}\tau.$$

Squaring and adding the above two equations, we get

$$\bar{\omega}^6 + \bar{Q}_1\bar{\omega}^4 + \bar{Q}_2\bar{\omega}^2 + \bar{Q}_3 = 0,$$

where

$$\bar{Q}_1 = \bar{l}^2 - \bar{f}^2 - 2\bar{m},$$
$$\bar{Q}_2 = \bar{m}^2 - 2\bar{l}\bar{n} + 2\bar{f}\bar{h} - \bar{g}^2,$$
$$\bar{Q}_3 = \bar{n}^2 - \bar{h}^2.$$

The sufficient conditions for the non-existence of a real number ω satisfying $D_E{*} = 0$ can be written as $\bar{\omega}^6 + \bar{Q}_1\bar{\omega}^4 + \bar{Q}_2\bar{\omega}^2 + \bar{Q}_3 > 0$, which can be transformed to

$$\bar{\omega}^6 + \bar{Q}_1\left[\bar{\omega}^2 + \frac{\bar{Q}_2}{2\bar{Q}_1}\right]^2 + \bar{Q}_3 - \frac{\bar{Q}_2{}^2}{4\bar{Q}_1} > 0.$$

Therefore, a sufficient condition for E^* to be stable is (a) $\bar{Q}_1 > 0$ and (b) $\bar{Q}_3 > \frac{\bar{Q}_2{}^2}{4\bar{Q}_1}$.

4.4. *Estimation of the length of delay to preserve stability at* E^*

In this section, we assume that in the absence of delays, E^* is locally asymptotically stable. Let $u(t)$, $v(t)$, and $w(t)$ be the respective linearized variables of this model. The system (Equation 8) becomes

$$\left.\begin{aligned} \frac{du}{dt} &= \bar{A}_1 u(t) + \bar{A}_2 v(t), \\ \frac{dv}{dt} &= \bar{B}_1 u(t) + \bar{B}_2 v(t) + \bar{B}_3 w(t), \\ \frac{dw}{dt} &= \bar{C}_1 v(t-\tau) + \bar{C}_2 w(t) + \bar{C}_3 w(t-\tau) \end{aligned}\right\}, \tag{19}$$

where $\bar{A}_1 = \frac{\alpha_1 P^* Z^*}{(K_1+P^*)^2} - \frac{rP^*}{K}$, $\bar{A}_2 = -\frac{\alpha_1 P^*}{K_1+P^*}$, $\bar{B}_1 = \frac{\alpha_2 K_1 Z^*}{(K_1+P^*)^2}$, $\bar{B}_2 = \frac{\gamma_1 Z^* F^*}{K_2+Z^*}$, $\bar{B}_3 = -\frac{\gamma_1 Z^*}{K_2+Z^*}$, $\bar{C}_1 = \frac{\gamma_2 K_2 F^*}{(K_2+Z^*)^2}$, $\bar{C}_2 = -\mu_3$, $\bar{C}_3 = \frac{\gamma_2 Z^*}{K_2+Z^*}$.

Let $\bar{u}(L)$, $\bar{v}(L)$, and $\bar{w}(L)$ be the Laplace transform of $u(t)$, $v(t)$, and $w(t)$, respectively. Taking the Laplace transformation of system (Equation 19), we have

$$(L-\bar{A}_1)\bar{u}(L)=\bar{A}_2\bar{v}(L)+u(0),$$
$$(L-\bar{B}_2)\bar{v}(L)=\bar{B}_1\bar{u}(L)+\bar{B}_3\bar{w}(L)+v(0),$$
$$(L-\bar{C}_2)\bar{w}(L)=\bar{C}_1e^{-L\tau}K_1+\bar{C}_1e^{-L\tau}\bar{v}(L)+\bar{C}_3e^{-L\tau}K_2+\bar{C}_3e^{-L\tau}\bar{w}(L)+w(0),$$

where $K_1(L)=\int_{-\tau}^{0}e^{-Lt}v(t)dt, K_2(L)=\int_{-\tau}^{0}e^{-Lt}w(t)dt$.

Let $\bar{F}(L)=L^3+\bar{l}L^2+\bar{m}L+\bar{n}-e^{-L\tau}(\bar{f}L^2+\bar{g}L+\bar{h})$.

Then, E^* is locally asymptotically stable if $Re\bar{F}(i\bar{v}_0)=0$ and $Im\bar{F}(i\bar{v}_0)>0$ i.e.

$$-\bar{l}\bar{v}_0^2+\bar{n}=-\bar{f}\bar{v}_0^2\cos\bar{v}_0\tau+\bar{h}\cos\bar{v}_0\tau+\bar{g}\,\bar{v}_0\sin\bar{v}_0\tau,$$
$$-\bar{v}_0^3+\bar{m}\,\bar{v}_0>\bar{f}\bar{v}_0^2\sin\bar{v}_0\tau-\bar{h}\sin\bar{v}_0\tau+\bar{g}\,\bar{v}_0\cos\bar{v}_0\tau,$$

where $\bar{v}_0$ be the smallest positive root of $Re(i\bar{v}_0)=0$.

To get an estimation on the length of delay, we use the following conditions,

$$-\bar{l}v^2+\bar{n}=-\bar{f}v^2\cos v\tau+\bar{h}\cos v\tau+\bar{g}\,v\sin v\tau, \tag{20}$$

$$-v^3+\bar{m}\,v>\bar{f}v^2\sin v\tau-\bar{h}\sin v\tau+\bar{g}\,v\cos v\tau. \tag{21}$$

Therefore, E^* will be stable if above inequality holds at $v=\bar{v}_0$, where $\bar{v}_0$ is the first positive root of Equation 20. We shall now estimate an upper bound v_+ of v_0, which would be independent of τ. Then, we estimate τ so that (Equation 21) holds for all $0\le v\le v_+$, and hence in particular at $\bar{v}=v_0$.

Maximizing $-\bar{f}v^2\cos v\tau+\bar{h}\cos v\tau+\bar{g}v\sin v\tau$, subject to $|\sin v\tau|\le 1, |\cos v\tau|\le 1$, we obtain

$$-\bar{l}v^2+\bar{n}\le\bar{f}v^2+\bar{h}+\bar{g}v. \tag{22}$$

Thus, the unique positive solution of $(\bar{f}+\bar{l})v^2+\bar{g}v-(\bar{n}-\bar{h})=0$ is denoted by v_+. Hence, if $v_+=\frac{-\bar{g}+\sqrt{\bar{g}^2+4(\bar{f}+\bar{l})(\bar{n}-\bar{h})}}{2(\bar{f}+\bar{l})}$, then from above equation we have $v_0\le v_+$ which indicates that v_+ is independent of τ. Then, we estimate τ so that (Equation 21) holds for all values $0\le v\le v_+$; now, rearranging (Equation 21) by $|\sin v\tau|\le\tau v$ and $|1-\cos v\tau|\le\frac{1}{2}\tau^2v^2$, we have

$$\frac{\bar{g}}{2}\tau^2v^2+(\bar{h}-\bar{f}\,v^2)\tau+(\bar{m}-\bar{g}-\bar{v}^2)>0. \tag{23}$$

Thus, Equation 21 will be satisfied if

$$A_\theta\tau^2+B_\theta\tau+C_\theta>0,\text{ where }A_\theta=\frac{\bar{g}}{2}v_+^2, B_\theta=\bar{h}-\bar{f}v_+^2, C_\theta=\bar{m}-\bar{g}-v_+^2.$$

Then, the Nyquist criterion holds for $0\le\tau\le\tau_+$, where $\tau_+=\frac{1}{2A_\theta}(-B_\theta+\sqrt{B_\theta^2+4A_\theta C_\theta})$ and τ_+ gives an estimate for the length of delay for which stability is preserved. Thus, we are now in a position to state the following theorem.

THEOREM 5 *If there exist τ in $0\le\tau\le\tau_+$ such that $A_\theta\tau^2+B_\theta\tau+C_\theta>0$, then τ_+ is the maximum value (length of delay) of τ for which E^* is asymptotically stable.*

5. Bifurcation analysis in the presence of discrete delay

Let us consider $\tau \neq 0$ and assume $\lambda_1 = \mu + \mathbf{i}\nu$ in Equation 18. Then, separating the real and imaginary parts, we get a system of transcendental equations as

$$\begin{aligned}&\mu^3 - 3\mu\nu^2 - \bar{l}(\mu^2 - \nu^2) + \bar{m}\mu + \bar{n}\\ &= \bar{f}(\mu^2 - \nu^2)\cos\nu\tau e^{-\mu\tau} + 2f\bar{\nu}\mu\sin\nu\tau e^{-\mu\tau} + \bar{g}\mu\cos\nu\tau e^{-\mu\tau} + \bar{h}\cos\nu\tau e^{-\mu\tau} + \bar{g}\nu\sin\nu\tau e^{-\mu\tau},\end{aligned} \tag{24}$$

$$\begin{aligned}&-\nu^3 + 3\mu^2\nu + 2\bar{l}\mu\nu + \bar{m}\nu\\ &= -\bar{f}(\mu^2 - \nu^2)\sin\nu\tau e^{-\mu\tau} + 2\bar{f}\nu\mu\cos\nu\tau e^{-\mu\tau} - \bar{g}\mu\sin\nu\tau - \bar{h}\sin\nu\tau e^{-\mu\tau} + \bar{g}\nu\cos\nu\tau e^{-\mu\tau}.\end{aligned} \tag{25}$$

Let us consider λ_1, and hence μ and ν are functions of τ. We want to know the change of stability of E^* which will occur at the values of $\tau = \hat{\tau}$ for $\mu = 0$ and $\nu \neq 0$.

Then, the above two equations become

$$\left.\begin{aligned}-\bar{l}\nu^2 + \bar{n} &= -\bar{f}\nu^2\cos\nu\hat{\tau} + \bar{h}\cos\nu\hat{\tau} + \bar{g}\nu\sin\nu\hat{\tau}\\ -\nu^3 + \bar{m}\nu &= \bar{f}\nu^2\sin\nu\hat{\tau} - \bar{h}\sin\nu\hat{\tau} + \bar{g}\nu\cos\nu\hat{\tau}\end{aligned}\right\}, \tag{26}$$

Now, eliminating $\hat{\tau}$, we have

$$\nu^6 + (\bar{l}^2 - \bar{f}^2 - 2\bar{m})\nu^4 + (\bar{m}^2 - 2\bar{l}\bar{n} + 2\bar{f}\bar{h} - \bar{g}^2)\nu^2 + \bar{n}^2 - \bar{h}^2 = 0. \tag{27}$$

From (Equation 26), we get $\hat{\tau}_n = \frac{1}{\hat{\nu}}tan^{-1}\{\frac{\hat{\nu}(\bar{f}\hat{\nu}^4 - \bar{f}\bar{m}\hat{\nu}^2 - \bar{h}\hat{\nu}^2 - \bar{l}\bar{g}\hat{\nu}^2 - \bar{h}\bar{m} - \bar{n}\bar{g})}{\bar{g}\hat{\nu}^4 - \bar{g}\bar{m}\hat{\nu}^2 + \bar{f}\bar{n}\hat{\nu}^2 - \bar{h}\bar{l}\hat{\nu}^2 + \bar{f}\bar{l}\hat{\nu}^4 - \bar{h}\bar{n}}\} + \frac{n\pi}{\hat{\nu}}$. The smallest $\hat{\tau}_n$ is given by $n = 0$ and we take it as $\hat{\tau}_n = \frac{1}{\hat{\nu}}tan^{-1}\{\frac{\hat{\nu}(\bar{f}\hat{\nu}^4 - \bar{f}\bar{m}\hat{\nu}^2 - \bar{h}\hat{\nu}^2 - \bar{l}\bar{g}\hat{\nu}^2 - \bar{h}\bar{m} - \bar{n}\bar{g})}{\bar{g}\hat{\nu}^4 - \bar{g}\bar{m}\hat{\nu}^2 + \bar{f}\bar{n}\hat{\nu}^2 - \bar{h}\bar{l}\hat{\nu}^2 + \bar{f}\bar{l}\hat{\nu}^4 - \bar{h}\bar{n}}\}$. In order to establish the Hopf bifurcation at $\tau = \hat{\tau}$, we need to show that $\frac{d\mu}{d\tau} \neq 0$ at $\tau = \hat{\tau}$. We differentiate Equations 24 and 25 with respect to τ and setting $\tau = \hat{\tau}$, $\mu = 0$ and $\nu = \hat{\nu}$, we get

$$\left.\begin{aligned}L_1\frac{d\mu}{d\tau}(\hat{\tau}) + M_1\frac{d\nu}{d\tau}(\hat{\tau}) = X_1\\ -M_1\frac{d\mu}{d\tau}(\hat{\tau}) + L_1\frac{d\nu}{d\tau}(\hat{\tau}) = Y_1\end{aligned}\right\}, \tag{28}$$

where

$$\begin{aligned}L_1 &= -3\hat{\nu}^2 + \bar{m} - \bar{g}\cos\hat{\nu}\hat{\tau} - 2\bar{f}\hat{\nu}\sin\hat{\nu}\hat{\tau} + \hat{\tau}(\bar{h}\cos\hat{\nu}\hat{\tau} + \bar{g}\hat{\nu}\sin\hat{\nu}\hat{\tau} - \bar{f}\hat{\nu}^2\cos\hat{\nu}\hat{\tau}),\\ M_1 &= -2\bar{l}\hat{\nu} + 2\bar{f}\hat{\nu}\cos\hat{\nu}\hat{\tau} - \bar{g}\sin\hat{\nu}\hat{\tau} + \hat{\tau}(\bar{h}\sin\hat{\nu}\hat{\tau} - \bar{g}\hat{\nu}\cos\hat{\nu}\hat{\tau} - \bar{f}\hat{\nu}^2\sin\hat{\nu}\hat{\tau}),\\ X_1 &= \bar{f}\hat{\nu}^3\sin\hat{\nu}\hat{\tau} - \bar{h}\hat{\nu}\sin\hat{\nu}\hat{\tau} + \bar{g}\hat{\nu}^2\cos\hat{\nu}\hat{\tau},\\ Y_1 &= \bar{f}\hat{\nu}^3\cos\hat{\nu}\hat{\tau} - \bar{h}\hat{\nu}\cos\hat{\nu}\hat{\tau} - \bar{g}\hat{\nu}^2\sin\hat{\nu}\hat{\tau}.\end{aligned}$$

Solving (Equation 28),

we get $\frac{d\mu}{d\tau}(\hat{\tau}) = \frac{L_1X_1 - M_1Y_1}{L_1^2 + M_1^2}$,

where $\frac{d\mu}{d\tau}(\hat{\tau})$ has the same sign as that of $L_1X_1 - M_1Y_1$.

Substituting the values of L_1, M_1, X_1, and Y_1 and using (Equation 26), we get

$$L_1X_1 - M_1Y_1 = \hat{\nu}^2[3\hat{\nu}^4 + 2(\bar{l}^2 - \bar{f}^2 - 2\bar{m})\hat{\nu}^2 + (\bar{m}^2 - 2\bar{l}\bar{n} + 2\bar{f}\bar{h} - \bar{g}^2)].$$

Let $H(z) = z^3 + B_1z^2 + B_2Zz + B_3$,

where $B_1 = \bar{l}^2 - \bar{f}^2 - 2\bar{m}$, $B_2 = \bar{m}^2 - 2\bar{l}\bar{n} + 2\bar{f}\bar{h} - \bar{g}^2$, $B_3 = (\bar{n}^2 - \bar{h}^2)$ which is left side of (Equation 27) with $\hat{v}^2 = z$. Then, $H(\hat{v}^2) = 0$ and we note that

$$\frac{d\mu}{d\tau}(\hat{\tau}) = \frac{\hat{v}^2}{L_1^2 + M_1^2}\frac{dH}{dz}(\hat{v}^2). \tag{29}$$

Hence, we can describe the criteria for the preservation of stability (instability) geometrically as follows:

(1) If the polynomial $H(z)$ has no positive roots, there is no change of stability.

(2) If $H(z)$is decreasing (increasing) at all its positives roots, stability (instability) is preserved.

We note the following facts:

(i) For the existence of $\hat{v}$, $H(z)$ must have at least one positive real root.

(ii) Since $H(z)$ is cubic in z ,

$lim_{z\to\infty} H(z) = \infty.$

(iii) If $H(z)$ has a unique positive root, then it must increase at that point to satisfy (ii).

(iv) If H(z) has two or three distinct positive real roots, then it must decrease at one root and increase at other; hence, (ii) is not satisfied.

(v) If $B_3 < 0$, then $H(z)$ has only one root.

(vi) If $B_1 > 0, B_3 > 0$, and $B_2 < 0$, then (i) will be satisfied. Now, if $B_1 > 0, B_3 > 0$, and $B_2 < 0$, then the minimum of $H(z)$ will exist at $z_{min} = \frac{-B_1 + \sqrt{B_1^2 - 3B_2}}{3}$ and (i) will be satisfied if $H(z_{min}) > 0$, i.e.

$$2B_1^3 - 9B_1B_2 + 27B_3 > 2(B_1^2 - 3B_2)^{3/2}, \tag{30}$$

or $2B_1(B_1^2 - 3B_2) + 27B_3 - 3B_1B_2 > 2(B_1^2 - 3B_2)^{3/2}$.

Since $27B_3 - 3B_1B_2 > 27B_3$ (since $B_1 > 0$, $B_3 > 0$, and $B_2 < 0$), and $B_1^2 - 3B_2 > B_1^2$, hence $2B_1(B_1^2 - 3B_2) + 27B_3 - 3B_1B_2 > 27B_3 + 2B_1^3$.

Thus, for Equation 30 to hold, it is sufficient that $27B_3 + 2B_1^3 > 2(B_1^2 - 3B_2)^{3/2} \Rightarrow B_2 > \frac{1}{3}\left[B_1^2 - \left(\frac{27B_3 + 2B_1^3}{2}\right)^{2/3}\right]$.

Now, we can state the following theorem.

THEOREM 6 *If $B_1 > 0$, $B_3 > 0$, and relation (Equation 30) hold, then the stable positive equilibrium E^* remains stable for all $\tau > 0$.*

6. Numerical simulations

In this section, we focus our attention on the occurrence and termination of the fluctuating plankton population. We begin with a parameter set (see Table 2, Chatterjee & Pal, 2011) for which the existence condition of the coexistence equilibrium point E^* is satisfied and the coexistence equilibrium point $E^* = (0.40, 0.10, 0.54)$ is locally asymptotically stable in the form of a stable focus with eigenvalues $-0.8003, -0.0129 \pm i0.2064$ (cf. Figure 1).

6.1. Effect of r

We have observed that the system shows oscillatory behavior around the positive interior equilibrium E^* for increasing the value of $r = 1.6$ with eigenvalues $-1.2305, 0.0013 \pm i0.0.2057$ (cf. Figure 2). Decreasing the value of r from 1.2 to 0.3 with the same set of parametric values in Table 2, the system

Table 2. A set of parametric values

Parameter	Definition	Value
r	Constant intrinsic growth rate for phytoplankton	1.2
K	Carrying capacity for phytoplankton	0.5
α_1	Maximal zooplankton ingestion rate	2
α_2	Maximal zooplankton conversion rate	1.5
γ_1	Maximal planktivorous fish ingestion rate	1
γ_2	Maximal planktivorous fish conversion rate	0.8
μ_1	Mortality rate of the phytoplankton	0.04
μ_2	Mortality rate of the zooplankton	0.06
μ_3	Mortality rate of the planktivorous fish	0.08
K_1	Half saturation constant for phytoplankton	0.8
K_2	Half saturation constant for zooplankton	0.9

shifts to planktivorous fish free equilibrium $E_1 = (0.0250, 0.0766, 0)$ in the form of a stable focus with eigenvalues $-0.0173, -0.0026 \pm i0.1188$ (cf. Figure 3). Figure 4(a–c) depicts the different steady-state behaviors of phytoplankton, zooplankton, and planktivorous fish in the system (Equation 1) for the parameter r. Here, we see a Hopf bifurcation point at r_c^0=1.583292 (denoted by a red star (*H*)) with first Lyapunov coefficient being -0.2999282. Clearly, the first Lyapunov coefficient is negative. This means that a stable limit cycle bifurcates from the equilibrium, when it loses stability.

6.2. Effects of K

For K = 0.8, leaving all other parameters unaltered, the system exhibits oscillation around the positive interior equilibrium E^* with eigenvalues $-0.8599, 0.0069 \pm i0.2370$ (cf. Figure 5). Figure 6(a–c) depicts the different steady-state behaviors of phytoplankton, zooplankton, and planktivorous fish in the system (Equation 1) for the parameter K. Here, we see a Hopf bifurcation point at K^* = 0.711977 (denoted by a red star (*H*)) with first Lyapunov coefficient being -0.3651997. Clearly, the first

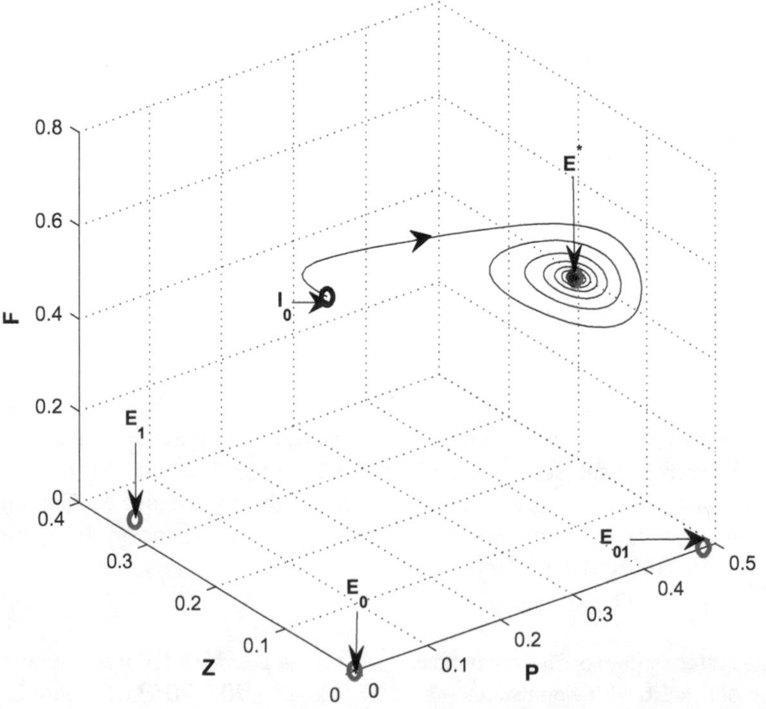

Figure 1. The equilibrium point E^* is stable for the parametric values as given in Table 2.

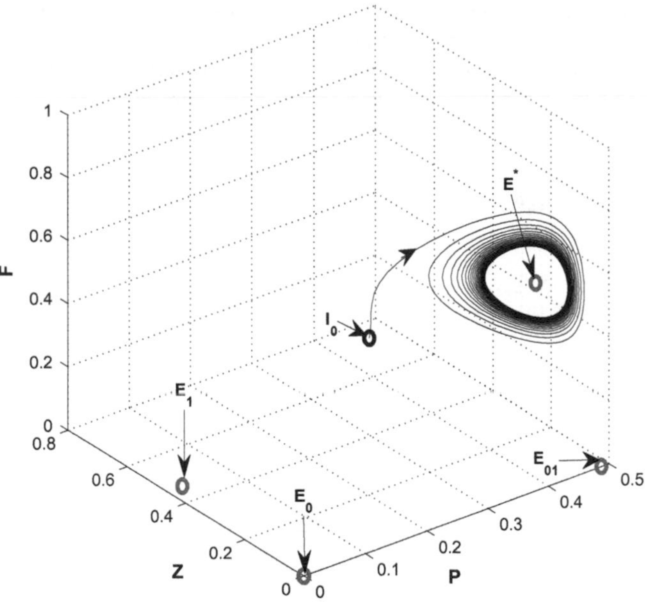

Figure 2. The figure depicts oscillatory behavior around the positive interior equilibrium point E^* of system (Equation 1) for increasing r, from 1.2 to 1.6 with other parametric values as given in Table 2.

Lyapunov coefficient is negative. This means that a stable limit cycle bifurcates from the equilibrium, when it loses stability.

6.3. Effects of μ_3

Keeping the other parameters fixed and decreasing the value of μ_3 from 0.08 to 0.02, the system exhibits oscillatory behaviors around the positive interior equilibrium E^* with eigenvalues $-1.0833, 0.0002 \pm i0.1084$ (cf. Figure 7). But the system shifts to planktivorous fish free equilibrium

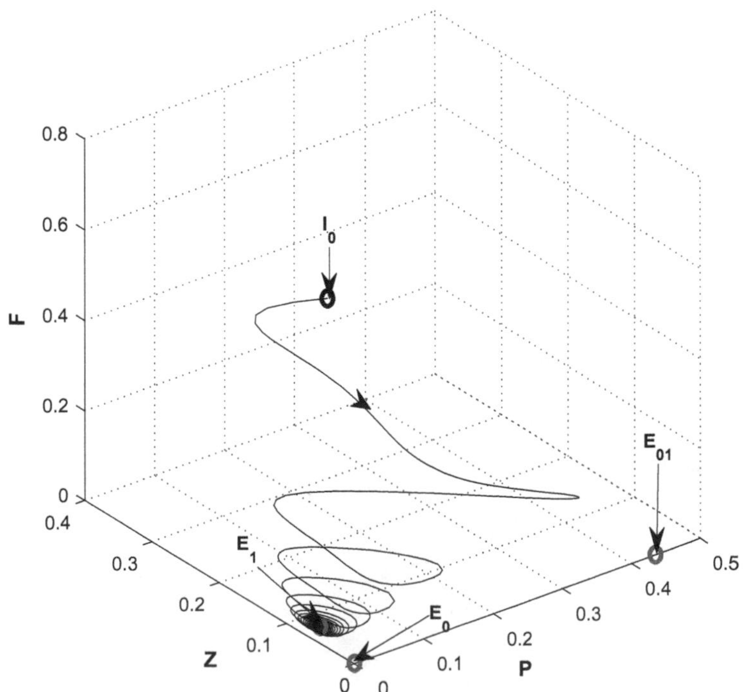

Figure 3. The figure depicts stable behavior around the planktivorous fish free equilibrium point E_1 of system (Equation 1) for decreasing r, from 1.2 to 0.3 with other parametric values as given in Table 2.

Figure 4. (a) The figure depicts different steady-state behaviors of phytoplankton for the effect of *r*. (b) The figure depicts different steady-state behaviors of zooplankton for the effect of *r* with other parametric values as given in Table 2. (c) The figure depicts different steady-state behaviors of planktivorous fish for the effect of *r* with other parametric values as given in Table 2.

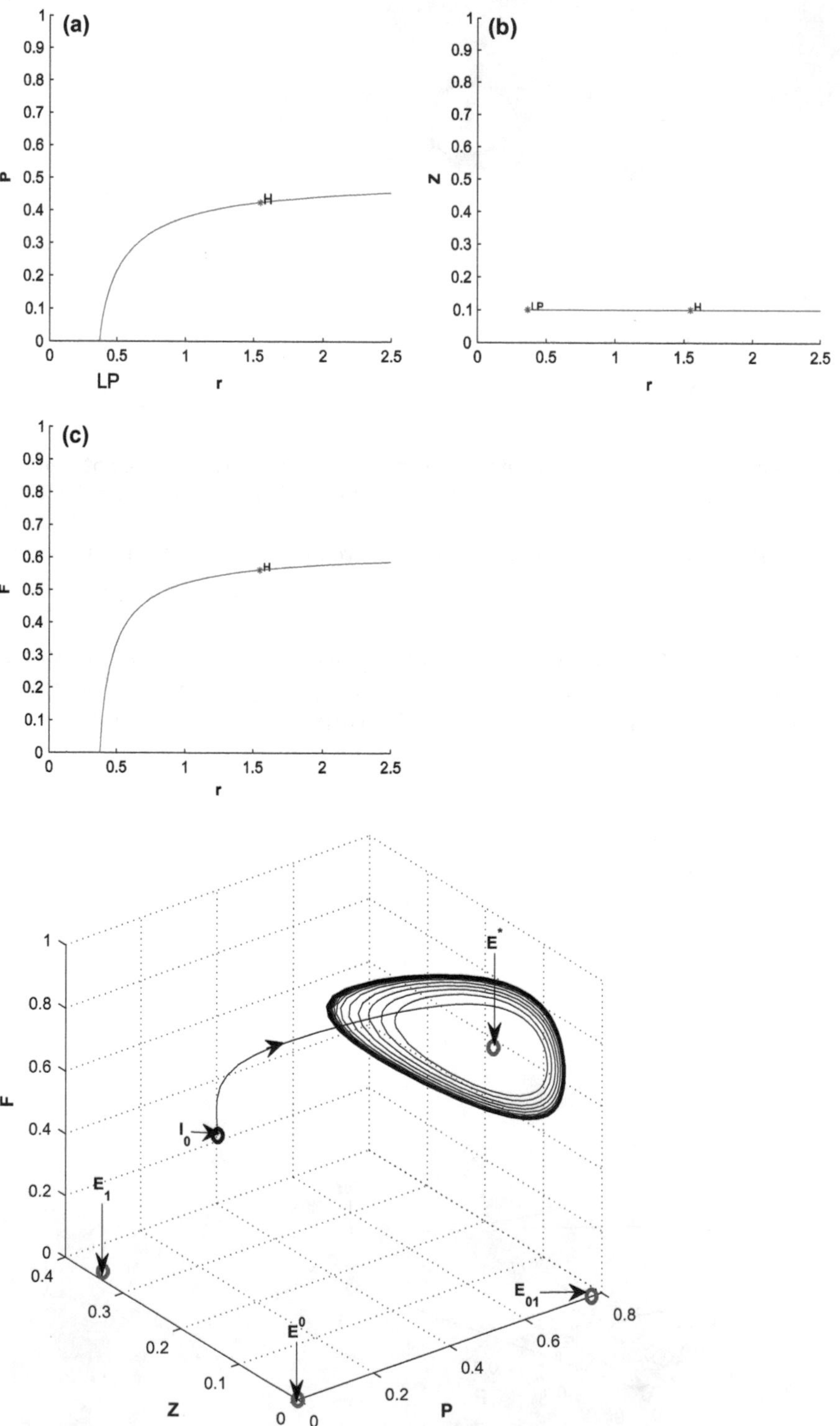

Figure 5. The figure depicts oscillatory behavior around the positive interior equilibrium point E^* of system (Equation 1) for increasing *K*, from 0.5 to 0.8 with other parametric values as given in Table 2.

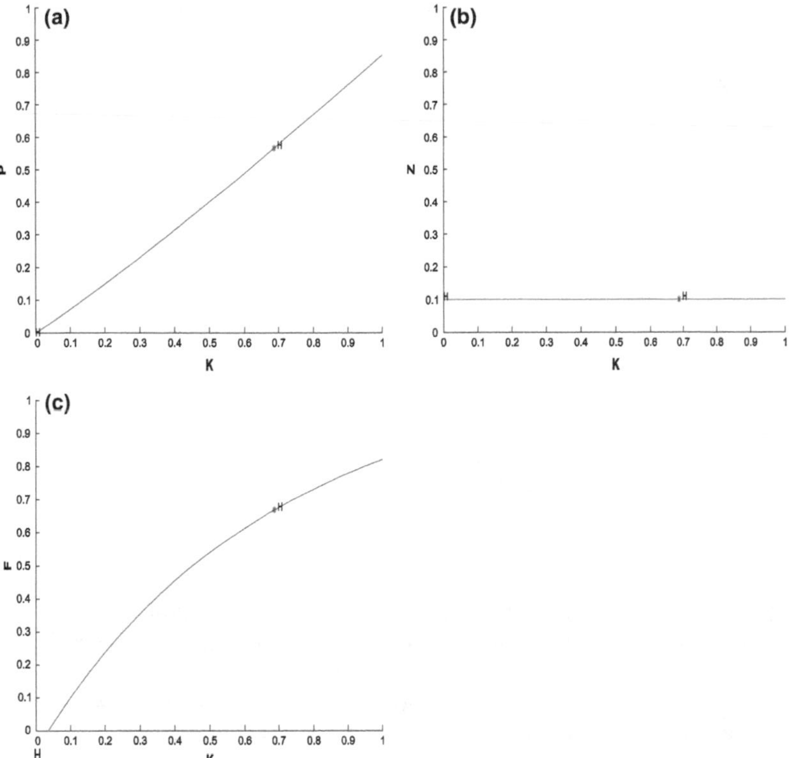

Figure 6. (a) The figure depicts different steady-state behaviors of phytoplankton for the effect of K. (b) The figure depicts different steady-state behaviors of zooplankton for the effect of K with other parametric values as given in Table 2. (c) The figure depicts different steady-state behaviors of planktivorous fish for the effect of K with other parametric values as given in Table 2.

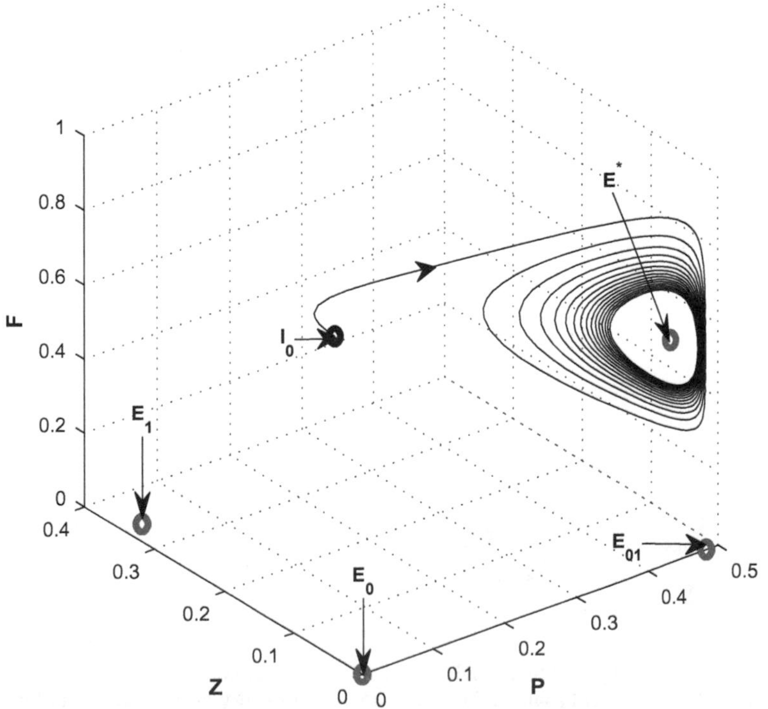

Figure 7. The figure depicts oscillatory behavior around the positive interior equilibrium point E^* of system (Equation 1) for decreasing μ_3, from 0.08 to 0.02 with same set of parametric values as use in Table 2.

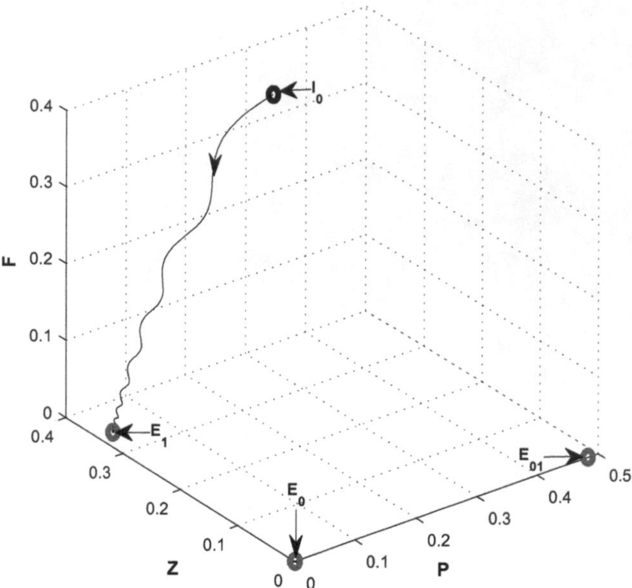

Figure 8. The figure depicts stable behavior around the planktivorous fish free equilibrium point E_1 of system (Equation 1) for increasing μ_3, from 0.08 to 0.3 with same set of parametric values as use in Table 2.

$E_1 = (0.0250, 0.3437, 0)$ with eigenvalues $-0.0289, -0.0080 \pm i0.2516$, when μ_3 increase from 0.08 to 0.3 with the same set of parametric values as in Table 2 (cf. Figure 8).

6.4. Combined effect of r and μ_3

It has already been shown that the system depicts oscillatory behavior around the positive interior equilibrium E^* for $r = 1.6$. If the mortality rate of planktivorous fish μ_3 in increased from 0.08 to 0.12, the

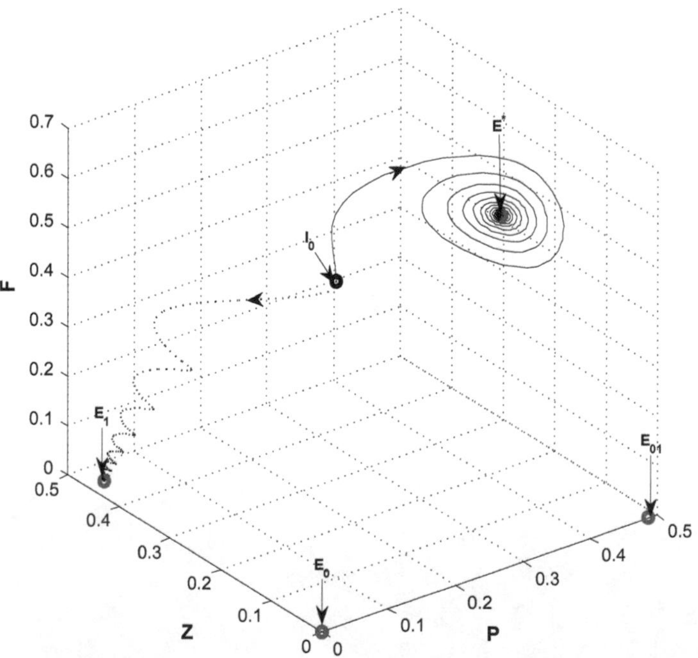

Figure 9. The figure depicts asymptotically stable behavior at positive interior equilibrium point E^* of system (Equation 1) for $r = 1.6$ and $\mu_3 = 0.12$ with other parametric values as given in Table 2 (Blue solid line). The trajectory (Black dotted line) shows stable behavior around the planktivorous fish free equilibrium point E_1 of system (Equation 1) for $r = 1.6$ and $\mu_3 = .3$ with other parametric values as given in Table 2.

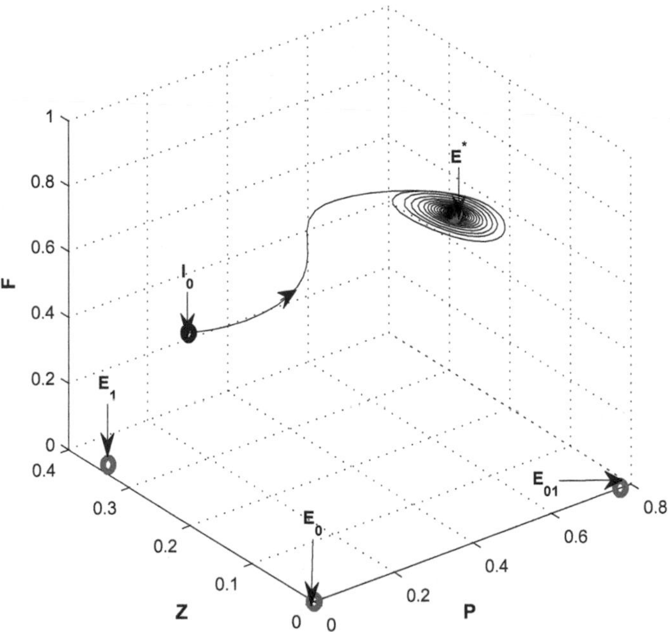

Figure 10. The figure depicts stable behavior around the positive interior equilibrium point E^* of system (Equation 1) for $K = 0.8$ and $\mu_3 = 0.12$ with other parametric values as given in Table 2.

system becomes asymptotically stable at positive interior equilibrium $E^* = (0.3868, 0.1588, 0.5591)$ in the form of a stable focus with eigenvalues $-1.0114, -0.0107 \pm i0.2431$ (cf. Figure 9). But if the mortality rate of planktivorous fish μ_3 is increased from 0.12 to 0.3, the system shifts to planktivorous fish free equilibrium $E_1 = (0.025, 0.4625, 0)$ with eigenvalues $-0.0284, -0.0104 \pm i0.2918$ (cf. Figure 9).

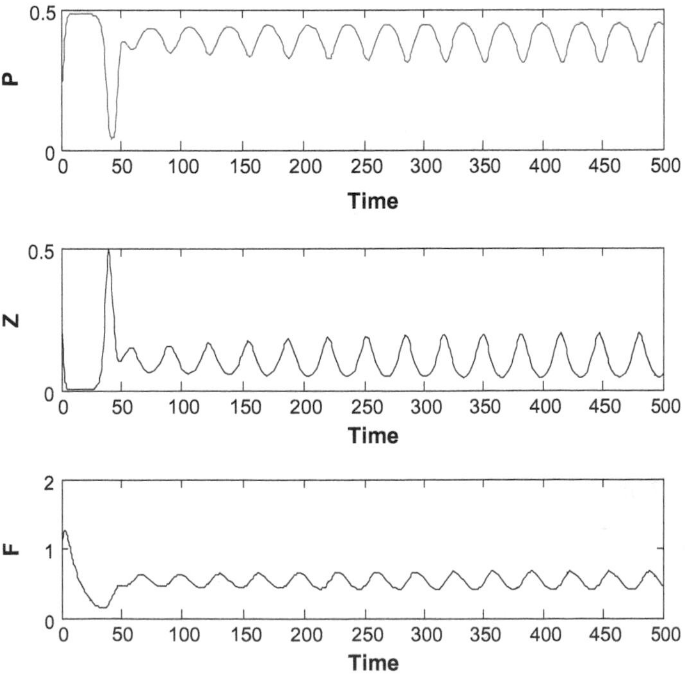

Figure 11. The figures depict oscillatory behavior of phytoplankton, Zooplankton, and planktivorous fish population for $\tau = 1$ with same set of other parametric values as given in Table 2.

6.5. Combined effect of K and μ_3

It has already been shown that the system depicts oscillatory behavior around the positive interior equilibrium E^* for $K = 0.8$ (cf. Figure 5). But the system shifts to asymptotically stable behavior at $E^* = (0.5960, 0.1590, 0.7282)$ in the form of a stable focus with eigenvalues $-0.6461, -0.0065 \pm i0.2876$ for increasing the value of μ_3 from 0.08 to 0.12 (cf. Figure 10).

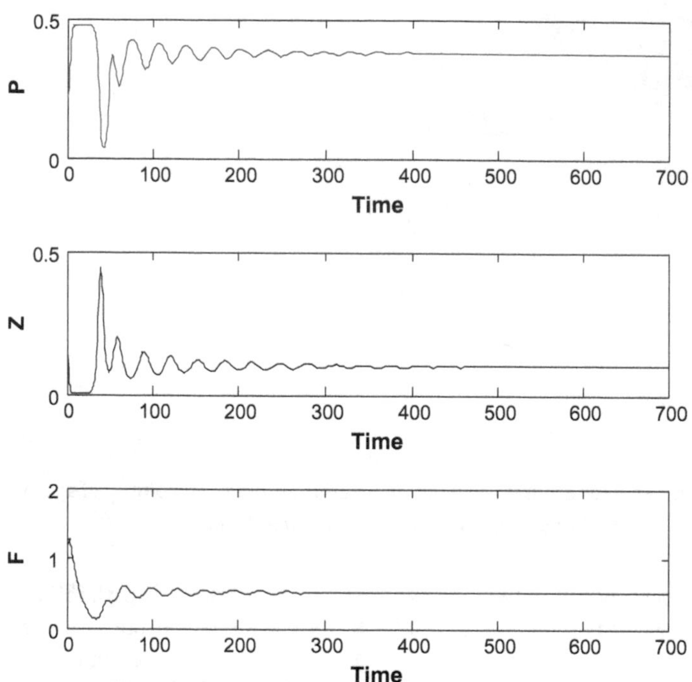

Figure 12. The coexistence equilibrium is locally asymptotically stable for $\tau = 1$ and $r = 1$.

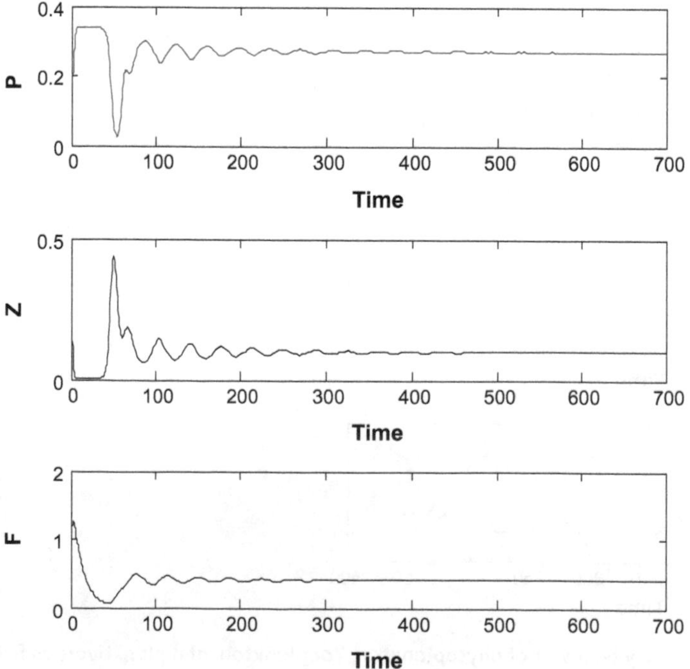

Figure 13. The coexistence equilibrium is locally asymptotically stable for $\tau = 1$ and $K = 0.35$.

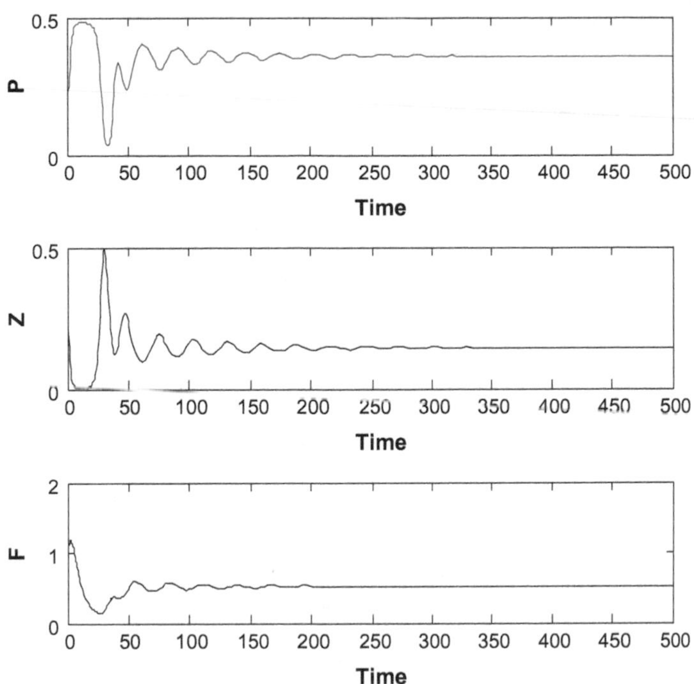

Figure 14. The coexistence equilibrium is locally asymptotically stable for $\tau = 1$ and $\mu_3 = 0.11$.

6.6. Effects of τ

Now, we are to introduce gestation delay, τ, in system (Equation 1). Keeping the other parameters fixed and increasing the value of the delay parameter τ from 0 to 1, we observe that the solution of

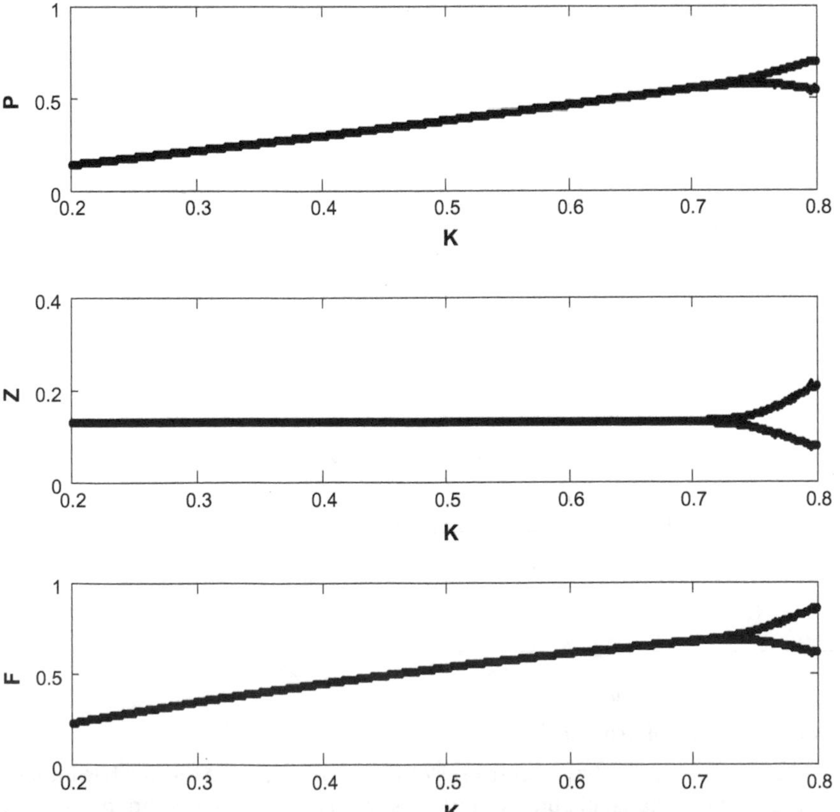

Figure 15. The bifurcation diagram of all the populations with *K* as the bifurcation parameter.

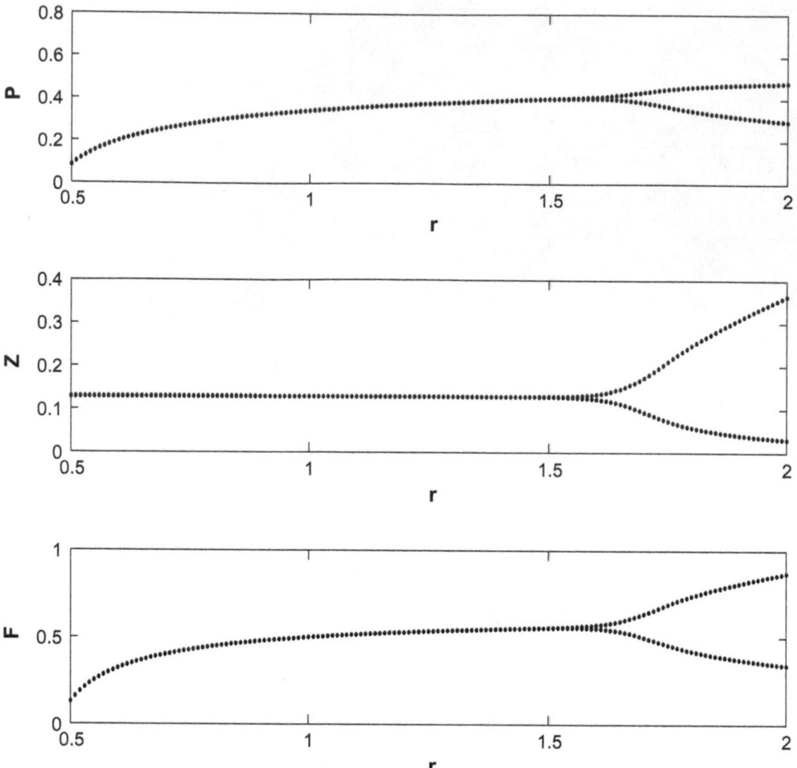

Figure 16. The bifurcation diagram of all the populations with *r* as the bifurcation parameter.

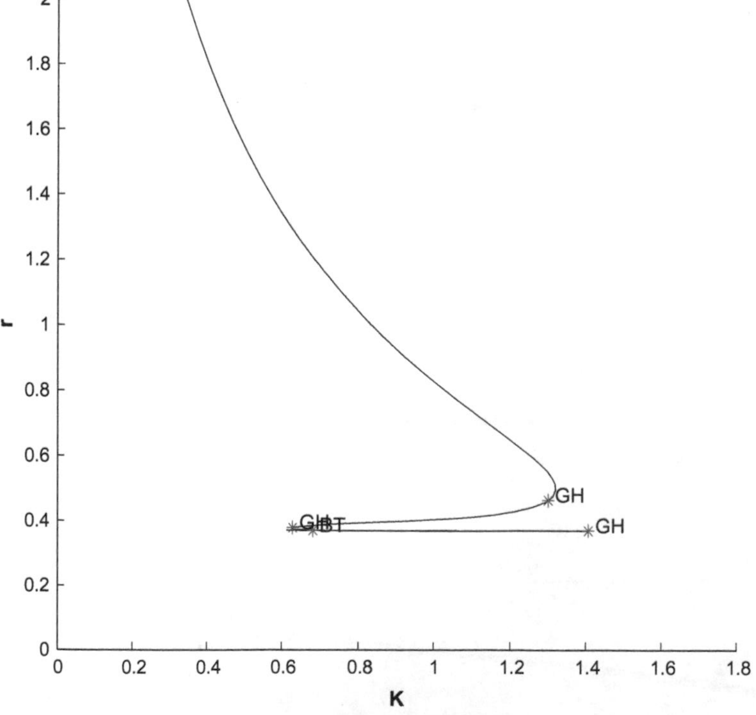

Figure 17. The *K*–*r* two parameters bifurcation diagram.

system (Equation 8) changes from stable behavior to oscillatory behavior around the positive interior equilibrium E^* (cf. Figure 11) for same set of parametric values as in Figure 1. We see that decreasing the value of r from 1.2 to 1, the system shifts to stable behavior around the positive interior

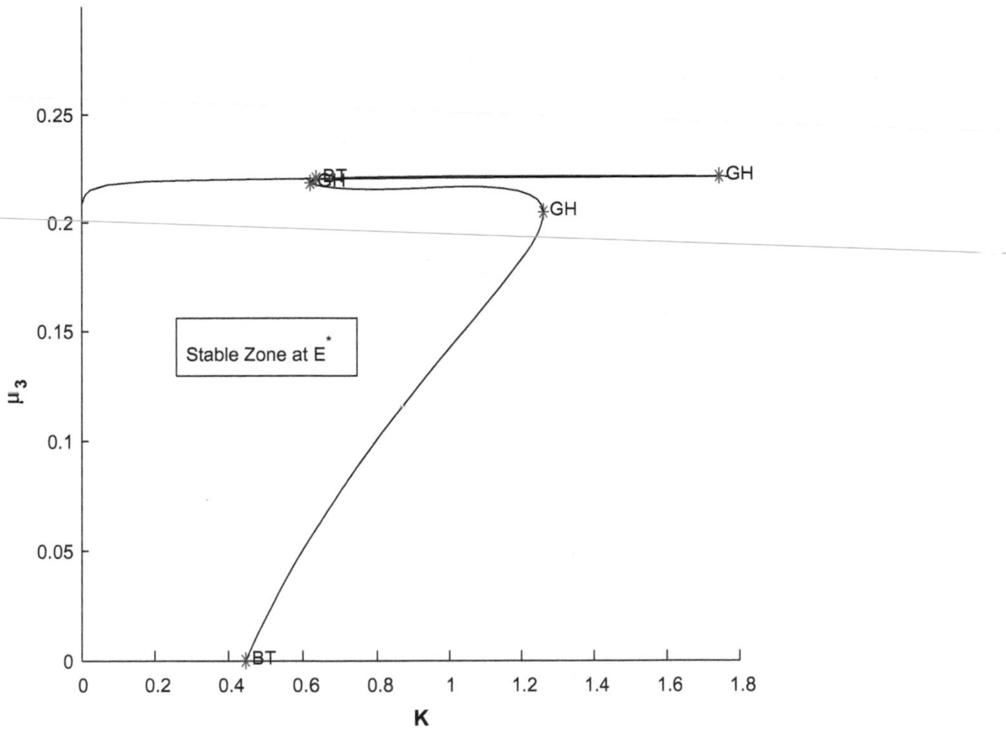

Figure 18. The $K - \mu_3$ two parameters bifurcation diagram.

equilibrium (cf. Figure 12). Similar cases happen when decreasing the carrying capacity and increasing mortality rate of planktivorous fish from 0.5 to 0.35 and 0.08 to 0.11, respectively, (cf. Figure 13 and cf. Figure 14) for same set of parametric values as in Figure 11.

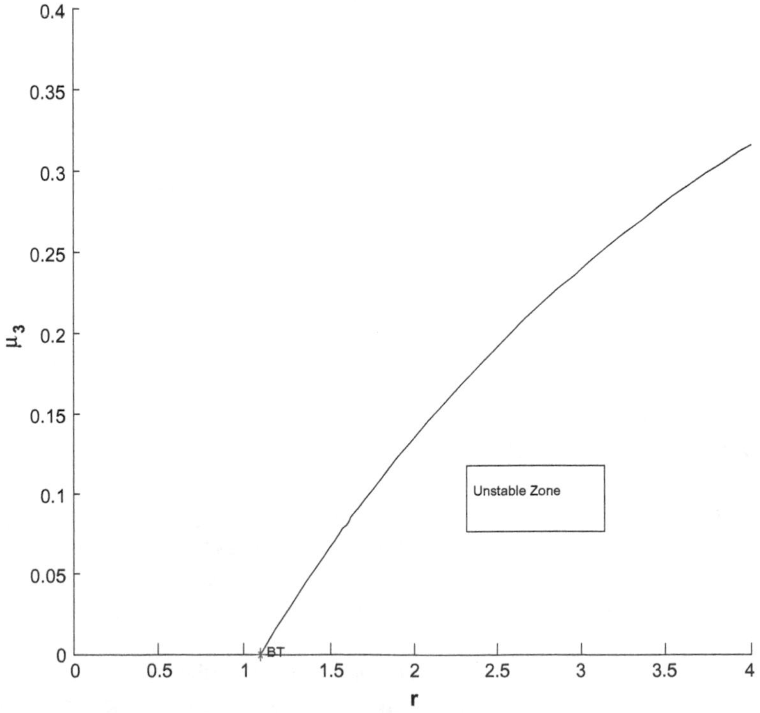

Figure 19. The $r - \mu_3$ two parameters bifurcation diagram.

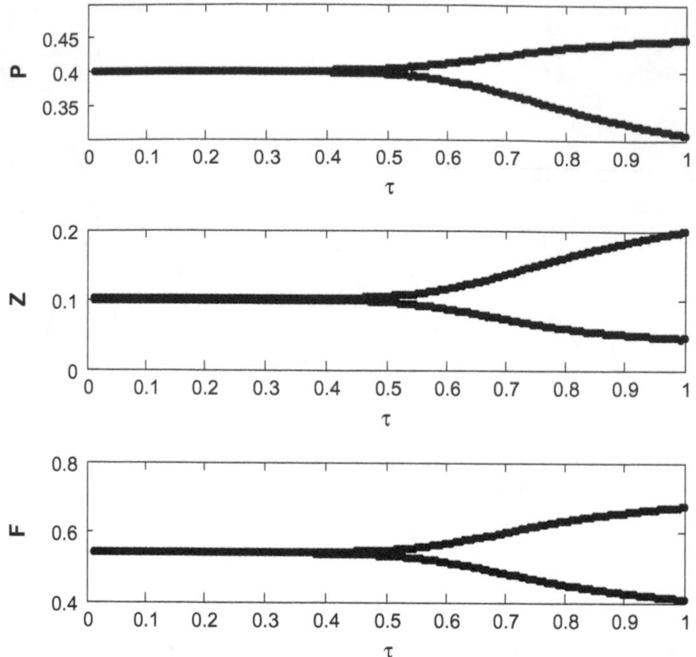

Figure 20. The bifurcation diagram for τ with all parametric values as given in Table2.

6.7. Hopf bifurcation

For a clear understanding of the dynamical changes due to change in constant nutrient input, K, a bifurcation diagram is plotted with this parameter as the bifurcation parameter with other three species (cf. Figure 15). We have also plotted another single parameter bifurcation diagram for r (cf. Figure 16). Now, we have plotted a two parameters bifurcation diagrams, K–r, $K - \mu_3$ and $r - \mu_3$ (cf. Figures 17–19). Here, we see a generalized Hopf (GH) point, where the first Lyapunov coefficient (l_1) vanishes but the second lyapounov coefficient is non-zero. The bifurcation point separates branches of sub and supercritical Andronov–Hopf bifurcations in the parameter plain. The Bogdanov–Takens (BT) point is the common point for the limit point curve and the curve corresponding to equilibria. Actually, at BT point, the system has an equilibrium with a double zero eigenvalues. For nearby parameter values, the system has two equilibria (a saddle and a non-saddle) which collide and disappear via a saddle-node bifurcation. The non-saddle equilibrium undergoes an Andronov–Hopf bifurcation, generating a limit cycle. Finally, a bifurcation diagram is plotted with τ as the bifurcation parameter with other three species (cf. Figure 20).

7. Discussion

A phytoplankton–zooplankton–fish interaction model is considered with nutrient uptake functions. Firstly, the model is studied analytically and the threshold conditions for the existence and stability of various steady states are worked out (see Table 1). It is observed that for constant intrinsic growth rate r, there is a chance for the population to fluctuate. But further increase in the value of constant intrinsic growth rate r may lead to extinction of planktivorous fish population.

It is observed that the system shifts to oscillatory behavior in presence of high value of carrying capacity. We have also observed that there is a chance of extension of planktivorous fish population for high value of mortality rate of planktivorous fish. Our study indicates that low value of mortality rate of planktivorous fish may lead to oscillatory behavior around the positive interior equilibrium.

Next, we have established that the system remains locally asymptotically stable and all solutions approach toward E^* in presence of discrete delay whenever its magnitude lies below some threshold condition. Further, we have also observed the maximum value (length of delay) of τ (i.e. τ_+) for which a locally asymptotically stable interior equilibrium E^* will remain asymptotically stable, where

$\tau_+ = \frac{1}{2A_\theta}(-B_\theta + \sqrt{B_\theta^2 + 4A_\theta C_\theta})$ (see Theorem 5). Moreover, we analyzed conditions for bifurcation of the positive interior equilibrium. We observed that a positive interior equilibrium remains stable if $B_1 > 0$ and $B_3 > 0$ for all $\tau > 0$ (see Theorem 6). Further, numerical simulations demonstrate the following conclusions:

(a) The system exhibits dynamic instability due to higher gestation delay.

(b) Low levels of constant intrinsic growth rate r induce stability around the positive interior equilibrium in presence of gestation delay.

(c) Low levels of carrying capacity of phytoplankton K help the system (Equation 8) to approach stability at E^* in presence of gestation delay.

(d) High value of mortality rate of planktivorous fish prevents instability behavior in presence of gestation delay.

Throughout the article of this model (analytically and numerically), an attempt has been made to search for a suitable way to control the system and maintain a stable coexistence between all the species. It has been observed that to control the fracturing population and to maintain stability around the coexistence equilibrium, we have to balance the rate of carrying capacity of phytoplankton, mortality rate of planktivorous fish population, and constant intrinsic growth rate.

Acknowledgements
The authors would like to thank the referees very much for their valuable comments and suggestions.

Funding
The research is partially supported by University Grants Commission, New Delhi [grant number MRP-MAJ-MATH-2013-609].

Author details
Samares Pal[1]
E-mail: samaresp@yahoo.co.in
Anal Chatterjee[2]
E-mail: chatterjeeanal172@gmail.com
ORCID ID: http://orcid.org/0000-0002-8792-0031
[1] Department of Mathematics, University of Kalyani, Kalyani 741235, India.
[2] Department of Mathematics, Sheikhpara A.R.M. Polytechnic, Sheikhpara 742308, India.

References

Beretta, E., Bischi, G. I., & Solimano, F. (1991). Stability in chemostat equations with delayed nutrient recycling. *Journal of Mathematical Biology, 28*, 99–111. doi:10.1007/BF00171521

Biktashev, V. N., Brindley, J., & Horwood, J. W. (2003). Phytoplankton blooms and fish recruitment rate. *Journal of plankton Research, 25*, 21–33. doi:10.1093/plankt/25.1.21

Bischi, G. I. (1992). Effects of time lags on transient characteristics of a nutrient cycling model. *Mathematical Biosciences, 109*, 151–175. doi:10.1016/0025-5564(92)90043-V

Chatterjee, A., & Pal, S. (2011). Effect of dilution rate on the predictability of a realistic ecosystem model with instantaneous nutrient recycling. *Journal of Biological Systems, 19*, 629–650. doi:10.1142/S021833901100410X

Cushing, J. M. (1977). *Integrodifferential equations and delay models in population dynamics* (Lecture notes in biomathematics, Vol. 20). Berlin: Springer-Verlag. doi:10.1007/978-3-642-93073-7

Das, K., & Ray, S. (2008). Effect of delay on nutrient cycling in phytoplankton–zooplankton interactions in estuarine system. *Ecological Modelling, 215*, 69–76. doi:10.1016/j.ecolmodel.2008.02.019

Freedman, H. I., Erbe, L. H., & Rao, V. S. H. (1986). Three species food chain models with mutual interference and time delays. *Mathematical Biosciences, 80*, 57–80. doi:10.1016/0025-5564(86)90067-2

Freedman, H. I., & Rao, V. S. H. (1983). The trade-off between mutual interference and time lags in predator-prey systems. *Bulletin of Mathematical Biology, 45*, 991–1004. doi:10.1007/BF02458826

Freedman, H. I., So, J., & Waltman, P. (1989). Coexistence in a model of competition in the chemostat incorporating discrete delay. *SIAM Journal on Mathematical Analysis, 49*, 859–870. doi:10.1137/0149050

Gazi, N. H., & Bandyopadyay, M. (2006). Effect of time delay on a detritus-based ecosystem. *International Journal of Mathematics and Mathematical Sciences*, Article ID: 25619, 1–28. doi:10.1155/IJMMS/2006/25619

Gazi, N. H., & Bandyopadhyay, M. (2008). Effect of time delay on a harvested predator-prey model. *Journal of Applied Mathematics and Computing, 26*, 263–280. doi:10.1007/s12190-007-0015-2

Gopalsamy, K. (1984). Delayed responses and stability in two-species systems. *The Journal of the Australian Mathematical Society Series B Applied Mathematics, 25*, 473–500. doi:10.1017/S0334270000004227

Gopalsamy, K. (1992). *Stability and oscillations in delay differential equations of population dynamics* (Mathematics and its applications, Vol. 74). Dordrecht: Kluwer. doi:10.1007/978-94-015-7920-9

Khare, S., Misra, O. P., Singh, C., & Dhar, J. (2011). Role of delay on planktonic ecosystem in the presence of a toxic producing phytoplankton. *International Journal of Differential Equations*, Article ID: 603183, doi:10.1155/2011/603183

Kuang, Y. (1993). *Delay differential equations with applications in popular dynamics*. New York, NY: Academic Press.

Liu, S., Beretta, E., & Breda, D. (2010). Predator-prey model of Beddington–DeAngelis type with maturation and gestation delays. *Nonlinear Analysis: Real*

World Applications, 11, 4072–4091. doi:10.1016/j.nonrwa.2010.03.013

Maity, A., Patra, B., & Samanta, G. P. (2007). Persistence and stability in a ratio-dependent predator-prey system with delay and harvesting. *Natural Resource Modeling, 20*, 575–600. doi:10.1111/j.1939-7445.2007.tb00221.x

Medvinsky, A. B., Rusakov, A. V., Bobyrev, A. E., Burmensky, V. A., Kriksunov, A. E., Nurieva, N. I., & Gonik, M. M. (2009). A conceptual mathematical model of aquatic communities in lakes Naroch and Myastro (Belarus). *Biophysics, 54*, 90–93. doi:10.1134/S0006350909010151

Mukhopadhyay, B., & Bhattacharyya, R. (2008). Role of gestation delay in a plankton–fish model under stochastic fluctuations. *Mathematical Biosciences, 215*, 26–34. doi:10.1016/j.mbs.2008.05.007

Mukhopadhyay, B., & Bhattacharyya, R. (2011). A stage-structured food chain model with stage dependent predation: Existence of codimension one and codimension two bifurcations. *Nonlinear Analysis: Real World Applications, 12*, 3056–3072. doi:10.1016/j.nonrwa.2011.05.007

Pal, S., & Chatterjee, A. (2012). Role of constant nutrient input and mortality rate of planktivorous fish in plankton community ecosystem with instantaneous nutrient recycling. *Canadian Applied Mathematics Quarterly, 20*, 179–207.

Rehim, M., & Imran, M. (2012). Dynamical analysis of a delay model of phytoplankton–zooplankton interaction. *Applied Mathematical Modelling, 36*, 638–647. doi:10.1016/j.apm.2011.07.018

Ruan, S. (1995). The effect of delays on stability and persistence in plankton models. *Nonlinear Analysis, Theory, Methods and Applications, 24*, 575–585. doi:10.1016/0362-546X(95)93092-I

Scheffer, M. (1991). Fish and nutrients interplay determines algel biomass: A minimal model. *Oikos, 62*, 271–282. Retrieved from http://www.jstor.org/stable/3545491

Scheffer, M., Rinaldi, S., & Kuznetsov, A. Y. (2000). Effects of fish on plankton dynamics: A theoretical analysis. *Canadian Journal of Fisheries and Aquatic Sciences, 57*, 1208–1219.

Singh, A., & Gakkhar, S. (2012). Analysis of delayed toxin producing phytoplankton–zooplankton system. *International Journal of Modeling and Optimization, 2*, 677–680. doi:10.7763/IJMO.2012.V2.208

Upadhyay, R. K., Kumari, N., & Rai, V. (2009). Wave of chaos in a diffusive system: Generating realistic patterns of patchiness in planktonfish dynamics. *Chaos, Solitons and Fractals, 40*, 262–276. doi:10.1016/j.chaos.2007.07.078

Yongzhen, P., Min, G., & Changguo, L. (2011). A delay digestion process with application in a three-species ecosystem. *Communications in Nonlinear Science and Numerical Simulation, 16*, 4365–4378. doi:10.1016/j.cnsns.2011.03.018

Permissions

All chapters in this book were first published in CM, by Cogent OA; hereby published with permission under the Creative Commons Attribution License or equivalent. Every chapter published in this book has been scrutinized by our experts. Their significance has been extensively debated. The topics covered herein carry significant findings which will fuel the growth of the discipline. They may even be implemented as practical applications or may be referred to as a beginning point for another development.

The contributors of this book come from diverse backgrounds, making this book a truly international effort. This book will bring forth new frontiers with its revolutionizing research information and detailed analysis of the nascent developments around the world.

We would like to thank all the contributing authors for lending their expertise to make the book truly unique. They have played a crucial role in the development of this book. Without their invaluable contributions this book wouldn't have been possible. They have made vital efforts to compile up to date information on the varied aspects of this subject to make this book a valuable addition to the collection of many professionals and students.

This book was conceptualized with the vision of imparting up-to-date information and advanced data in this field. To ensure the same, a matchless editorial board was set up. Every individual on the board went through rigorous rounds of assessment to prove their worth. After which they invested a large part of their time researching and compiling the most relevant data for our readers.

The editorial board has been involved in producing this book since its inception. They have spent rigorous hours researching and exploring the diverse topics which have resulted in the successful publishing of this book. They have passed on their knowledge of decades through this book. To expedite this challenging task, the publisher supported the team at every step. A small team of assistant editors was also appointed to further simplify the editing procedure and attain best results for the readers.

Apart from the editorial board, the designing team has also invested a significant amount of their time in understanding the subject and creating the most relevant covers. They scrutinized every image to scout for the most suitable representation of the subject and create an appropriate cover for the book.

The publishing team has been an ardent support to the editorial, designing and production team. Their endless efforts to recruit the best for this project, has resulted in the accomplishment of this book. They are a veteran in the field of academics and their pool of knowledge is as vast as their experience in printing. Their expertise and guidance has proved useful at every step. Their uncompromising quality standards have made this book an exceptional effort. Their encouragement from time to time has been an inspiration for everyone.

The publisher and the editorial board hope that this book will prove to be a valuable piece of knowledge for researchers, students, practitioners and scholars across the globe.

List of Contributors

Craig George Leslie Hopf and Gurudeo Anand Tularam
Mathematics and Statistics, Griffith Sciences (ENV), Environmental Futures Research Institute (EFRI), Griffith University, Brisbane, Australia

Renu Chugh and Preety Malik and Vivek Kumar
Department of Mathematics, KLP College, Rewari, India

G. Nagamani and S. Ramasamy
Department of Mathematics, Gandhigram Rural Institute - Deemed University, Gandhigram, Tamil Nadu, 624 302 India

Kewal Krishan Nailwal
Department of Mathematics, A.P.J. College of Fine Arts, Jalandhar, Punjab 144001, India

Deepak Gupta
Department of Mathematics, M. M. University, Mullana, Ambala, Haryana, India

Sameer Sharma
Department of Mathematics, D.A.V. College, Jalandhar, Punjab 144008, India

Firdous A. Shah
Department of Mathematics, University of Kashmir, South Campus, Anantnag 192 101, Jammu and Kashmir, India

R. Abass and J. Iqbal
Department of Mathematical Sciences, BGSB University, Rajouri 185234, Jammu and Kashmir, India

Bui Van Dinh
Department of Mathematics, Le Quy Don Technical University, No. 236 Hoang Quoc Viet Road, Hanoi, Vietnam

Lung-Hui Chen
Department of Mathematics, National Chung Cheng University, 168 University Rd. Min-Hsiung, Chia-Yi County, 621, Taiwan

Michail Zak
California Institute of Technology, 9386 Cambridge street, Cypress, CA 90630, USA

Juergen Geiser
Department of Electrical Engineering and Information Technology, Ruhr University of Bochum, Universitätsstrasse 150, D-44801 Bochum, Germany

S. Kartal
Faculty of Education, Department of Mathematics, Nevsehir Haci Bektas Veli University, Nevsehir 50300, Turkey

Amir T. Payandeh Najafabadi
Department of Mathematical Sciences, Shahid Beheshti University, G.C. Evin, Tehran, 1983963113, Iran

Dan Z. Kucerovsky
Department of Mathematics and Statistics, University of New Brunswick, Fredericton, New Brunswick, Canada, E3B 5A3

Ying Wang
College of Science, Sichuan University of Science and Engineering, 643000 Zigong, China

Samares Pal
Department of Mathematics, University of Kalyani, Kalyani 741235, India

Anal Chatterjee
Department of Mathematics, Sheikhpara A.R.M. Polytechnic, Sheikhpara 742308, India

Craig George Leslie Hopf and Gurudeo Anand Tularam
Mathematics and Statistics, Griffith Sciences (ENV), Environmental Futures Research Institute (EFRI), Griffith University, Brisbane, Australia

Renu Chugh and Preety Malik and Vivek Kumar
Department of Mathematics, KLP College, Rewari, India

G. Nagamani and S. Ramasamy
Department of Mathematics, Gandhigram Rural Institute - Deemed University, Gandhigram, Tamil Nadu, 624 302 India

Kewal Krishan Nailwal
Department of Mathematics, A.P.J. College of Fine Arts, Jalandhar, Punjab 144001, India

Deepak Gupta
Department of Mathematics, M. M. University, Mullana, Ambala, Haryana, India

Sameer Sharma
Department of Mathematics, D.A.V. College, Jalandhar, Punjab 144008, India

Firdous A. Shah
Department of Mathematics, University of Kashmir, South Campus, Anantnag 192 101, Jammu and Kashmir, India

R. Abass and J. Iqbal
Department of Mathematical Sciences, BGSB University, Rajouri 185234, Jammu and Kashmir, India

Bui Van Dinh
Department of Mathematics, Le Quy Don Technical University, No. 236 Hoang Quoc Viet Road, Hanoi, Vietnam

Lung-Hui Chen
Department of Mathematics, National Chung Cheng University, 168 University Rd. Min-Hsiung, Chia-Yi County, 621, Taiwan

Michail Zak
California Institute of Technology, 9386 Cambridge street, Cypress, CA 90630, USA

Juergen Geiser
Department of Electrical Engineering and Information Technology, Ruhr University of Bochum, Universitätsstrasse 150, D-44801 Bochum, Germany

S. Kartal
Faculty of Education, Department of Mathematics, Nevsehir Haci Bektas Veli University, Nevsehir 50300, Turkey

Amir T. Payandeh Najafabadi
Department of Mathematical Sciences, Shahid Beheshti University, G.C. Evin, Tehran, 1983963113, Iran

Dan Z. Kucerovsky
Department of Mathematics and Statistics, University of New Brunswick, Fredericton, New Brunswick, Canada, E3B 5A3

Ying Wang
College of Science, Sichuan University of Science and Engineering, 643000 Zigong, China

Samares Pal
Department of Mathematics, University of Kalyani, Kalyani 741235, India

Anal Chatterjee
Department of Mathematics, Sheikhpara A.R.M. Polytechnic, Sheikhpara 742308, India

Denise S Casselli
School of Dentistry, Campus of Sobral, Federal University of Ceará, Rua Coronel Estanislau Frota s/n, Sobral CE 62010-560, Brazil

Marcela G Borges
School of Dentistry, Federal University of Uberlândia, Av. Pará 1720, Bloco 4L, Uberlândia MG 38400-902, Brazil

Murilo S Menezes
School of Dentistry, Federal University of Uberlândia, Av. Pará 1720, Bloco 4L, Uberlândia MG 38400-902, Brazil

André L Faria-e-Silva
School of Dentistry, Federal University of Sergipe, Rua Cláudio Batista s/n, Aracaju SE 49060-100, Brazil

Siripong Mahaphasukwat
Graduate school, Tokyo Institute of Technology, 4259 Nagatsuta, Midori-ku, Yokohama 226-8503, Japan

Kazumasa Shimamoto
Graduate school, Tokyo Institute of Technology, 4259 Nagatsuta, Midori-ku, Yokohama 226-8503, Japan

Shota Hayashida
Graduate school, Tokyo Institute of Technology, 4259 Nagatsuta, Midori-ku, Yokohama 226-8503, Japan

Yu Sekiguchi
Precision and Intelligence Laboratory, Tokyo Institute of Technology, 4259 Nagatsuta, Midori-ku, Yokohama 226-8503, Japan

Chiaki Sato
Precision and Intelligence Laboratory, Tokyo Institute of Technology, 4259 Nagatsuta, Midori-ku, Yokohama 226- 503, Japan

Index

Printed in the USA
CPSIA information can be obtained
at www.ICGtesting.com
JSHW051413221024
72173JS00006B/1352